Ground Penetrating Radar

Dedicated
to
Professor Matteo Pastorino
An Inspiring Teacher of Science and Humanity

SCIENCES

Waves, Field Director – Frédérique de Fornel

Instrumentation, Sensors and Measurements,
Subject Head – Cyril Lupi

Ground Penetrating Radar

From Theoretical Endeavors to Computational Electromagnetics, Signal Processing, Antenna Design and Field Applications

Coordinated by
Mohammed Serhir
Dominique Lesselier

iSTE

WILEY

First published 2024 in Great Britain and the United States by ISTE Ltd and John Wiley & Sons, Inc.

ISTE Ltd
27-37 St George's Road
London SW19 4EU
UK

www.iste.co.uk

John Wiley & Sons, Inc.
111 River Street
Hoboken, NJ 07030
USA

www.wiley.com

Library of Congress Control Number: 2023943503

British Library Cataloguing-in-Publication Data
A CIP record for this book is available from the British Library
ISBN 978-1-78945-157-3

ERC code:
PE8 Products and Processes Engineering
PE8_3 Civil engineering, architecture, maritime/hydraulic engineering, geotechnics, waste treatment

Contents

Preface

Mohammed SERHIR[1] **and Dominique LESSELIER**[2]

[1] *Université Paris-Saclay, CentraleSupélec, SONDRA, Gif-sur-Yvette, France*

[2] *Université Paris-Saclay, CNRS, CentraleSupélec, Laboratoire des Signaux et Systèmes, Gif-sur-Yvette, France*

The present volume, *Ground Penetrating Radar: From Theoretical Endeavors to Computational Electromagnetics, Signal Processing, Antenna Design and Field Applications*, proposed within the Waves field of the Sciences series, is in the subject "Instrumentation, Sensors and Measurements", since it covers these three areas of interest. All the above elements are illustrated in this book, emphasis being on how to meet the constraints and how to investigate the ground penetrating radar (GPR) functioning accompanied by a wealth of experiments in properly controlled conditions. Yet, for the sake of completeness and in order to keep in touch with most GPR research, electromagnetic modeling and microwave imaging go far beyond, and they involve complex challenges, as illustrated by the chapter titles of this book.

The material, as proposed in the book, is comprised of eight closely related chapters. The essential aspects of each contribution are highlighted in this preface, however, we must emphasize that each chapter may not do justice to all of the aspects studied at length by the authors. Above all, the information is provided in a pedagogical way, within the framework of the key objectives of the Sciences series.

Chapter 1 is entitled "Electromagnetic Imaging of GPR Data: Theory and Experiments". Its aim is to propose a careful analysis of tomographic approaches, with a strong focus on qualitative and linearized approaches, since those are quite handy in the field of study, due to their reduced computational burden and the fact that they produce quick yet still quite sound solutions to the end-user. Then, the proposed algorithms are tested on experimental data acquired in controlled conditions, and their pros and cons are cleverly illustrated. In terms of modeling, the approach is based on a Green's formulation, scalarized according to the commonly made hypothesis of a transverse magnetic mode of polarization. If the analysis starts from a free space model, it quickly goes into the situation of a singly or multiply layered embedding as is necessary for GPR practices. The imaging strategy of choice is, as said above, qualitative. After considering the vast literature on methodologies of that kind, including recent ones by the authors based on discovering the boundaries of the targets and not sampling the whole space, the discussion is on the so-called linear sampling method (LSM), the orthogonality sampling method (OSM) and a more recently introduced method termed boundary inverse source and sparsity (B-IS). In the latter, the quest is of the boundaries of the target support, where equivalent sources are located whatever illuminations are assumed. Then, the sparsity comes into play as only a few non-zero pixels within a possibly vast domain of search are sought. The above is followed by a careful, substantive analysis of the different methodologies, and it is completed by a close look at linearized approaches, based on the Born approximation and the distorted Born approximation. The virtual experiments (VE) approach completes the previous analyses as it particularly enables a more general interpretation beyond LSM here. After a comprehensive discussion of regularization strategies, multistatic tomographic imaging is run on controlled laboratory data, concerning 3D landmines and pipes – those provided by Georgia Tech – and they show that the targets can indeed be identified with accuracy, even if the proposed solutions have been developed in a 2D setting, that is, with slices being superimposed at the end.

Chapter 2 is entitled "Nonlinear and Hybrid Inversion Techniques for Ground Penetrating Radar Imaging". It focuses on nonlinear and hybrid methods specifically developed within the GPR application framework to map the electromagnetic parameters of buried objects (in the hypothesis of frequency-independent parameters) from multifrequency–multi-view data. The approach is built within Green's framework. Once the problem is sketched properly into it, with the usual data and state integral operators, a nonlinearized deterministic method is proposed, based on an inexact Newton approach. The algorithm is discussed at length in the chapter, and argues that it is feasible to separate zones wherein an object is detected and nothing is detected. Then, attention is turned to hybrid inversion techniques, in which a qualitative delay-and-sum method is inserted in the above algorithm. In short, the qualitative method yields information on the presence or absence of discontinuities with respect to the background medium. Examples illustrate the algorithmic advances in a quasi-monostatic configuration where results of the qualitative method, the quantitative method and the hybridized method are compared with due detail, confirming the better quality of the latter.

Chapter 3 is entitled "Sensor Deployment in Subsurface GPR Imaging". The starting point encompasses a variety of approaches dedicated to imaging buried targets (including through-wall cases) to exemplify the challenges faced and the solutions to which those are amenable. Then, the reader is reminded that the so-called linearized imaging solutions lead to a significant reduction of complexity, which in the GPR situation, with very demanding conditions of exploitation on terrain, can be very useful, if run properly, and if accepted that only detection, localization and rough estimates of shape can be achieved. This is the focus of this chapter, which gives careful explanations of the pertinent direct operator in the Green formalism, the inversion of which is then discussed at length and ends up with a so-called regularized reconstruction operator containing everything, and prone to a migration scheme; its theoretical analysis is investigated in-depth and it is illustrated by a simple pedagogical example. Then, the key point is how we deploy the sensors by managing the usual trade-off of having the least data to feed the algorithm with the proper sampling (among other factors) that fits the inverse operator properties (the spectral ones in effect). A careful analysis blending mathematics and numerics follows, considering optimal frequency and space samplings, to devise a proper solution, also carefully compared to solutions in the literature. A host of numerical results follows in various configurations, aiming at depicting the best data in terms of resolution, the cases of free and half-space with sensors away or on the surface of separation being specifically managed. Overall, a detailed mathematical analysis is presented for further reference, aiming at the optimization of sensors' deployment, and a practitioner can thus, for example, imagine what they can get at best in a given situation, even if idealized for the sake of understanding by attentive linearization.

Chapter 4 is entitled "Ground Penetrating Radar Tomography for Cultural Heritage". It is about providing a non-invasive assessment of buried/hidden structures in the field of cultural heritage applications. Indeed, in archaeological surveys, underground GPR images enable us to map a given site at some degree in advance in order to properly plan out excavation activities as well as to get precious knowledge about hidden features of the sites that cannot be excavated. There are numerous examples in Italy, as well as in many other places, such as the well-preserved floors of Roman "domus", churches and a host of historical buildings. Moreover, GPR allows the performance of structural surveys to yield information useful for assessing the state of health of historical assets as well as for characterizing previous restoration actions. In this framework, a further enhancement of the GPR capabilities is by the means of tomographic data processing approaches based on proper models of electromagnetic propagation in complex scenarios, denoted as GPR tomography (GPRT) in the microwave domain. As analyzed previously, this should hopefully provide the end-user with a wealth of high-resolution images, more easily interpretable than standard radargrams. The chapter begins by reviewing the concepts underlying GPR and microwave tomography approaches with an emphasis on linear procedures and discusses effective implementations in detail. This enables the authors to get a sound basis for the practical investigations that follow. The latter are illustrated by several examples of GPRT surveys performed by the authors and colleagues at three famous archaeology sites. In the archaeological park of Paestum in Campania, rather wide areas in the park and close to it had been left unexplored by classical means. GPRT surveys with frequencies around 200 MHz ensure deep and fast surveys, which have allowed us to obtain impressive maps of potential structures, all of which have been subsequently confirmed by proper excavations. For the Knossos Palace in Crete, the main concern was to further examine the possible drastic restoration (by Evans) in the early 20th century of its walls and pillars and how it was deployed, with the need for a sharp resolution now involving the use of the much higher central frequency of 2 GHz. The third case is Paphos' Tombs of the Kings in Cyprus, for which again earlier restorations were to be verified, especially for cracks and other fractures in the columns, with the potential to lead to further damage. This meant that again, GPRT was run in the 2 GHz range like in the previous case. To conclude, the new or at least additional generation of GPR systems mounted on unmanned aerial vehicles is considered, and their potential in the field of archaeological surveys is deliberated, in particular, how high permittivity of the underground can affect the subsurface probing itself if the antennas are operated in the GHz range, but a contrario enabling a rich surface mapping, as shown with a prototype tested at the archaeological parks of Paestum and Velia (and also in Campania).

Chapter 5 is entitled "The Full-Wave Radar Equation for Wave Propagation in Multilayered Media and Its Applications". Its primary objective is to present and explain the far- and near-field formulations for full-wave modeling of radar data. Examples are given to illustrate the main features and demonstrate the potential or proven uses in the field. The far-field formulation assumes that the antenna is far enough away and that the field over the antenna aperture is homogeneous. On the other hand, the near-field formulation is a more sophisticated model that applies when the scattered field distribution depends on the medium's distance and properties. The near-field model generalizes the far-field approach using the superposition principle, decomposing the field distribution into a sum of basic distributions using a set of point sources and receivers. Both formulations are developed within the framework of a planar-stratified lossy dielectric medium, and the spectral and spatial Green's functions are explained for both. In both the far- and near-field models, the antenna characteristic functions do not depend on the medium distance or its properties. In fact, the far-field model is a particular case of the near-field model, specifically the generalized model with $n = 1$, where n represents the number of source and receiver points. Retrieving the antenna characteristic functions becomes more complex as n increases. When $n = 1$, the retrieval is analytical, but for $n > 1$, it is numerical. An analytical inverse estimation is still being sought. Remote sensing is performed by properly minimizing an objective function. The first application illustrated is soil moisture mapping using quad and drone, with many examples of interest given. The second application is gesture recognition within a given scene or through various materials, with examples provided. The authors also envision using this technology for breathing pattern recognition, which is still in its infancy regarding antenna engineering and processing. The authors conclude that the far-field methodology is well-developed, but further analysis of the near-field approach is necessary, particularly in regards to calibration issues. They propose a wealth of references for further reading on these aspects.

Chapter 6 is entitled "Assessment of Flexible Pavements by GPR: 20 Years of R&D in France". It is clearly focused on a specific application, yet this application is essential if we wish to maintain roads in good condition at not too high a price, and GPR appears as the perfect tool for this challenge (even if interest is shown in other tools, like gammadensimetry and a host of mechanical devices, as underlined also by the authors). The contribution first proposes a neat and comprehensive analysis of the state of the art of road pavement inspections, with due, but not privileged emphasis on the development of solutions in France where a strong effort has been underway these last two decades and is on a high level nowadays. The first part is devoted to GPR systems and acquisition, as existing, with the peculiar situation of rapid diagnostic being underlined. Then, the reader will go on to find processing and interpretation, proposing a wealth of results to illustrate the main points. Afterward, complements on other non-destructive testing are provided, and a due analysis of their pros and cons is carried out. Then, possibly at the core of this chapter, a wealth of novel approaches is proposed, to enable better evaluations rather than via conventional processing, which may fail unless a stringent array of hypotheses on the pavement is fulfilled. These approaches, discussed in detail, come under the name of high-resolution time-delay algorithms. Yet full-waveform inversion also appears to have high potential, as underlined by the authors, and many investigators, with the promises of artificial intelligence which are specifically mentioned. Tomographic synthetic aperture radar is also shown to be worthwhile to further investigate, as it is delivering fast and highly accurate maps. This is one of the conclusions of the chapter, among many others.

Chapter 7 is entitled "GPR for Tree Roots Reconstruction under Heterogeneous Soil Conditions". Indeed, the main idea is that we cannot analyze – save exceptions – root tree systems in the way needed for civil engineering to optimize the natural environment and biomass estimation with destructive means, unless a single experiment (e.g. excavation) is attempted for comparison. GPR then appears to be a good and possibly cheap way forward for obtaining further knowledge on those systems with the least amount of hassle. The contribution, after an elaborate discussion of root systems and their environment, develops a processing framework that is valid for complex soil conditions and a variety of root architectures, and it shows how it applies using specifically tailored antennas to map a root system, with both experiments in the laboratory and the field to illustrate the advances and to pinpoint open challenges. This chapter does not emphasize sophisticated solvers, that is, FD-TD which is the (agile powerful) workhorse chosen to obtain the simulated data that are helping the authors, and which uses rather classical processing tools, to remove the clutter, estimate the soil permittivity, and in the end produce well-focused images quickly. This pragmatic yet in-depth investigated approach should speak to many end-users in the GPR community. The key role of tree roots in nature and their climatic evolution should appeal to many readers, particularly since it was Leonardo Da Vinci who was one of the very first to have hypothesized the architecture of tree roots in a mathematical fashion.

Chapter 8, entitled "Sounding Radars and Ground Penetrating Radars Designed for the Exploration of our Solar System – Focus on Planet Mars", beautifully ends the volume, as it takes us to the fascinating environments of planets and asteroids, and investigates the development and usage of sounding radars and GPRs as essential elements of instrumental payload in almost all planetary missions. To enable the reader to grasp the present advances, first an overview of past, current and future radars developed for planetary exploration is proposed, illustrated by carefully chosen examples showing the breadth of the efforts made so far. Then, because there are evidently many constraints that impact the design and technical performances of a radar consequently, data interpretation is heavily weighed on. After a proper analysis of these issues, the author focuses on the characterization of the Martian subsurface, which has already been envisaged by radars both in orbit and on the ground. The detail of the modes of operation, studies carried out from the collected data, and lessons concerning the diagnoses of subsurface structures are appraised in succession. It is particularly seen that, among other conclusions, a qualitative understanding of the subsurface configuration is rather easily obtained from processed GPR images, yet quantitative analyzes are needed to support the attempted identification of subsurface materials and in-depth understandings of the environment. However, the wide variety of sounders operating within different frequency ranges does provide lots of complementary results and yields significant inputs to tough scientific questions, while knowledge of buried geological structures is indeed essential to understanding the mechanisms of formation and the evolution of the sites studied. Then, the development of two radars to probe the Martian subsurface is discussed: Electromagnetic Investigation of the SubSurface (EISS) and

Water Ice Subsurface Deposit Observation on Mars (WISDOM): designed for the purpose of the ExoMars mission. As of today, the first has been “descoped”, so insistence is on the second, as programmed later. However, the work led around them is useful in any case as underlined. In the context of Martian exploration, the WISDOM GPR frequency step 500 MHz–3 GHz radar has been developed to meet the needs for characterizing the subsoil of Mars, as part of the ExoMars Rover Payload Instrument Suite, this again taking place with severe constraints linked to the spatial context (very limited mass, volume and consumption). Thanks to the great results obtained by soundings radars (on Mars, the Moon and Titan) and GPRs (on Mars and the Moon), it is then concluded in the last section about a series of radars that are currently being developed for the future missions that will characterize the subsurface of other bodies of our solar system.

To conclude this preface, we will present a host of questions that still arise today and are certainly not fully answered within the present volume. Yet, in our opinion, as conveners, this book encompasses many tough issues already raised, and this is what makes the analysis a good read both on an educational level as well as on a research level, for theoreticians, numericians, as well as engineers, and practitioners. We can mention: (i) multi-sensors and true three-dimensional modeling, the consistency of Green’s scalar and dyadic formulations of the electromagnetic fields (those may be the basis of solution methods, the mathematical analysis of which can then go quite in-depth and provide further understanding of the imaging itself, since, for example, FD-TD is very powerful, but may seem less amenable to use whenever inverting the signals themselves) and/or related quantities (the well-known S parameters as the most relevant ones); (ii) operating environments since combining dispersive, heterogeneous and in effect heavily uncertain cases; (iii) optimized sparse acquisitions in order to save the precious time of the handler on terrain; (iv) (almost) real-time signal processing and image production with a view to end-user decision-making; and (v), it goes without saying that the fact that any signal is produced/collected by real antenna structures that are also claiming for careful optimizations within the context of the applications envisaged (a set of those applications is discussed in the several chapters as proposed, yet they are certainly not unique), and all of these open questions are recognized as not exhaustive. In addition, deep learning in the present context could have a full volume devoted to it, and is hinted to at least by several contributors, since GPR is a clear case of epistemic and aleatoric uncertainty as is now discussed in the deep learning community under various guises. As for the need to smartly mix possibly sparse and certainly erroneous experimental data with, in principle, perfectly simulated data yet completed in a proper way by cleverly built errors (not reduced to additive white Gaussian noises as too often in truth), while building up the needed networks, it remains, to say the least, a demanding issue.

Finally, as coordinators of the present volume, we would like to say that it would not be without the sophisticated and timely contributions of the authors, nor without the impetus given to us by Dr. Frédérique de Fornel, who is cleverly managing the Waves field of the Sciences series. It is true that a number of other contributors in the field of GPR could have provided their research in this book, but we are certainly not pretending to be exhaustive, mainly in the hope that this volume not only focuses on the right tools in the demanding field of GPR at the highest level of research, but that it also attracts students and young scientists to this field, as it claims many new developments in harmony with concrete and indeed the increasingly urgent demands of society in a wide range of fields.

June 2023

1

Electromagnetic Imaging of GPR Data: Theory and Experiments

Michele AMBROSANIO[1,3], Martina Teresa BEVACQUA[2,3], Tommaso ISERNIA[2,3] and Vito PASCAZIO[1,3]

[1] *Dipartimento di Ingegneria, Università degli Studi di Napoli "Parthenope", Naples, Italy*

[2] *Dipartimento di Ingegneria dell'Informazione delle Infrastrutture e dell'Energia Sostenibile, Università Mediterranea di Reggio Calabria, Italy*

[3] *Consorzio Nazionale Interuniversitario per le Telecomunicazioni (CNIT), Parma, Italy*

Ground penetrating radar (GPR) represents a powerful technology able to investigate non-accessible scenarios in a non-invasive and non-destructive way, as witnessed by the numerous and different applications, which include demining, lunar explorations, archaeology, geology and civil engineering (Persico 2014; Sehrir et al. 2017; Aboudourib et al. 2020). However, standard GPRs provide a user-dependent map of the underground and subsurface targets, known as "radargram", whose interpretation requires firm human expertise and does not allow us to infer about the nature of the targets buried below the soil, unless some a priori information is available.

In this respect, tomographic approaches can play a very relevant role (Ambrosanio et al. 2019). These techniques aim at providing images regarding the variations of dielectric permittivity and electrical conductivity profiles of the scene

---

For a color version of all the figures in this chapter, see www.iste.co.uk/serhir/radar.zip.

Ground Penetrating Radar,
coordinated by Mohammed SERHIR and Dominique LESSELIER.

under investigation, which are more reliable and readable than those achieved by using standard GPR data processing. Image formation requires an accurate modeling of the scattering phenomena to properly take into account the complex interactions among sources, receivers, ground and buried objects. As a result, the development of robust and reliable recovery algorithms still has some issues related to both modeling errors as well as the presence of uncertainties in the data. Indeed, the underlying inverse scattering problem (ISP) is both nonlinear and ill-posed (Bertero and Boccacci 1998; Colton and Kress 1998).

Nowadays, most of the tomographic approaches present in the literature can be classified into three main categories (Ambrosanio 2019). The first class includes qualitative methods, which aim at detecting the scatterers hosted in the imaging domain (ID) and retrieving only a limited amount of information, such as scatter location and shape. The second ones, known as approximated methods, adopt some of the approximations of the scattering phenomena, which allow us to simplify the mathematical model and to reduce the computational complexity. Lastly, the nonlinear quantitative methods aim at retrieving both electrical and morphological properties of the ID, by tackling the problem in its full nonlinearity, without involving any approximation. A further class of imaging strategies includes hybrid approaches, which are usually based on a first qualitative step that aims at providing some rough information, that is then exploited in the next quantitative steps to avoid false solutions and to allow for reliable and effective recoveries.

In this respect, this chapter aims at providing to the research community an overview of all the main aspects on tomographic approaches in the framework of the underground exploration of the soil by GPR technologies, with particular attention given to qualitative and linearized approaches. Indeed, they can be relevant in GPR prospecting due to their simplicity and reduced computational burden.

The chapter is organized as follows. In sections 1.1 and 1.2, the equations modeling scattering problem underlying tomographic approaches is reported and extended to the case of planarly layered media. Section 1.3 introduces the basics of qualitative methods, in particular the linear sampling method (LSM), the orthogonality sampling method (OSM) and the boundary retrieval through inverse source and sparsity (B-IS) for the half-space case. In section 1.4, linearized methods are discussed, while section 1.5 is focused on regularization techniques. Finally, in section 1.6 experimental imaging reconstructions are discussed and reported from (Ambrosanio 2021).

1.1. Inverse scattering problem: mathematical formulation

In ISPs, an object-of-interest (OI) is located inside an ID Ω, which is then contained in the problem domain. The problem domain is immersed in a background medium $(\varepsilon'_b, \sigma_b, \mu_0)$ whose electrical properties are known and usually assumed to be homogeneous. It is worth noting that, for the aim of this chapter, only non-magnetic targets and background will be considered. The ID is surrounded by a boundary Γ of general shape and size depending on the employed imaging setup on which the measuring probes are located. In all of the following applications, a multi-view–multistatic system will be employed, that is, an imaging system made up of more transmitters and receivers in which the samples of scattered fields are collected at receiver locations as per each transmitting antenna.

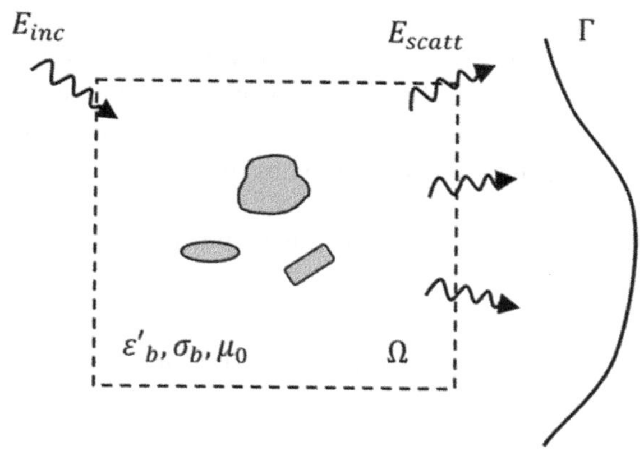

Figure 1.1. *Problem statement. The picture illustrates the incident electric field E_{inc} and the scattered electric field E_{scatt} collected at receivers locations on the curve Γ. The background is a non-magnetic medium $(\varepsilon'_b, \sigma_b, \mu_0)$ and the targets are in located the imaging domain Ω*

The whole scenario has been sketched in Figure 1.1. In general, it is possible to define the complex permittivity as:

$$\varepsilon(\vec{r}\,) = \varepsilon'(\vec{r}\,) - j\,\frac{\sigma(\vec{r}\,)}{2\pi f \varepsilon_0}, \qquad [1.1]$$

in which ε' and σ are the relative permittivity and conductivity, respectively, f is the operating frequency, $\vec{r}$ is the position vector identifying each point inside the ID, j refers to the imaginary unit and $\varepsilon_0 = 8.854 \times 10^{-12}$ F/m is the permittivity value in the void and represents a universal constant. Thus, by adopting the definition in equation [1.1], it is possible to introduce the contrast function "$\chi(\vec{r}\,)$":

$$\chi(\vec{r}\,) = \frac{\varepsilon_x(\vec{r}\,)-\varepsilon_b}{\varepsilon_b}, \qquad [1.2]$$

in which $\varepsilon_b = \varepsilon'_b - j\frac{\sigma_b}{2\pi f\varepsilon_0}$.

In this framework, two different electromagnetic scattering problems can be identified: the first problem, also known as *forward problem*, consists of evaluating the scattered field due to the object of interest inside the ID, assuming as known quantities both the incident field "$\vec{E}_{inc}(\vec{r}\,)$" and the contrast function "$\chi(\vec{r}\,)$" that encodes the targets properties; the second problem, also referred to as *inverse problem*, aims at estimating the unknown contrast function "$\chi(\vec{r}\,)$" from the samples of the scattered field "$\vec{E}_{sct}(\vec{r}\,)$" collected on Γ and from the illuminations "$\vec{E}_{inc}(\vec{r}\,)$". The two problems are strictly related to each other; as a matter of fact, a whole class of inversion procedure exploits the forward problem as a fundamental step for obtaining an estimate of the contrast function by means of an iterative strategy.

In order to treat these scattering problems properly, it is convenient to introduce the two equations:

$$\begin{aligned}\vec{E}^{(v)}_{tot}(\vec{r},\omega) &= \vec{E}^{(v)}_{inc}(\vec{r},\omega) + \int_\Omega \bar{\bar{G}}_i(\vec{r},\vec{r}\,',\omega)\chi(\vec{r}\,',\omega)\vec{E}^{(v)}_{tot}(\vec{r}\,',\omega)d\vec{r}\,, \\ &= \vec{E}^{(v)}_{inc}(\vec{r},\omega) + \mathcal{A}_i[\chi\vec{E}^{(v)}_{tot}],\quad \vec{r}\in\Omega,\quad v=1,\ldots,N_v,\end{aligned} \qquad [1.3]$$

$$\begin{aligned}\vec{E}^{(v)}_{scat}(\vec{r},\omega) &= \int_\Omega \bar{\bar{G}}_e(\vec{r},\vec{r}\,',\omega)\chi(\vec{r}\,',\omega)\vec{E}^{(v)}_{tot}(\vec{r}\,',\omega)d\vec{r}\,' = \\ &= \mathcal{A}_e[\chi\vec{E}^{(v)}_{tot}](\vec{r},\omega),\quad \vec{r}\in\Gamma,\quad v=1,\ldots,N_v,\end{aligned} \qquad [1.4]$$

in which $\omega = 2\pi f$, where "(v)" represents the "V" illuminations of the multi-view–multistatic microwave system whose incident field impinges on the scatterers located in the ID, "$\vec{E}^{(v)}_{inc}$, $\vec{E}^{(v)}_{tot}$, $\vec{E}^{(v)}_{scatt}$" refer to the incident, total and scattered field at the vth illumination, respectively, while "$\bar{\bar{G}}_i$" and "$\bar{\bar{G}}_e$" are the so-called internal and external dyadic Green functions of electric type that describe the relation between the contrast function and the field generated both inside the ID Ω and the scattered field measured along the boundary Γ. These functions define the two operators introduced previously, also known as internal and external integral radiating operators "$\mathcal{A}_i$" and "$\mathcal{A}_e$" or domain and data operators, respectively. It is easy to note that such operators are bilinear related to the pair $(\chi, \vec{E}_{tot})$. These bilinear operators are defined as $\mathcal{A}_i: X \times T \to S_i$ and $\mathcal{A}_e: X \times T \to S_e$, with $X \subset L^\infty(\Omega)$ being the subspace of the possible contrast functions, $T \subset L^2(\Omega)$ being a proper subspace for the total electric field inside the object, and $S_i \subset L^2(\Omega)$ and $S_e \subset L^2(\Gamma)$ being two subspaces for the scattered field inside and outside the ID.

Rewriting equation [1.3] by exploiting the operator notation as:

$$\vec{E}_{tot}(\vec{r},\omega) = \left(I - A_i^{\chi}\right)^{-1}\left[\vec{E}_{inc}(\vec{r},\omega)\right], \tag{1.5}$$

and substituting it in the data equation [1.4], we obtain:

$$\vec{E}_{sct}(\vec{r},\omega) = A_e\left[\chi(\vec{r},\omega)\left(I - A_i^{\chi}\right)^{-1}\left[\vec{E}_{inc}(\vec{r},\omega)\right]\right], \tag{1.6}$$

where "I" is the identity operator and the superscript "-1" denotes the inverse operator. Note that although the data and domain operators themselves are linear, the relation between the scattered field and the contrast function is nonlinear. The inversion of the operator in [1.6] is a mathematical representation of the nonlinearity associated with the ISP. Such an issue is handled by casting the ISP as an optimization algorithm over variables representing the unknown electromagnetic properties of the objects to be retrieved, and it occurs because of the information related to the multi-view–multistatic system.

It is worth observing that equations [1.3.] and [1.4] can be rewritten in a different form by introducing the quantity:

$$\vec{W}(\vec{r},\omega) = \chi(\vec{r},\omega) \cdot \vec{E}_{tot}(\vec{r},\omega), \tag{1.7}$$

which is known as *contrast source*. In the following, and for the scope of this chapter, the case of transverse magnetic (TM) mode will be considered. Under this assumption, the vectorial equations [1.3] and [1.4] become scalar, since only one component of the electric field is expected.

1.2. Inverse scattering problem in case of planarly-layered media

In section 1.1, a brief introduction on the mathematical statement for the ISP in a homogeneous, unbounded background medium was considered. In this section, a more suitable formulation of the concepts described in the theory of the planarly-stratified media will be dealt with.

Among all the quantities introduced previously, the main actors are still the "incident field" in the ID, the "scattered field" collected at receiver locations and the internal and external Green's functions.

For the aims proposed in this chapter, the derivation of the two-dimensional (2D) planarly-three-layered medium (known as "slab") will be considered, whose case also includes the planarly-two-layered one. These formulations belong to the framework of the simplified 2D intra-wall imaging (IWI) and GPR applications, respectively. In the following, the ID will be assumed as buried in medium 2.

To evaluate the electric field generated by a filamentary source located in a simplified 2D planarly-layered medium, the case of a single plane-wave in medium 1 impinging on the planar interface of a slab will be considered. Finally, by exploiting the Sommerfeld's (1964) identity and integrating on all the spatial frequencies, a closed-form for the incident field generated by a filamentary source in medium 1 and observed in medium 2 will be obtained.

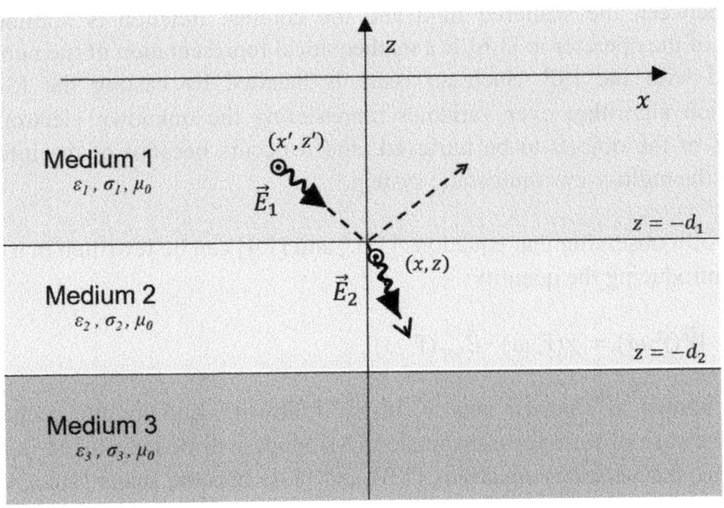

Figure 1.2. *Sketch of a planarly-three-layered medium*

For a generic 2D TM wave, the scalar electric field for a three-layered medium with source located in (x',z') and field observed in (x,z) can be written as:

$$E_{2z}(k_x,x',z,z',\omega) = A_2 e^{jk_{1z}z'} e^{jk_x x'} \left[e^{-jk_{2z}z} + R_{23}^{TM} \cdot e^{jk_{2z}(2d_2+z)}\right], \qquad [1.8]$$

in which $k_{iz} = \sqrt{\omega^2\mu_0\epsilon_i - k_x^2}$, with $i,j = \{1,2\}$ and $\Im[k_{iz}] < 0$, d_2 being the depth of the related interface, and A_2 is defined as:

$$A_2 = \frac{A_1 \cdot T_{12}^{TM} \cdot e^{j(k_{1z}-k_{2z})d_1}}{1 - R_{21}^{TM}R_{23}^{TM}e^{jk_{2z}2(d_2-d_1)}}, \quad [1.9]$$

in which T_{ij}^{TM} and R_{ij}^{TM}, $i,j = \{1,2,3\}$, refers to the transmission and reflection coefficients at interface between media i and j for the TM mode, respectively, A_1 is the amplitude of the incident field and d_1 is the depth of the first interface (between mediums 1 and 2).

By choosing the reference coordinate system as making the x axis coinciding with the interface between mediums 1 and 2 (i.e. $z = -d_1 = 0$) and assuming $d = d_2 - d_1$ and unitary amplitude for the incident field in medium 1 ($A_1 = 1$), then equation [1.8] becomes:

$$E_{2z}(k_x, x', z, z', \omega) = T_{12}^{TM}\frac{1}{1 - R_{21}^{TM}R_{23}^{TM}e^{jk_{2z}2d}} \cdot \left[e^{j(k_{1z}z' - k_{2z}z)} + R_{23}^{TM}e^{jk_{1z}z'}e^{jk_{2z}(2d+z)}e^{jk_xx'}\right], \quad [1.10]$$

where $T_{12}^{TM} = \frac{2}{k_{2z}+k_{1z}}$ and $R_{ij}^{TM} = \frac{k_{iz}-k_{jz}}{k_{iz}+k_{jz}}$, $i,j = \{1,2,3\}$.

Finally, by exploiting the inverse Fourier transform for equation [1.10], we get:

$$E_{2z}(x, x', z, z', \omega) = \int_{-\infty}^{+\infty} T_{12}^{TM}\frac{1}{1 - R_{21}^{TM}R_{23}^{TM}e^{jk_{2z}2d}} \cdot \left[e^{j(k_{1z}z' - k_{2z}z)} + R_{23}^{TM}e^{jk_{1z}z'}e^{jk_{2z}(2d+z)}\right] \cdot e^{jk_x(x-x')}dk_x . \quad [1.11]$$

Thanks to the reciprocity, it is quite easy to evaluate the external Green function by replacing the observation point (x, z) with the source location and the probe location (x', z') with the receivers (Catapano et al. 2004). Finally, the internal Green's function can be written as (Chew 1995):

$$g_{int}(k_x, x', z, z', \omega) = M_2e^{-jk_{2z}|z-z'|} + M_2R_{21}e^{-jk_{2z}(z+z')} + +M_2R_{23}e^{jk_{2z}(z+z'+2d)} + M_2R_{21}R_{23}e^{-jk_{2z}|z-z'|}e^{jk_{2z}2d}, \quad [1.12]$$

which, by adding and subtracting the quantity $M_2R_{21}R_{23}e^{jk_{2z}2d}e^{-jk_{2z}|z-z'|}$, making some calculations and exploiting the inverse Fourier Transform along the k_x direction results in:

$$g_{int}(x,x',z,z',\omega) = -\frac{j}{4}H_0^{(2)}\left(k_{2z}\sqrt{\Delta x^2+\Delta z_-^2}\right)$$

$$-\frac{j}{2\pi}\int_{-\infty}^{+\infty}\frac{R_{23}M_2R_{21}}{k_{2z}}e^{jk_{2z}2d}\cos k_{2z}\Delta z_- e^{jk_x\Delta x}dk_x + \qquad [1.13]$$

$$-\frac{j}{4\pi}\int_{-\infty}^{+\infty}\frac{M_2}{k_{2z}}\left[R_{21}e^{-jk_{2z}(\Delta z_+)} + R_{23}e^{jk_{2z}(\Delta z_+ + jk_{2z}2d)}\right]e^{jk_x\Delta x}dk_x,$$

where $\Delta x = x - x'$, $\Delta z_- = z - z'$ and $\Delta z_+ = z + z'$ (Crocco and Soldovieri 2003). It is easy to derive the previous functions for the case of a two-layered medium (GPR) since it is sufficient to impose $R_{23} = 0$.

In particular, for the three-dimensional (3D) half-space case, it is possible to evaluate analytically the expression of the Green's functions by exploiting similar calculations, which drive into the formulas proposed in Lo Monte et al. (2010) and Lo Monte et al. (2012) that, for the sake of brevity, will not be dealt with in this chapter.

1.3. Solution strategies

The electromagnetic ISP defined by the previous equations is intrinsically nonlinear in the relationship between data and unknowns. However, features that are quite attractive for any inversion approach are the requirement of being fast, computationally effective and reliable from the point of view of reconstructions, trying to exploit as little a priori information as possible. Moreover, such approaches should also avoid the occurrence of false solutions, also known as "local minima" (Isernia et al. 2001).

Nowadays, all the approaches present in the literature for the solution of EIS problems can be classified into three main categories, which are as follows:

– *qualitative methods*, where only a limited amount of information is provided, such as scatterer location and shape;

– *linearized methods*, which include famous approximations such as *physical optics* and *Born approximation* (BA), or even high-order approximate models;

– *nonlinear optimization methods*, in which an approximate solution is obtained by means of an iterative process starting from an initial guess. This class will be not dealt with.

The following sections will be focused on the case of qualitative methods and linear approximations, considering some of the most adopted approaches in practical GPR scenarios.

1.4. Qualitative methods

To overcome the difficulties related to nonlinearity and ill-posedness of the inverse problem underlying GPR imaging as well as the limited number of measurements, many studies have been focused on the introduction of solution methods for the corresponding inverse obstacle problem, whose scope is to recover only the morphological information about the targets, namely their supports, giving up their electromagnetic properties (Cakoni and Colton 2006). Such methods are known as *qualitative* methods and are different from *quantitative* methods, which aim at retrieving both electrical and morphological properties of the region of interest (Cakoni and Colton 2006). Indeed, the latter tackle the problem in its full nonlinearity, without involving any approximation and allow to widen the class of retrievable objects considerably. Unfortunately, they are characterized by a higher computational burden and long processing time. On the other hand, qualitative methods are characterized by simplified mathematical models and lower computational burden. Part of the discussion about qualitative methods reported below can be found detailed in the recent paper from the same authors of the chapter (Bevacqua 2019).

Among qualitative methods, it is worth mentioning sampling methods that include the LSM (Colton 2003) (and the related factorization method; Kirsch 2008) as well as the OSM (Potthast 2010). The main idea is that of sampling the investigation domain in an arbitrary grid of points and computing in each point an indicator function whose energy will assume different values depending on whether the sampled point belongs or not to the scatter (see, for instance, Figure 1.3). Although both LSM and OSM belong to the class of sampling methods, they exhibit different features. For instance, the LSM indicator function is computed by solving an auxiliary linear problem, whose solution involves the adoption of a regularization technique. In a different fashion, the OSM does not require the solution of a linear problem, as the indicator function is just related to the evaluation of a scalar product. Moreover, the OSM seems to be more flexible with respect to the measurement configurations.

By taking advantage from recent results in area of compressive sensing and sparsity promoting techniques (Donoho 2006), a new qualitative method has been very recently introduced in literature which, provided some conditions hold true, is able to retrieve the boundary of unknown targets rather than their support (Bevacqua 2017, 2018). Such a method exploits the equivalence theorem in case of dielectric obstacles and takes advantage from the particular distribution assumed by the induced currents in case of perfect electric conductors (Bevacqua 2017). Differently from sampling methods, it does not sample the investigation domain in a grid of points, and it does not require the selection of a fixed threshold to discriminate between points inside and outside the targets. It just retrieves the boundary of the targets (see, for instance, Figure 1.3). However, if compared with LSM and OSM, it shows a heavier computational burden.

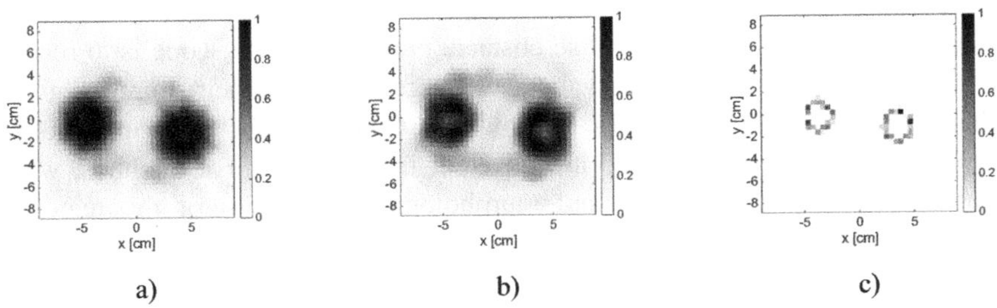

Figure 1.3. *Normalized indicators against experimental data: the TwinDielTM target. 8GHz and number of measurements M = 36. (a) LSM, (b) OSM and (c) B-IS. Extracted from Bevacqua (2019)*

1.4.1. *Linear Sampling Method*

The LSM is one of the most popular qualitative methods to retrieve objects' support from the measurements of the corresponding scattered field data (Colton 2003). In detail, it consists of sampling the region under test into an arbitrary grid of points and solving for each point the so-called *far-field integral* equation (FFE) given by:

$$\int_{\Gamma} E_{sct}{}^{\infty}(\hat{r}_m, \hat{r}_t)\ \xi(\vec{r}_s, \hat{r}_t)\ d\hat{r}_t = G_e{}^{\infty}(\hat{r}_m, \vec{r}), \qquad [1.14]$$

wherein ξ represents the unknown function, $\vec{r}_s$ is the sampling point and $E_{sct}{}^{\infty}$ is the scattered far-field pattern measured on a closed circle curve Γ in the far zone of the scatterers in direction $\hat{r}_m$ when a unit amplitude plane wave impinges from direction

$\hat{r}_t$. $G_e{}^{\infty}$ is the Green's function in the far-field zone. Despite the linearity of [1.14], the evaluation of the solution ξ is not straightforward due to the compactness of the operator at the left-hand side, which implies ill-posedness in its inversion (Bertero 1998). Hence, a regularization technique is required. An effective choice is that of solving it by adopting the Tikhonov regularization (Catapano et al. 2007). Then, the estimation of the unknown support is pursued by evaluating the l_2 norm (i.e. the "energy") of the unknown function ξ with respect to $\hat{r}_t$ given by:

$$I_{LSM}(\vec{r}_s) = \|\xi(\vec{r}_s, \hat{r}_t)\|_2^2 = \int_\Gamma |\xi(\vec{r}_s, \hat{r}_t)|^2 d\hat{r}_t. \quad [1.15]$$

The above defined indicator depends on the sampling point $\vec{r}_s$ and assumes a finite value when the sampling point belongs to the unknown object, while it blows up elsewhere (Catapano et al. 2007). As such, by selecting a fixed threshold, the indicator [1.15] allows the discrimination between points inside and outside the scatterers and finally retrieve the support of the target (Catapano et al. 2007). It is worth underlining that the computational burden is low, as the solution of [1.14] only involves the evaluation of the SVD of the measured data. The method has been also extended to the case of subsurface imaging (Catapano et al. 2008).

Note that, in case of a lossy background, there is the need of a modified indicator due to the attenuation of the field radiated by a buried electric dipole according to a factor which depends on the amount of losses in the host medium. Then, the energy of the LSM solution is normalized to the energy of the known term of the FFE. As a matter of fact, the field to be matched in the LSM equation becomes progressively less intense when the distance between the sampling point and the measurement domain grows. Interestingly, the presence of losses is not the only case, wherein the modified far-field equation can play a useful role. For example, for a given measurement domain and a given domain under test, in case the directions of the test dipole (whose polarization is chosen equal to the transmitting one) and the measurement dipoles do not match, the different amount of polarization mismatches, which is experienced for different depths implying very different energies for the right-hand term (Catapano 2008).

1.4.2. *Orthogonality sampling method*

The OSM has been recently introduced in literature in case of free space (Potthast 2010). It consists of computing the *reduced scattered field* from the far-field pattern $E_{sct}{}^{\infty}(\hat{r}_m, \hat{r}_t)$. Such a reduced field can be computed by evaluating the scalar product between the far-field and the Green's function in far-field zone, that is (Potthast 2010):

$$E_{sct}{}^{red}(\vec{r}_s, \hat{r}_t) = \int_{\Gamma} E_{sct}{}^{\infty}(\hat{r}_m, \hat{r}_t)\, e^{jk_b \vec{r}_s \cdot \hat{r}_m}\, d\hat{r}_m = \frac{1}{\gamma} \langle E_s{}^{\infty}, G_e{}^{\infty} \rangle_{\Gamma} \qquad [1.16]$$

where $\langle \cdot,\cdot \rangle$ denotes the scalar product and γ is a constant for a fixed frequency (Bevacqua 2020). In a nutshell, equation [1.16] tests the *orthogonality* relation between the far-field pattern and the Green function in far zone (apart from a constant). Then, the OSM indicator $I_{OSM}(\vec{r}_s)$ is defined as:

$$I_{OSM}(\vec{r}_s) = \left\| E_{sct}{}^{red}(\vec{r}_s, \hat{r}_t) \right\|_2^2 = \int_{\Gamma} \left| E_{sct}{}^{red}(\vec{r}_s, \hat{r}_t) \right|^2 d\,\hat{r}_t \qquad [1.17]$$

and achieves large values when the far-field pattern approaches the one of the background Green's function for the considered sampling point (Bevacqua 2020). Again, by selecting a fixed threshold, it is possible to discriminate sampling points belonging to the support (Bevacqua 2020).

The method has been also extended to the case of near-field (Akinci 2018b) subsurface imaging (Akinci 2018a). However, further investigations are required to better understand its physical interpretation in case of a non-homogeneous scenario.

1.4.3. *Shape reconstruction via joint sparsity based inverse source and equivalence principles*

To deal with inverse obstacle problems, we can also take advantage from the joint solution of a number of related *inverse source* problems, which consist of recovering the induced currents from the knowledge of the measured far-field pattern. The strategy is obviously suggested by the fact that the support of the contrast sources W induced in the unknown object is exactly equal to support of the target whatever E_{inc}. Note that inverse source problems are linear (see equation [1.4]) but are still (severely) ill-posed, so that some additional care is needed.

In this respect, note that when a generic electromagnetic field is propagating in the space in presence of dielectric obstacles, it induces inside the obstacles some currents, which are expected to be different from zero inside the target support and become in turn sources of the scattered field data. However, by virtue of the equivalence theorem (Franceschetti 2013) (which is a key point of the approach), the measured scattered field can be eventually conceived as the one radiated by some equivalent surface currents, that is, electric W_s and magnetic W_{smx} and W_{smy} surface currents, which can play the role of the original ones.

The main advantage of considering the equivalent currents rather than the original ones is that they are distributed over the boundary of target support. As a result, they can be identified by only few non-zero pixels in the investigation domain and, hence, can be considered sparse with respect to the pixel basis representation. Then, a possible idea for their retrieval is that of looking for the sparsest currents distributions that contemporarily satisfy the measured data. To this end, sparsity promoting approaches can be exploited (Donoho 2006).

Interestingly, the property of sparsity displayed by the equivalent currents holds true, regardless of the direction of illumination $\hat{r}_t$ or the kind of measurements configuration. To enforce this joint sparsity among the different scattering experiments, an auxiliary variable I_{BIS} can be defined as the common upper bound on the amplitude of the equivalent currents for the different scattering illumination conditions. Accordingly, the problem is recast as (Bevacqua 2017, 2018):

$$\min \|I_{BIS}(\vec{r})\|_1 \qquad [1.18a]$$

$$s.t. \quad \left\|E_s^{\infty} - \mathcal{A}_e[W_s] - \mathcal{A}_{mx}[W_{smx}] - \mathcal{A}_{my}\left[W_{smy}\right]\right\|_2 \leq \delta, \qquad [1.18b]$$

$$|W_s(\vec{r}, \hat{r}_t)| \leq I_{BIS}(\vec{r}),$$

$$|W_{smx}(\vec{r}, \hat{r}_t)| \leq I_{BIS}(\vec{r}), \quad \forall \hat{r}_t \in \Gamma \qquad [1.18c]$$

$$\left|W_{smy}(\vec{r}, \hat{r}_t)\right| \leq I_{BIS}(\vec{r}),$$

where $\|\cdot\|_1$ is the l_1 norm, and $\mathcal{A}_{mx}$ and $\mathcal{A}_{my}$ are the integral radiation operators which relate the magnetic surface currents to the scattered electric fields.

The variable I_{BIS} does not depend on the direction of illumination $\hat{r}_t$, but only on the position $\vec{r}$ in the investigation domain. Moreover, it is expected to assume non-zero values only on the boundary of target support, so that it will act as a "boundary" indicator. For this reason, the method is called B-IS. As such, differently from LSM and OSM, there is no need for a thresholding to retrieve the target support from the indicator map, as the approach directly retrieves the boundary of the targets.

Note that the approach can be also applied to the case of perfect electrical conductors. In fact, in such a case, the induced currents W exist only on the boundary of Σ and there is no need to introduce equivalent surface currents. Then, the problem [1.7] is further simplified (Bevacqua 2017, 2018).

The B-IS method belongs to the class of the joint sparsity promoting procedures. Note, the joint sparsity used herein is not enforced by means of the mixed norm $l_1 - l_2$ as in Kowalski (2009) and Fornasier (2008), but through the auxiliary variable I_{BIS}.

1.4.4. *Applicability and limitations of the three methods*

In LSM, the far-field equation [1.14] can be interpreted as an attempt to synthetize induced currents such that their radiating part is focused and/or circularly symmetric with respect to the given sampling point (Catapano 2007; Di Donato 2011). As such, it correctly works as long as the object is convex, and the nonradiating component of the induced currents is negligible (Catapano 2007). Usually, when the product of the maximum amplitude of the contrast χ times the maximum size of the target is large, which corresponds to an increasing amount of nonradiating sources, it does not guarantee reliable reconstructions.

The OSM is based on a test of orthogonality that allows us to compute the reduced scattered field. As far as its physical interpretation is concerned, it has been recently argued in Bevacqua (2020) that the reduced scattered field is related to the radiative part of the currents. As such, similarly to LSM, the OSM indicator does not take into account the nonradiating part of the currents, so that, again, limitations are expected with increasing electrical dimensions of the scatterer.

While both the LSM and OSM act on volumetric currents that are located on all points belonging to the support of the target, B-IS aims at jointly solving several inverse source problems, wherein one is looking for currents which (under some reasonable hypotheses) are located on the boundary of target support. Notably, the optimization problem [1.18] does not aim at retrieving exactly the equivalent currents but rather to find a common upper bound to their amplitudes. Also note the possible (superficial) nonradiating currents do not seem to play any role in the process, while the possible nonradiating volumetric currents are filtered out by the sparsity regularization. The B-IS method can be understood as more robust with respect to the presence of nonradiating currents than LSM and OSM. Moreover, the upper bound in [1.18] can be understood as a trade-off between sparsity promoting and data fitting. As a result, B-IS will be able to retrieve at least the convex-hull of the targets. Remarkably, probably because of the use of the l_1 norm rather than the l_0 pseudo-norm, we can also retrieve the shape of some non-convex target (Bevacqua 2017). As a result, B-IS is expected to fail when the boundary of the target is not sparse (for instance in case of a ragged boundary) or when, according to the sparsity theory (Donoho 2006), the number of measurements is not sufficiently large with respect to the degree of sparsity of the target contour.

Concerning the requirements about the kind of measured data, the LSM needs to correctly discretize the integral operator in [1.14] by collecting a sufficient number of data in multi-view–multistatic measurement configuration. As a result, even if the product of the maximum amplitude of the contrast χ times the maximum size of the target is not large; the undersampling of the measured data can compromise the reconstructions. On the other side, the OSM requires an accurate discretization of the integral [1.16] over the measurements curve and, hence, a sufficient number of measurements. However, differently from LSM, OSM can compensate the low amount of data by considering other kind of experiments different from multi-view–multistatic ones. In fact, OSM as well as B-IS exhibit wide flexibility with respect to processed data and can offer the possibility of combining information arising from frequency diversity without the need of a posteriori merging the single results. To this end, the OSM indicator can be easily modified by integrating over a given frequency band (Potthast 2010; Bevacqua 2020), and the B-IS indicator function implicitly defined by [1.18c] can be redefined as the upper bound to the different sources at the different frequencies (Bevacqua 2017).

Another important difference among the methods regards the computational burden. The OSM indicator does not involve the solution of an ill-posed problem but only the evaluation of a scalar product. This implies more robustness with respect to measurement errors on the scattered data and a very low computation burden. On the contrary, the LSM indicator function is computed by solving an auxiliary linear problem, whose solution involves the adoption of a regularization technique. However, the main computational effort of the method is the evaluation of the SVD of measured data, whose dimension is only dictated by the number of scattering experiments. This task must be carried out only once, since the kernel of the linear system is the same for all sampling points. With respect to the LSM and the OSM, B-IS exhibits a heavier computational burden, which drastically increases with the number of unknowns, since an l_1 norm optimization problem has to be solved. On the other side, the problem underlying the B-IS belongs to the class of convex programming problems, whose single optimal solution can be reached by any off-the-shelf local optimization procedure.

In terms of reconstruction accuracy, both the LSM and OSM require the selection of a fixed threshold to discriminate between points inside and outside the targets. In a different fashion, the B-IS does not provide an indicator map monotonically increasing outside the actual support, but directly looks for the boundary of the target. As such, it is expected to be more accurate.

1.5. Linearized methods

As a countermeasure to nonlinearity, linearized methods allow a very easy implementation and require a limited amount of computer memory and computational time, even though they suffer from several limitations induced by the adopted approximated model. These limitations can be more pronounced in the case of GPR surveys where data are collected only in reflection mode, especially in the case of single-frequency data processing.

The most popular approximation of the tomographic model [1.5] is the BA (Lo Monte et al. 2010; Ambrosanio et al. 2015). In BA, the tomographic problem is linearized as the auxiliary unknown total field inside the object is assumed equal to the known incident field. This hypothesis is satisfied in case of weak scatterers, that is, for objects whose internal characteristics are very close to the ones of the hosting medium, and/or for objects whose dimension is very small in terms of the wavelength in the hosting medium. When such assumptions do not hold true, BA can be properly exploited to perform target location. This can be also explained by considering that, in case of far-field measurements and incident plane wave, the BA operator is very closely related to the one of Kirchhoff approximation (KA) (Soldovieri 2008). The KA, or physical optics approximation, can be adopted in case of electrically large perfect electric conductors. Under such a linear approximation, the solution algorithm is formulated as the inversion of a linear integral operator that relates the scattered data to a distribution unknown function that is directly related to the illuminated boundary of the object. Therefore, the inversion allows the retrieval of target support boundary of the object.

Another relevant linearized approach has been introduced in the last few years within the virtual scattering experiments (VE) framework (Bevacqua et al. 2016). The approach is based on a proper linear combination of the original incident (in Ω) and corresponding scattered fields (on Γ) to introduce new (virtual) experiments. In this framework, a novel approximation to effectively linearize the scattering problem, even in the case of non-weak scattering regime, is introduced.

1.5.1. *Born approximation*

In the BA, the unknown total field inside the object is approximated with the incident field, that is, the object is considered negligible and therefore such a hypothesis is fully satisfied when the internal characteristics of the scatterers are close to those of the external hosting medium and its dimensions are small compared to the wavelength in the considered medium. Under these assumptions, equation [1.6] becomes:

$$E_{sct}^{(v)} = \mathcal{A}_e\left[\chi E_{inc}^{(v)}\right] = \mathcal{L}^{(v)}[\chi], \quad v = 1,\ldots,N_v. \qquad [1.19]$$

The operator "$\mathcal{L}^{(v)}$", incorporating $E_{inc}^{(v)}$ in $\mathcal{A}_e$, is a linear mapping between data $E_{scat}^{(v)}$ and unknowns χ. It is worth pointing out that the linearization of model [1.6] into [1.19] does not overcome the ill-posedness of the considered inverse problem due to the finite dimensional nature of the scattered field. The order of the finite dimensional nature reduces further when accounting the aspect-limited nature of the GPR configuration. The more critical the aspect-limited is, the lower the order of the finite dimension and the more critical the capability to recover profiles. Hence, to improve the reconstruction performance of the tomographic approach, we can collect more information by using multifrequency data, since monochromatic data may be too poor to guarantee good results, as proven by the spectral coverage of the multi-view–multistatic multifrequency configuration (Leone and Soldovieri 2003; Ambrosanio and Pascazio 2015).

In choosing a multifrequency approach some attention must be paid in defining the unknown of the problem, since the contrast function does not change with the illumination, but it changes with frequency. Nevertheless, the fact that the contrast function changes with the frequency would imply solving more monochromatic ISP, but since the dependence on the frequency in the complex permittivity of the media and buried objects is supposed to be as:

$$\varepsilon(\vec{r},\omega) = \varepsilon'(\vec{r}\,) - j\frac{\sigma(\vec{r}\,)}{\omega\varepsilon_0}, \qquad [1.20]$$

in which $\varepsilon'(\vec{r}\,)$ and $\sigma(\vec{r}\,)$ represent the permittivity and conductivity of the second medium and buried targets, with the hypothesis that they are independent from the frequency. Keeping in mind what has been said previously, it is possible to exploit the multifrequency data to improve the amount of independent information to obtain better reconstructions. By means of simple changes in the expression of the contrast function, it is possible to recover the unknown profile at a certain frequency (e.g. the maximum one) by using all of the bandwidth information. In this framework, the unknown contrast function can be written as (Bucci et al. 2001):

$$\chi(\vec{r},\omega) = \frac{1}{2}\left(1+\frac{\overline{\omega}}{\omega}\right)\overline{\chi}(\vec{r}\,) + \frac{1}{2}\left(1-\frac{\overline{\omega}}{\omega}\right)\overline{\chi}^*(\vec{r}\,) = \mathcal{B}_\omega[\overline{\chi}], \qquad [1.21]$$

in which $\overline{\chi}(\vec{r}) = \chi(\vec{r}, \overline{\omega})$ is the value of the contrast function at the selected frequency and the mathematical symbol "*" represents the conjugate operation and $\mathcal{B}_\omega[\cdot]$ represents a linear operator. Therefore, by looking at [1.21], by knowing the contrast function at the frequency $\overline{\omega}$, it is possible to evaluate the contrast function at any frequency in the selected bandwidth.

Therefore, if we consider N_f working frequencies spanning a total bandwidth $(\omega_{max} - \omega_1)$ and assume as reference the maximum frequency in the selected bandwidth, which means to assume that $\chi(\omega_{\max}) = \chi(\overline{\omega}) = \overline{\chi}$, then starting from equation [1.19] for the single-frequency case and moving toward a multifrequency approach, it is possible to write the system of equations in a compact form for the multifrequency case:

$$E_{sct}^{MF}(\vec{r}) = \mathcal{A}_e^{MF}[\mathcal{B}_\omega[\overline{\chi}(\vec{r})] \cdot E_{inc}^{MF}(\vec{r})] = \mathcal{L}^{MF}[\overline{\chi}]. \quad [1.22]$$

which generalizes the BA to the multifrequency case.

1.5.2. *Distorted Born approximation*

Conversely from the linearization introduced in the previous section via the BA, such a linearization can also be performed in another scenario different from the homogeneous background, which yields the so-called distorted Born approximation (DBA). In this case, the linearization is performed about an assigned reference permittivity function (possibly inhomogeneous) and the actual internal field is approximated by the unperturbed field. This approximation is connected to the fact that in several applications, a reference dielectric permittivity, which represents a priori information about the unknown profile, can be available, so the problem can be recast as the reconstruction of the difference between the actual and the reference profiles from the scattered field data (Devaney and Oristaglio 1983; Leone et al. 1999).

Regarding the class of retrievable functions, only a limited class of permittivity profiles can be reconstructed within the DBA even if the maximum deviation from the true profile with respect to the reference profile is small. Moreover, the number and variability of these profiles can depend on the reference permittivity. However, the DBA allows the increase in the class of retrievable profile with respect to the case of simple BA.

Thus, with respect to the linear model shown in equation [1.19], the main difference consists of updating the Green's function for the specific inhomogeneous reference medium and the possible boundary conditions, and consequently also the related incident field. Unfortunately, such a function is usually not available in closed form except for special canonical geometries: in all the other cases, numerical evaluation can be exploited to obtain an estimate.

Therefore, equation [1.19] can be rewritten (in the single-frequency case) as:

$$E_{sct}^{(v)} = A_e^{bck}\left[\chi E_{bck}^{(v)}\right] = \mathscr{L}_{bck}^{(v)}[\chi], \quad v = 1,\ldots,N_v. \qquad [1.23]$$

1.5.3. *A linear approach for quantitative subsurface imaging based on VE*

In order to avoid nonlinear optimization for quantitative imaging, a linear data-driven approximation has been recently developed in Crocco et al. (2012) and successively extended to GPR prospecting in Bevacqua et al. (2016). The approach has been introduced as a new scattering approximation based on the physical meaning of LSM, but a more general understanding and interpretation can be also given with respect to the framework of VE (Di Donato et al. 2015).

The linear approximation relies on a proper linear combination of the original incident in the investigation domain and the corresponding scattered fields on the measurement domain to introduce new (virtual) experiments. In case of the subsurface survey, both the VE design equation (that is the LSM equation) and the field approximation must be readapted by considering the presence of the interface between air and soil (Bevacqua et al. 2016).

The solution of the LSM is exploited to image the support indicator and choose P evenly spaced points, known as a "pivot point" $\vec{r}_p$, to build the *multi-pivot* VE experiments (Di Donato et al. 2015). Indeed, the LSM design equation [1.14] can be understood as the attempt to recombine the original scattered field in order to fit at the measurement positions $\vec{r}_m \in \Gamma$, the field radiated in the air by an elementary source placed in the soil in the considered sampling point $\vec{r}_s$.

Then, for each VE, when the pivot point $\vec{r}_p$ is inside the scatterer's support, the virtual total field $\mathcal{E}_{tot}$ can be conveniently approximated by means of the following expression:

$$\mathscr{E}_{tot}(\vec{r},\vec{r}_p) \approx \mathscr{E}_{inc}(\vec{r},\vec{r}_p) + LP\left[G_i(\vec{r},\vec{r}_p)\right], \qquad [1.24]$$

where $\mathcal{E}_{inc}$ is the virtual incident field as defined in the VE framework (Bevacqua et al. 2016). Moreover, at the right-hand side, the second addendum is a low pass version of the internal Green's function. This operation avoids field singularity in the pivot point for the scattered field. It is worth underlining that this approximation takes implicitly into account the nature of the scatterer through the weighting function used to recombine the original incident field (Bevacqua et al. 2016). Moreover, in the field approximation, the unknown $\mathcal{E}$ is approximated in each $\vec{r} \in \Omega$ by considering the field G_i radiated in the soil by an elementary source placed in the same medium.

The above linear framework has a limited range of validity, as it is based on an approximation of the internal field, as well as on the validity range of the design equation, that is the FFE (Cakoni 2006). This notwithstanding exploiting the FFE as pre-processing step ensures the possibility to cope with a class of dielectric profiles (in terms of dimension and contrast values) widely exceeding the BA, as shown in Bevacqua et al. (2016). Such a property is related to the nature of the field approximation [1.5], that is *scatterer-aware* through the recombination of the data provided by LSM coefficient.

The VE framework can be exploited not only to introduce the above linearized approach, but also, and, more interestingly for the subsurface imaging problem at hand to counteract the aspect limitation of data. Indeed, aspect limited data are not sufficient to achieve quantitative characterization of the target even when the model assumptions are fulfilled, like, for example, the use of the BA in the case of weak scatterers or even assuming the exact knowledge of the internal field (the latter being an ideal case).

In this respect, the field conditioning enforced in each VE allows us also to predict the value of the field in locations other than the measurement points, so that we can exploit a number of additional equations corresponding to "fictitious measurements" (Di Donato and Crocco 2015). Accordingly, let us consider a fictitious curve Γ^* (complementary to Γ), which surrounds the investigation domain and allows us to locate measurement points in the soil. The value of the VE measured scattered field at these fictitious locations would be given by $G_i(\vec{r}_r^{\,*}, \vec{r}_p)$, where $\vec{r}_r$ and $\vec{r}_r^{\,*}$ denote the actual and the fictitious measurement points, respectively, as fictitious measurements are placed in the lower half space. As mentioned above, the virtual data equation can be suitably rewritten as:

$$k_b^{\,2} \int_{\Omega} \tilde{G}_e(\vec{r}, \vec{r}\,')\, \chi(\vec{r}\,') \mathscr{E}_{tot}(\vec{r}\,', \vec{r}_p) d\vec{r}\,' = \begin{cases} \mathscr{E}_{sct}(\vec{r}, \vec{r}_p) \; if \;\; r = r_r \in \Gamma \\ G_i(\vec{r}, \vec{r}_p) \; if \;\; r = r_r^{\,*} \in \Gamma^* \end{cases}, \qquad [1.25]$$

wherein $\mathcal{E}_{sct}$ is the virtual scattered field. To take into account the dependence of the kernel of [1.25] from the different kinds of measurement (the actual ones on surface and the fictitious one in the underground), the Green's function is defined as:

$$\tilde{G}_e(\vec{r},\vec{r}') = \begin{cases} G_e(\vec{r},\vec{r}_p) \; if \; \vec{r} = \vec{r}_r \in \Gamma \\ G_i(\vec{r},\vec{r}_p) \; if \; \vec{r} = \vec{r}_r^{\,*} \in \Gamma^* \end{cases}, \qquad [1.26]$$

that stands for the external or the internal Green's function depending on the measurement at hand.

In Figure 1.4, a sketch of the fictitious receivers, placed over the fictitious curve denoted with Γ^*, is depicted. Details about the number of fictitious receivers and the selection of Γ^* can be found in Di Donato and Crocco (2015) and Bevacqua et al. (2016).

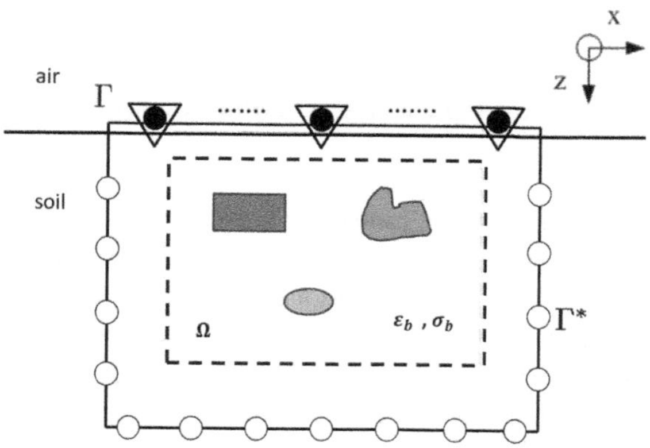

Figure 1.4. *The virtual measurements setup. The black and white circles refer to actual and fictitious measurements, respectively, while the white triangles denote the transmitters*

1.6. Regularization strategies

GPR tomography involves the solution of a severe ill-posed problem (Bertero and Boccacci 1998; Colton and Kress 1998) due to the filtering properties of the radiation operator (Bucci et al. 2001). The need for facing the low ratio between data and unknowns in the imaging problem drives the need for exploiting some regularization strategies to make the solution stable. Regularization techniques consist of adding some a priori information to the problem at hand to restore the well posedness. Such a priori information cannot be derived from the data or from the properties of the mapping which relates data and unknowns, but it is something strictly related to the expected properties of the contrast function χ to be retrieved. The effectiveness of the different regularizations will depend on how much the unknown scenario "fits" the regularization model.

The simplest form to enforce some a priori information is to include a regularizing term in the cost functional to be minimized when dealing with the inverse problem. Indeed, such an additive term will allow us to construct from the beginning a solution both compatible with the data (within the experimental error) and which exhibits the specific physical features. The additional information which can be exploited and/or enforced can be quite different. For instance, a requirement on the energy of the solution could be imposed, that is, a minimum l_2 energy requirement, which yields to the case of the well-known Tikhonov regularization technique (Colton and Kress 1998), or even some piecewise constant behavior on the contrast function (Bevacqua et al. 2016) as well as some physical bounds imposed on the permittivities and conductivities (e.g. positive permittivity and conductivities). The main issue in such a kind of regularization strategies is related to the choice of the regularization parameters, that is, the weights assigned at each regularization term, and no simple and general rules exist to perform such a choice in an optimal way.

While adding a penalty term is probably the most popular way of enforcing the regularization, there are of course other ways. Among the others, a possible strategy amounts to introduce proper statistical tools, such as in Pascazio and Ferraiuolo (2003), in which a Markov random field approach based on maximum a posteriori estimation method is proposed, which shows better performance also when compared with conventional techniques like Tikhonov regularization.

Another alternative consists of keeping the dimensionality of the unknowns space low by adopting a suitable (convenient) basis to represent the unknown function. Such a strategy is usually referred to as "regularization by projection", as the unknown function is represented by means of a suitable set of basis functions $\{\phi_p\}_{p=1}^{P}$ as follows:

$$\chi(\cdot) = \sum_{p=1}^{P} c_p \, \phi_p(\cdot) , \qquad [1.27]$$

in which the c_p are coefficients of the unknown contrast and $(\cdot)$ is the independent variable (e.g. space and frequency). In a continuous space, the truncation index P should be infinite, but since we are dealing with a discrete version of the contrast function, in case of lack of any a priori information and additional forms of regularization, this number must be lower than the essential dimension of data (i.e. the degrees of freedom) and can be determined by means of an analytic or numeric criterion (Bucci and Franceschetti 1987, 1989).

As far as the choice of the basis functions is concerned, different options can be used: among them, the most adopted ones are pixel basis, spectral harmonics and Wavelet decomposition (Mallat 1999). In particular, the Wavelet basis offers the possibility to decompose the unknown profile into two sets of coefficients, namely coarse and detail coefficients, thus allowing intrinsic multiresolution representation of the unknown.

A similar idea is adopted in multiresolution regularization techniques (Bucci et al. 2001; Salucci et al. 2017). These regularization techniques allow us to accommodate the representation coefficients for the contrast unknowns in a nonuniform fashion within the investigated domain, thus allowing us to focus the processing only in those regions where the scatterers are located and to adopt a variable degree of resolution with different depths. This feature is quite interesting in GPR imaging as it permits a coarser resolution for the deeper parts of the investigated domain and preserves a higher number on unknowns for the shallower regions (Salucci et al. 2017).

In order to restore the well-posedness of the problem, another interesting opportunity is offered by the *sparsity*-based regularization techniques, which are based on the assumption that the unknown can be represented through a limited number of non-zero coefficients in a convenient basis. Such techniques guarantee that an accurate retrieval of the unknowns is possible even for a number of data much lower than the overall number of basis coefficients, but sufficiently larger than the number of non-zero elements. The most widely used sparsity based methods are probably the l_p-regularized methods, which estimate the regression coefficients by minimizing a least square objective function as (Estatico et al. 2015):

$$\Phi_p(\chi) = \|E_{sct} - \mathcal{L}[\chi]\|_{l_2}^2 + \lambda\|\chi\|_{l_p}^p , \qquad [1.28]$$

where the constant $\lambda > 0$ is the regularization parameter, $\mathcal{L}$ represents the linearized GPR operator relating data (the scattered field samples E_{scat}) to the unknowns (the contrast function χ) and "p" is a parameter in the interval [0, 1].

With the advent of *compressive sensing*, a great attention has been paid to [1.28] by the signal processing community when $p = 1$. The theory of compressive sensing has shown that under sparsity and incoherence assumptions, solving the l_1-minimization problem is equivalent to solving the l_0-minimization problem (i.e.

equation [1.28]) when $p = 0$ (Figueiredo et al. 2007). In this framework, different iterative algorithms have been proposed that are based on linear programming or interior-point methods (Tan et al. 2011).

Recently, wider sparsity concepts have been explored by the research community, which are known as group-sparsity, block-sparsity or joint-sparsity. In these cases, the non-zero elements are arranged in clusters that allow us to achieve better reconstruction results than just treating the signal as being arbitrarily sparse (Sun et al. 2017; Bevacqua and Isernia 2018).

Finally, a binary regularization can be eventually considered in case of homogeneous or step wise constant unknown objects. It implies an additive term to be considered in the relevant cost functional to be minimized. The aim of this additive term is to make the punctual value of the unknown function only belong to a given finite alphabet of values $\{X_m\}$ (Catapano 2004). The values $\{X_m\}$ may be either known or unknown, depending on the available a priori information.

1.7. An example of multistatic tomographic imaging for GPR prospecting: the Gergia Tech case

In this section, BA method equipped with sparsity regularization is applied in case of an experimental multistatic GPR system developed at the Georgia Institute of Technology. The discussion and the results which follow come from the paper by Ambrosanio (2021).

The system is composed of a linear array of resistive-vee antennas (Kim and Scott 2004) made of two transmitters and four receivers arranged as in Figure 1.5. Each pair of contiguous receivers (R_1-R_2, R_2-R_3, R_3-R_4) is 12-cm spaced, while the two transmitters (T_1-T_2) are located at a 48-cm distance from each other. Thus, this arrangement of the antennas provides eight bistatic pairs spacing from 12 to 96 cm in 12 cm increments.

The GPR scanned area extends for a 1.8 × 1.8 m^2 region with the scanning system located above the surface of the ground at a constant height of 27.8 cm. The linear array of transmitters and receivers moves above the investigated area in a stepped fashion with a spatial sampling step of 2 cm. Thus, the investigated surface is discretized into a grid of 91 points by 91 points. Each time the array stops in a location, it collects data from the eight bistatic pairs by means of 401 equally spaced frequency points from 60 MHz to 8.06 GHz (i.e. a frequency step equal to 20 MHz). After collection, the data are calibrated and stored (more details regarding the data calibration are reported in Counts et al. (2007)). Note that, as far as the GPR perspective, the highest frequencies in the employed 8-GHz bandwidth are not useful due to the attenuation in the soil, while this bandwidth is efficient for the air-target test case (Ambrosanio et al. 2019).

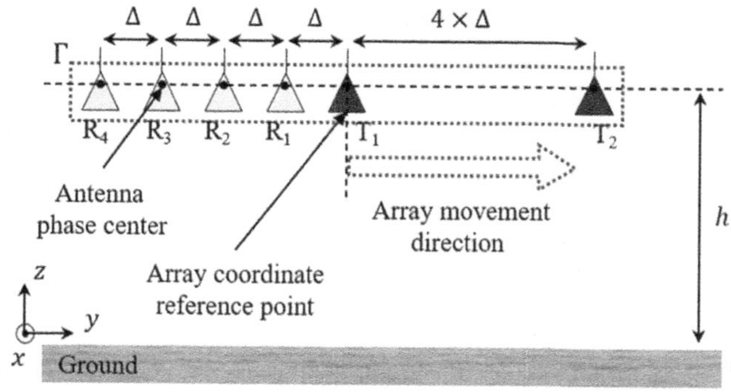

Figure 1.5. *The Georgia Tech GPR system. The spacing between contiguous receivers is Δ = 12 cm and the height from the ground is h = 27.8 cm*

The calibrated data (made available by Professor Waymond Scott[1]) require a proper pre-processing step before running the tomographic imaging procedures. The ideal clutter removal procedure consists of subtracting from the total field the one due to the clutter sources. As a matter of fact, a key factor in clutter removal techniques is the muting of the air–soil interface, as well as the estimation of its roughness (Cmielewski et al. 2007; Lo Monte et al. 2010; Temlioglu and Erer 2016; Cristofani et al. 2018).

In the following, a proper pre-processing of the data has been adopted prior to inversion. Details about the exploited strategy can be found in Ambrosanio (2021).

1.7.1. *Landmines data set*

A first experimental data set is composed of different targets that include metal spheres and a variety of anti-tank (AT) and anti-personnel (AP) dielectric mines of different sizes. Target depths range from a few centimeters up to 30 cm and the distance is to be intended from the surface of the sand to the top of the target. The ground medium is a damp, compacted sandy soil. Figure 1.6 shows a picture of the targets involved in the scenario under test and their spatial locations and depths. Both plastic AT as well as AP mines are present, together with other various objects.

1 https://waymond-scott.ece.gatech.edu/multistaticbeamformingdata/multistaticbeamformingdata/.

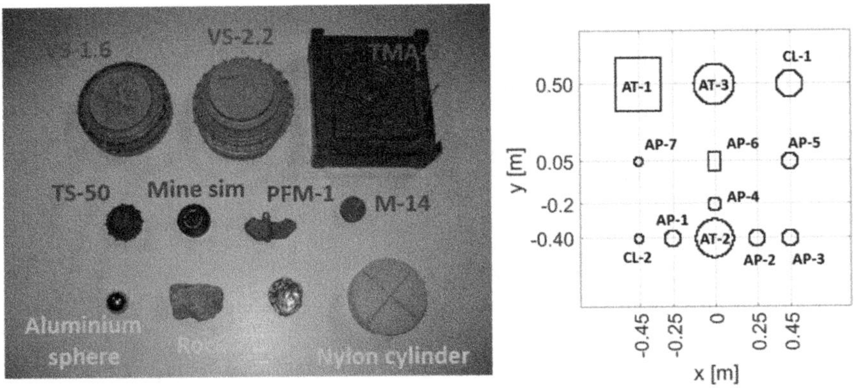

a) b)

Figure 1.6. *Targets employed for the mine data sets (a) and their spatial locations (b). The objects in red refer to anti-tank (AT) plastic mines, blue refer to anti-personnel (AP) mines and the dark yellow refer to other objects. The depths of the buried targets are as follows: AT-1 (TMA-5): 12 cm; AT-2 (VS-1.6): 11.5 cm; AT-3 (VS-2.2): 13 cm; AP-1 (TS-50): 1.5 cm; AP-2 (TS-50): 1.5 cm; AP-3 (TS-50): 2 cm; AP-4 (Mine simulant): 2 cm; AP-5 (TS-50): 1.5 cm; AP-6 (PFM1): 2 cm; AP-7 (M-14): 1.5 cm; CL-1 (Nylon cylinder): 10.5 cm; CL-2 (Aluminum sphere): 11.5 cm*

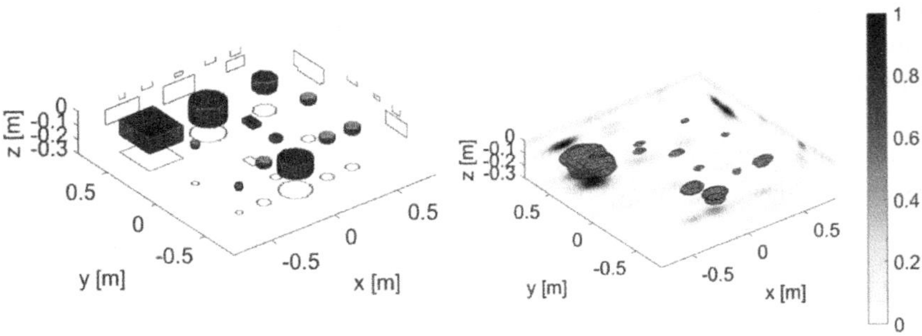

a) b)

Figure 1.7. *Reference scenario with AT and AP plastic mines and other clutter objects (a) and its recovery via multifrequency sparse BA approach (b). These results were obtained by processing 45 equally spaced frequencies in the range [0.8, 4] GHz*

To evaluate the quality of GPR recoveries, Figure 1.7(b) reports a 3D estimate of the buried targets support via exploiting sparse multifrequency BA method, while Figure 1.7(a) shows its reference. Regarding the frequency selection, 45 equally spaced frequencies in the range [0.8, 4] GHz were employed. Note that, the highest frequencies are processed as they are not useful due to the attenuation in the soil. Figure 1.7(b) shows the iso-surfaces of target supports and the colored slices representing the projections on each plane. For the sake of clarity, some slices of the retrieved profiles are reported in Figure 1.8.

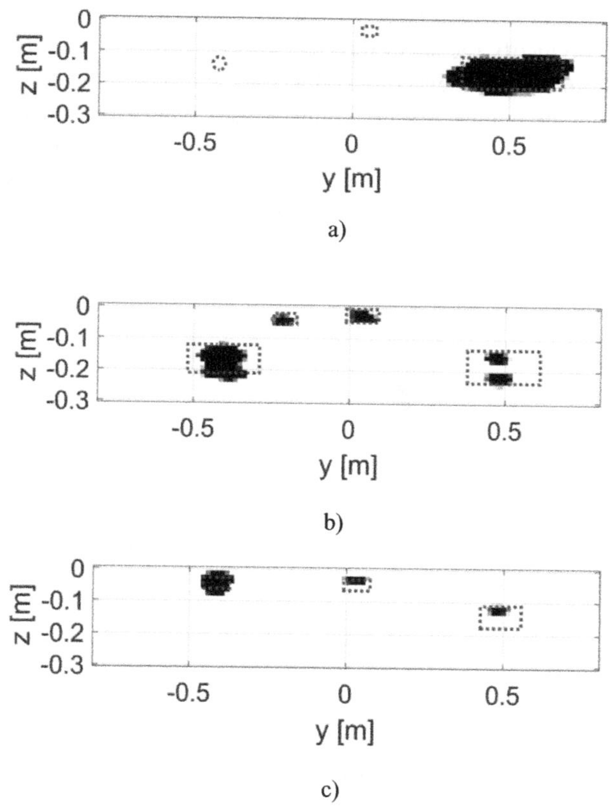

Figure 1.8. *Some slices of retrieved objects supports (i.e. black areas) in Figure 1.7(b): slices at (a)* $x = -0.45$ *m; (b)* $x = 0$ *m; (c)* $x = +0.45$ *m*

Since the aim of the analysis is to provide targets support, we can assume the contrast constant with respect to the frequency, thus reducing the computational burden and memory requirement for the recovery procedure. It is worth noting that the inversion was carried out via a 2.5D procedure, that is, the GPR data corresponding to each slice are processed via a 2D algorithm despite the 3D geometry, and then each reconstructed slice along the x axis is merged with the others to cover the whole ID.

The investigated area is discretized into 33 × 108 × 24 voxels and all the slices were normalized by the largest magnitude in the whole 3D image. As expected, larger targets with higher contrasts have stronger responses: indeed, the TMA-5 AT plastic mine has a bigger retrieved support rather than other objects like the VS-1.6 and VS-2.2, notwithstanding their size. The strong contribution of the TMA-5 mine to the data might also justify the small recovery for the VS-2.2, which is worse than the one of the VS-1.6 despite their comparable size, and this might be related to its farther distance from the biggest objects in the domain. Regarding other objects like the metal sphere and the M-14 AP plastic mine, they are not visible in Figure 1.7(b) due to their small size, depth and/or composition. For similar reasons, the big nylon cylinder is not properly imaged, since its dielectric permittivity is close to that of the soil.

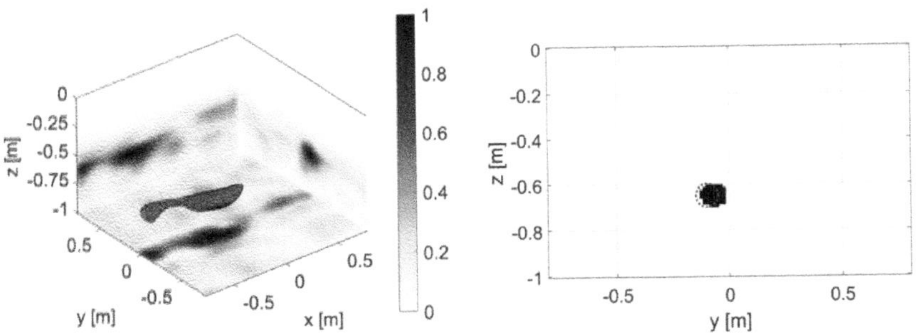

a) b)

Figure 1.9. *Support recovery of a 10 cm diameter plastic pipe buried at 58 cm (a) and its cross-section recovery at x = –50 cm. (b) A picture of the reference is reported in Counts et al. (2007)*

1.7.2. *Buried pipe*

Another example is composed of a hollow 10 cm diameter plastic pipe buried at 58 cm depth in the same sandy soil employed for the data set of landmines one.

The very accurate results of the recovery algorithm based on the use of the multifrequency BA method and iterative shrinkage thresholding algorithm (Beck and Teboulle 2009) with the same settings as described previously are reported in Figure 1.9. The threshold exploited to create the iso-surface in the image was fixed at 0.5. The figure shows clearly the location and the size of the buried pipe, as proven in (Counts et al. 2007).

1.8. References

Aboudourib, A., Serhir, M., Lesselier, D. (2020). A processing framework for tree-root reconstruction using ground-penetrating radar under heterogeneous soil conditions. *IEEE Transactions on Geoscience and Remote Sensing*, 59(1), 208–219.

Akinci, M.N. (2018a). An efficient sampling method for cross-borehole GPR imaging. *IEEE Geoscience and Remote Sensing Letters*, 15(12), 1857–1861.

Akinci, M.N. (2018b). Improving near-field orthogonality sampling method for qualitative microwave imaging. *IEEE Transactions on Antennas and Propagation*, 66(10), 5475–5484.

Ambrosanio, M. and Pascazio, V. (2015). A compressive-sensing-based approach for the detection and characterization of buried objects. *IEEE Journal of Selected Topics in Applied Earth Observations and Remote Sensing*, 8(7), 3386–3395.

Ambrosanio, M., Bevacqua, M.T., Isernia, T., Pascazio. V. (2019a). Experimental multistatic imaging via the linear sampling method. In *IEEE International Geoscience and Remote Sensing Symposium*, Yokohama.

Ambrosanio, M., Bevacqua, M.T., Isernia, T., Pascazio, V. (2019b). The tomographic approach to ground-penetrating radar for underground exploration and monitoring: A more user-friendly and unconventional method for subsurface investigation. *IEEE Signal Processing Magazine*, 36(4), 62–73.

Ambrosanio, M., Bevacqua, M.T., Isernia, T., Pascazio, V. (2021). Performance analysis of tomographic methods against experimental contactless multistatic ground penetrating radar. *IEEE Journal of Selected Topics in Applied Earth Observations and Remote Sensing*, 14, 1171–1183.

Beck, A. and Teboulle, M. (2009). A fast iterative shrinkage-thresholding algorithm for linear inverse problems. *SIAM Journal on Imaging Sciences*, 2(1), 183–202.

Bevacqua, M.T. and Isernia, T. (2017). Shape reconstruction via equivalence principles, constrained inverse source problems and sparsity promotion. *Progress in Electromagnetics Research*, 158, 37–48.

Bevacqua, M.T. and Isernia, T. (2018). Boundary indicator for aspect limited sensing of hidden dielectric objects. *IEEE Geoscience and Remote Sensing Letters*, 15(6), 838–842.

Bevacqua, M.T. and Palmeri, R. (2019). Qualitative methods for the inverse obstacle problem: A comparison on experimental data. *Journal of Imaging*, 5(4), 47.

Bevacqua, M.T., Crocco, L., Di Donato, L., Isernia, T., Palmeri, R. (2016). Exploiting sparsity and field conditioning in subsurface microwave imaging of nonweak buried targets. *Radio Science*, 51(4), 301–310.

Bevacqua, M.T., Isernia, T., Palmeri, R., Akinci, M.N., Crocco, L. (2020). Physical insight unveils new imaging capabilities of orthogonality sampling method. *IEEE Transactions on Antennas and Propagation*, 68(5), 4014–4021.

Bucci, O.M. and Franceschetti, G. (1987). On the spatial bandwidth of scattered fields. *IEEE Transactions on Antennas and Propagation*, 35(12), 1445–1455.

Bucci, O.M. and Franceschetti, G. (1989). On the degrees of freedom of scattered fields. *IEEE Transactions on Antennas and Propagation*, 37(7), 918–926.

Bucci, O.M., Crocco, L., Isernia, T., Pascazio, V. (2001). Subsurface inverse scattering problems: Quantifying, qualifying and achieving the available information. *IEEE Transactions on Geoscience and Remote Sensing*, 39(11), 2527–2538.

Cakoni, F. and Colton, D. (2006). *Qualitative Methods in Inverse Scattering Theory*. Springer-Verlag, Berlin.

Catapano, I., Crocco, L., Isernia, T. (2004). A simple two-dimensional inversion technique for imaging homogeneous targets in stratified media. *Radio Science*, 39(1), 1–14.

Catapano, I., Crocco, L., Isernia, T. (2007). On simple methods for shape reconstruction of unknown scatterers. *IEEE Transactions on Antennas and Propagation*, 55(5), 1431–1436.

Catapano, I., Crocco, L., Isernia, T. (2008). Improved sampling methods for shape reconstruction of 3-D buried targets. *IEEE Transactions on Geoscience and Remote Sensing*, 46(10), 3265–3273.

Chew, W.C. (1995). *Waves and Fields in Inhomogeneous Media*. IEEE Press, Piscataway.

Cmielewski, O., Tortel, H., Litman, A., Saillard, M. (2007). A two-step procedure for characterizing obstacles under a rough surface from bistatic measurements. *IEEE Transactions on Geoscience and Remote Sensing*, 45(9), 2850–2858.

Colton, D. and Kress, R. (1998). *Inverse Acoustic and Electromagnetic Scattering Theory*, 2nd edition. Springer, Berlin.

Colton, D., Haddar, H., Piana, M. (2003). The linear sampling method in inverse electromagnetic scattering theory. *Inverse Problems*, 19, 105–137.

Counts, T., Gurbuz, A.C., Scott, W.R., McClellan, J.H., Kim, K. (2007). Multistatic ground-penetrating radar experiments. *IEEE Transactions on Geoscience and Remote Sensing*, 45(8), 2544–2553.

Cristofani, E., Becquaert, M., Lambot, S., Vandewal, M., Stiens, J., Deligiannis, N. (2018). Random subsampling and data preconditioning for ground penetrating radars. *IEEE Access*, 6, 26866–26880.

Crocco, L. and Soldovieri, F. (2003). GPR prospecting in a layered medium via microwave tomography. *Annals of Geophysics*, 46, 559–572.

Crocco, L., Catapano, I., Di Donato, L., Isernia, T. (2012). The linear sampling method as a way for quantitative inverse scattering. *IEEE Transactions on Antennas and Propagation*, 4(60), 1844–1853.

Devaney, A.J. and Oristaglio, M.L. (1983). Inversion procedure for inverse scattering within the distorted-wave Born approximation. *Physical Review Letters*, 51(4), 237.

Di Donato, L. and Crocco, L. (2015). Model-based quantitative cross-borehole GPR imaging via virtual experiments. *IEEE Transactions on Geoscience and Remote Sensing*, 53(8), 4178–4185.

Di Donato, L., Bevacqua, M.T., Isernia, T., Catapano, I., Crocco, L. (2011). Improved quantitative microwave tomography by exploiting the physical meaning of the linear sampling method. In *Proceedings of the 5th European Conference on Antennas and Propagation*, IEEE, 11–15 April.

Di Donato, L., Bevacqua, M.T., Crocco, L., Isernia, T. (2015). Inverse scattering via virtual experiments and contrast source regularization. *IEEE Transactions on Antennas and Propagation*, 63(4), 1669–1677.

Donoho, D. (2006). Compressed sensing. *IEEE Transactions on Information Theory*, 52(4), 1289–1306.

Estatico, C., Fedeli, A., Pastorino, M., Randazzo, A. (2015). A multifrequency inexact-Newton method in L^p Banach spaces for buried objects detection. *IEEE Transactions on Antennas and Propagation*, 63(9), 4198–4204.

Figueiredo, M.A., Nowak, R.D., Wright, S.J. (2007). Gradient projection for sparse reconstruction: Application to compressed sensing and other inverse problems. *IEEE Journal of Selected Topics in Signal Processing*, 1(4), 586–597.

Fornasier, M. and Rauhut, H. (2008). Recovery algorithms for vector-valued data with joint sparsity constraints. *SIAM Journal on Numerical Analysis*, 46(2), 577–613.

Franceschetti, G. (2013). *Electromagnetics: Theory, Techniques, and Engineering Paradigms*. Springer, New York.

Isernia, T., Pascazio, V., Pierri, R. (2001). On the local minima in a tomographic imaging technique. *IEEE Transactions on Geoscience and Remote Sensing*, 39(7), 1596–1607.

Kim, K. and Scott Jr., W.R. (2004). A resistive linear antenna for ground-penetrating radars. *Proceedings of SPIE – The International Society for Optical Engineering*, 5415.

Kirsch, A. and Grinberg, N.I. (2007). *The Factorization Method for Inverse Problems*. Oxford University Press.

Kowalski, M. (2009). Sparse regression using mixed norms. *Applied and Computational Harmonic Analysis*, 27(3), 303–324.

Leone, G. and Soldovieri, F. (2003). Analysis of the distorted Born approximation for subsurface reconstruction: Truncation and uncertainties effects. *IEEE Transactions on Geoscience and Remote Sensing*, 41(1), 66–74.

Leone, G., Persico, R., Pierri, R. (1999). Inverse scattering under the distorted Born approximation for cylindrical geometries. *Journal of the Optical Society of America A*, 16(7), 1779–1787.

Lo Monte, L., Erricolo, D., Soldovieri, F., Wicks, M.C. (2010). Radio frequency tomography for tunnel detection. *IEEE Transactions on Geoscience and Remote Sensing*, 48, 1128–1137.

Lo Monte, L., Soldovieri, F., Erricolo, D., Wicks, M.C. (2012). Imaging below irregular terrain using RF tomography. *IEEE Transactions on Geoscience and Remote Sensing*, 50(9), 3364–3373.

Mallat, S. (1999). *A Wavelet Tour of Signal Processing*. Elsevier Academic Press, Florida.

Mario, B. and Patrizia, B. (1998). *Introduction to Inverse Problems in Imaging*. Institute of Physics Publishing, Bristol and Philadelphia.

Pascazio, V. and Ferraiuolo, G. (2003). Statistical regularization in linearized microwave imaging through MRF-based MAP estimation: Hyperparameter estimation and image computation. *IEEE Transactions on Image Processing*, 12(5), 572–582.

Persico, R. (2014). *Introduction to Ground Penetrating Radar: Inverse Scattering and Data Processing*. John Wiley & Sons, New York.

Potthast, R. (2010). A study on orthogonality sampling. *Inverse Problems*, 26(7), 1–8.

Salucci, M., Poli, L., Anselmi, N., Massa, A. (2017). Multifrequency particle swarm optimization for enhanced multiresolution GPR microwave imaging. *IEEE Transactions on Geoscience and Remote Sensing*, 55(3), 1305–1317.

Serhir, M. and Lesselier, D. (2017). Wideband reflector-backed folded bowtie antenna for ground penetrating radar. *IEEE Transactions on Antennas and Propagation*, 66(3), 1056–1063.

Soldovieri, F., Brancaccio, A., Prisco, G., Leone, G., Pierri, R. (2008). A Kirchhoff-based shape reconstruction algorithm for the multimonostatic configuration: The realistic case of buried pipes. *IEEE Transactions on Geoscience and Remote Sensing*, 46(10), 3031–3038.

Sommerfeld, A. (1964). *Partial Differential Equations in Physics*. Academic Press, New York.

Sun, S., Kooij, B.J., Yarovoy, A.G. (2017). Linearized 3-D electromagnetic contrast source inversion and its applications to half-space configurations. *IEEE Transactions on Geoscience and Remote Sensing*, 55(6), 3475–3487.

Tan, X., Roberts, W., Li, J., Stoica, P. (2011). Sparse learning via iterative minimization with application to MIMO radar imaging. *IEEE Transactions on Signal Processing*, 59(3), 1088–1101.

Temlioglu, E. and Erer, I. (2016). Clutter removal in ground-penetrating radar images using morphological component analysis. *IEEE Geoscience and Remote Sensing Letters*, 13(12), 1802–1806.

2

Nonlinear and Hybrid Inversion Techniques for Ground Penetrating Radar Imaging

Valentina SCHENONE, Alessandro FEDELI, Matteo PASTORINO† and Andrea RANDAZZO

Department of Electrical, Electronic, Telecommunications Engineering and Naval Architecture, Università di Genova, Italy

The retrieval of the properties of buried structures is a fascinating topic which has a crucial importance in many applicative fields: from archaeology to civil engineering and geophysics. For the characterization and imaging of buried targets by means of electromagnetic waves, the possible use of ground penetrating radar (GPR) for such a task has emerged with very promising outcomes (Daniels 2004; Persico 2014). Despite the great benefits of GPR, like its non-invasiveness and practicality, the retrieval of targets' properties starting from GPR measurements needs the solution of an inverse scattering problem, which is in general nonlinear and affected by a strong level of ill-posedness (Pastorino 2010).

Compared to other applications of electromagnetic imaging, GPR prospecting poses additional issues mainly related to (i) the aspect-limited acquisition system, which means that measurements all around the target of interest are often impossible (Bucci et al. 2001), (ii) the presence of the air–soil interface, which decreases the

For a color version of all the figures in this chapter, see www.iste.co.uk/serhir/radar.zip.

Ground Penetrating Radar,
coordinated by Mohammed SERHIR and Dominique LESSELIER.

intensity of the scattered field at the probing points and complicates the problem due to the interactions with the antenna system (De Coster and Lambot 2019) and (iii) the significant presence of clutter (Solimene et al. 2014). Consequently, classical approaches for the treatment of electromagnetic inverse problems may fail in the application to GPR data, and specific inversion strategies need to be devised.

In the framework of GPR imaging, several methods can be found in the related scientific literature (Baussard et al. 2004; Breard et al. 2008; Catapano et al. 2015), where both two- and three-dimensional approaches were considered (Catapano et al. 2008; Gurbuz et al. 2014; Sun et al. 2017; Hajebi and Hoorfar 2020). However, many approaches are qualitative, that is, able to retrieve only some features of buried targets, like their position and/or shape (Bevacqua and Isernia 2018; Gennarelli et al. 2020; Ambrosanio et al. 2021).

The quantitative inversion of GPR data is more challenging since it requires us to deal with the nonlinearity of the problem (Pastorino and Randazzo 2018). However, the potential of quantitative GPR imaging clearly emerges when the identification of targets is concerned (Fedeli et al. 2017). This is why scientific research is continuously promoting novel ideas to address this kind of problem (Gurbuz et al. 2014; Salucci et al. 2017; Hajebi et al. 2018). Among quantitative imaging approaches, methods based on neural networks also show promising results (Chen et al. 2020; Liu et al. 2021; Wang et al. 2022), as well as methods that combine both neural-network processing and deterministic inversion techniques (Wei and Chen 2019; Fedeli 2021).

In this chapter, inversion techniques for obtaining maps of buried structures starting from GPR data are discussed, with particular reference to some recently proposed nonlinear and hybrid methods developed in nonconventional functional spaces (Fedeli et al. 2020; Schenone et al. 2022). The main theoretical aspects on which data inversion is based are outlined, providing some reconstruction examples.

The chapter has the following outline. Section 2.1 introduces the problem of nonlinear GPR imaging, establishing the bases for the definition of the related inverse problem in the presence of multifrequency data. Section 2.2 introduces quantitative inversion approaches for GPR imaging by taking the example of a nonlinear method based on an inexact-Newton scheme. The hybridization of this approach with a qualitative imaging technique is presented in section 2.3. Finally, a numerical test case for the assessment and comparison of the considered techniques is analyzed in section 2.4.

2.1. GPR imaging as a nonlinear inverse problem

Let us consider the standard GPR imaging configuration drafted in Figure 2.1, where a common-offset GPR device composed of a transmitting (Tx) and a receiving (Rx) antenna performs a series of quasi-monostatic measurements along its movement on a horizontal line of length l, at a height h from the air–soil interface. We assume that measurements with S different positions of Tx are available, that is, S *views*, and that for each view the Tx is an ideal line-current source that gives rise to a z-polarized transverse magnetic (TM) field. More precisely, in the sth view ($s = 1, \ldots, S$) the source is located on the xy plane at $\left(x_{Tx}^{(s)}, y_{Tx}^{(s)}\right)$ and is characterized by a current density $\mathbf{J}_{Tx}^{(s)}(x, y, f) = J_{Tx}^{(s)}(x, y, f)\hat{\mathbf{z}} = I(f)\delta\left(x - x_{Tx}^{(s)}\right)\delta\left(y - y_{Tx}^{(s)}\right)\hat{\mathbf{z}}$, $\delta(\cdot)$ being the Dirac delta function and I an arbitrary amplitude. Time-harmonic field quantities at the frequency f are considered, with an $e^{j2\pi ft}$ time dependence. The scattering problem is formulated in a two-layer environment that includes two half spaces:

– the upper half space, for $y > 0$, characterized by a dielectric permittivity ε_0 and zero electric conductivity (i.e. air, modeled as vacuum);

– the lower one, for $y < 0$, characterized by a dielectric permittivity ε_b and an electric conductivity σ_b (i.e. soil).

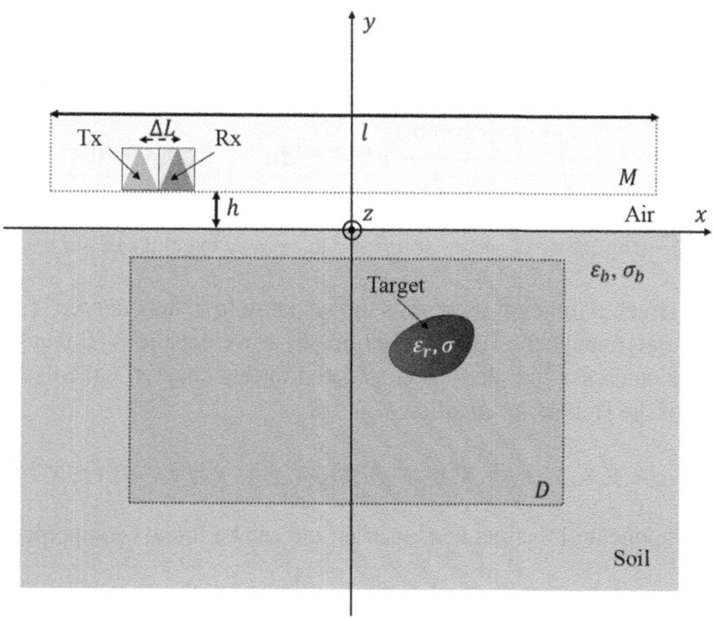

Figure 2.1. *Configuration of the problem considered in GPR imaging*

Let us suppose that no targets are present in the considered scenario. Under the above hypotheses, the electric field vector due to the radiation of the source at the sth position in a target-free scenario can be described as the *incident* field $\mathbf{E}_{inc}^{(s)}(x, y, f) = E_{inc,z}^{(s)}(x, y, f)\hat{\mathbf{z}}$. In particular, its z-component is the function that satisfies the scalar Helmholtz equation:

$$\nabla^2 E_{inc,z}^{(s)}(x, y, f) + k^2(x, y, f) E_{inc,z}^{(s)}(x, y, f) = j2\pi f \mu_0 J_{Tx}^{(s)}(x, y, f) \qquad [2.1]$$

where

$$k^2(x, y, f) = \begin{cases} k_0^2 & y \geq 0 \\ k_b^2 & y < 0 \end{cases} \qquad [2.2]$$

with $k_0^2 = (2\pi f)^2 \varepsilon_0 \mu_0$ and $k_b^2 = (2\pi f)^2 (\varepsilon_b - j\sigma_b / 2\pi f)\mu_0$.

As a result, we have

$$E_{inc,z}^{(s)}(x, y, f) = j2\pi f \mu_0 I(f) g_{HS}\left(x, y, x_{Tx}^{(s)}, y_{Tx}^{(s)}, f\right) \qquad [2.3]$$

where g_{HS} is the half-space Green's function given by Stinson (1976)

$$g_{HS}(x, y, x', y', f) = \frac{1}{2\pi j} \begin{cases} \displaystyle\int_{-\infty}^{+\infty} \frac{e^{-j\gamma_0 |y - y'|} + R e^{-j\gamma_0 (y + y')}}{2\gamma_0} e^{ju(x - x')} du & y \geq 0 \\ \displaystyle\int_{-\infty}^{+\infty} \frac{e^{j(\gamma_b y - \gamma_0 y')}}{\gamma_0 + \gamma_b} e^{ju(x - x')} du & y < 0 \end{cases} \qquad [2.4]$$

with $\gamma_0^2 = k_0^2 - u^2$, $\gamma_b^2 = k_b^2 - u^2$, and $R = (\gamma_0 - \gamma_b)/(\gamma_0 + \gamma_b)$.

NOTE.– The Green's function represents the response to a "unit source" (i.e. a Dirac delta) of the electromagnetic system described by wave equations (Balanis 2012). In a two-dimensional case like the one considered in this chapter, Green's function is the solution of the Helmholtz equation.

$$\nabla^2 g_{HS}(x, y, x', y', f) + k^2(x, y, f) g_{HS}(x, y, x', y', f) = \delta(x - x')\delta(y - y')$$

and the z-component of the incident electric field can be found by superposition as

$$E_{inc,z}^{(s)}(x,y,f) = j2\pi f\mu_0 \iint_{-\infty}^{\infty} J_{Tx}^{(s)}(x',y',f) g_{HS}(x,y,x',y',f) dx'dy'$$

which yields the result of equation [2.3].

Let us now consider the case in which some targets are present in the subsoil. For the sake of simplicity, we impose the additional hypothesis that such buried targets have homogeneous dielectric and geometric properties along the z axis, that is, that they are characterized by a dielectric permittivity $\varepsilon(x,y)$ and electric conductivity $\sigma(x,y)$. These targets are assumed to be contained in the so-called *investigation domain D*, which is the underground region of interest for the GPR imaging approach (Figure 2.1).

With z-polarized TM incident fields and uniform target properties along z, it can be shown that the electric field when targets are present (i.e. the *total* field) is also z-polarized and it can be expressed as $\mathbf{E}_{tot}^{(s)}(x,y,f) = E_{tot,z}^{(s)}(x,y,f)\hat{\mathbf{z}}$. In more detail, $E_{tot,z}^{(s)}(x,y,f)$ is the solution of equation

$$\nabla^2 E_{tot,z}^{(s)}(x,y,f) + k_t^2(x,y,f) E_{tot,z}^{(s)}(x,y,f) = j2\pi f\mu_0 J_{Tx}^{(s)}(x,y,f) \qquad [2.5]$$

where

$$k_t^2(x,y,f) = \begin{cases} k_0^2 & y \geq 0 \\ k_b^2 & y < 0, (x,y) \notin D \\ (2\pi f)^2 \left[\varepsilon(x,y) - \dfrac{j\sigma(x,y)}{2\pi f}\right]\mu_0 & y < 0, (x,y) \in D \end{cases} \qquad [2.6]$$

If we add the term $k^2(x,y,f)E_{tot,z}^{(s)}(x,y,f)$ to both sides of equation [2.5], we obtain

$$\begin{aligned} &\nabla^2 E_{tot,z}^{(s)}(x,y,f) + k_t^2(x,y,f)E_{tot,z}^{(s)}(x,y,f) + k^2(x,y,f)E_{tot,z}^{(s)}(x,y,f) \\ &= j2\pi f\mu_0 J_{Tx}^{(s)}(x,y,f) + k^2(x,y,f)E_{tot,z}^{(s)}(x,y,f) \end{aligned} \qquad [2.7]$$

which can also be written as

$$\begin{aligned} &\nabla^2 E_{tot,z}^{(s)}(x,y,f) + k^2(x,y,f)E_{tot,z}^{(s)}(x,y,f) \\ &= j2\pi f\mu_0 J_{Tx}^{(s)}(x,y,f) + [k^2(x,y,f) - k_t^2(x,y,f)]E_{tot,z}^{(s)}(x,y,f) \end{aligned} \qquad [2.8]$$

Introducing the term

$$J_{eq}^{(s)}(x,y,f) = \frac{k^2(x,y,f) - k_t^2(x,y,f)}{j2\pi f \mu_0} E_{tot,z}^{(s)}(x,y,f) \quad [2.9]$$

which is clearly equal to zero for $(x,y) \notin D$, the equation to be solved to find the z-component of the total electric field becomes

$$\nabla^2 E_{tot,z}^{(s)}(x,y,f) + k^2(x,y,f) E_{tot,z}^{(s)}(x,y,f) = j2\pi f \mu_0 \left[J_{Tx}^{(s)}(x,y,f) + J_{eq}^{(s)}(x,y,f) \right] \quad [2.10]$$

By exploiting the superposition as we did to compute the incident field from Green's function (see the note above), the following interesting result is obtained:

$$\begin{aligned} E_{tot,z}^{(s)}(x,y,f) &= j2\pi f \mu_0 \iint_{-\infty}^{\infty} \left[J_{Tx}^{(s)}(x',y',f) + J_{eq}^{(s)}(x',y',f) \right] g_{HS}(x,y,x',y',f) dx'dy' \\ &= E_{inc,z}^{(s)}(x,y,f) + j2\pi f \mu_0 \iint_D J_{eq}^{(s)}(x',y',f) g_{HS}(x,y,x',y',f) dx'dy' \end{aligned} \quad [2.11]$$

where the last step has been done by recalling that $J_{eq}^{(s)}(x',y',f) = 0$ for $(x',y') \notin D$.

If we define the scattered electric field as

$$\mathbf{E}_{scat}^{(s)}(x,y,f) = \mathbf{E}_{tot}^{(s)}(x,y,f) - \mathbf{E}_{inc}^{(s)}(x,y,f) \quad [2.12]$$

it turns out that its z component $E_{scat,z}^{(s)}(x,y,f) = E_{tot,z}^{(s)}(x,y,f) - E_{inc,z}^{(s)}(x,y,f)$, thanks to equation [2.11], is equal to

$$E_{scat,z}^{(s)}(x,y,f) = j2\pi f \mu_0 \iint_D J_{eq}^{(s)}(x',y',f) g_{HS}(x,y,x',y',f) dx'dy' \quad [2.13]$$

Considering that $J_{eq}^{(s)}$ is non-zero only inside D, it is easy to show from [2.9] that

$$\begin{aligned} J_{eq}^{(s)}(x,y,f) &= j2\pi f \left[\varepsilon(x,y) - \varepsilon_b - j\frac{\sigma(x,y) - \sigma_b}{2\pi f} \right] E_{tot,z}^{(s)}(x,y,f) \\ &= j2\pi f \left(\varepsilon_b - j\frac{\sigma_b}{2\pi f} \right) \chi(x,y,f) E_{tot,z}^{(s)}(x,y,f), (x,y) \in D \end{aligned} \quad [2.14]$$

where

$$\chi(x,y,f) = \frac{\varepsilon(x,y) - \varepsilon_b - j\frac{\sigma(x,y) - \sigma_b}{2\pi f}}{\varepsilon_b - j\frac{\sigma_b}{2\pi f}} \quad [2.15]$$

is the so-called *contrast function*. Combining [2.13] and [2.14], it results

$$E_{scat,z}^{(s)}(x,y,f) = -k_b^2 \iint_D \chi(x',y',f) E_{tot,z}^{(s)}(x',y',f) g_{HS}(x,y,x',y',f) dx'dy' \quad [2.16]$$

The goal of quantitative GPR imaging is to retrieve the dielectric properties of targets in the investigation domain D (i.e. $\varepsilon(x,y)$ and $\sigma(x,y)$, or the contrast function $\chi(x,y,f)$, for $(x,y) \in D$) starting from a set of measurements of the scattered electric field $E_{scat,z}^{(s)}(x,y,f)$ collected in the measurement domain M, which is the set of locations where field probes can be placed. This constitutes the *inverse problem* we have to deal with. In our particular case of quasi-monostatic measurement configuration, for each sth view, just one point is available in M, that is, the one where the Rx antenna is placed with coordinates $\left(x_{Rx}^{(s)}, y_{Rx}^{(s)}\right)$.

The use of equation [2.16] to find the dielectric properties of buried targets solving the inverse problem is not trivial at all. First, it should be noted that the double integral over D requires, in addition to the Green's function, the knowledge of $E_{tot,z}^{(s)}(x,y,f)$ for $(x,y) \in D$. This is somewhat tricky, since $E_{tot,z}^{(s)}$ in turn depends on the properties of the buried targets, here represented by $\chi(x,y,f)$. Consequently, the link between $E_{scat,z}^{(s)}$ and χ (or, equivalently, to ε and σ) is a *nonlinear* relationship. Second, in GPR imaging $E_{scat,z}^{(s)}$ is only available in an aspect-limited configuration since field measurements are possible only over the surface of the terrain or in boreholes (Jia et al. 2002; Di Donato and Crocco 2015; Estatico et al. 2015b, 2019). Third, the presence of the air–soil interface and possibly lossy soils cause an additional decrease in the available information (Bucci et al. 2001). These facts further complicate the solution of the inverse problem, which is severely ill-posed.

In order to get as much information as possible from the measured samples of the scattered field, a great help comes from the simultaneous exploitation of multifrequency data (Bucci et al. 2000). This strategy plays a key role in GPR prospecting (Estatico et al. 2015a). So, let us assume we have the following measured data available:

$$E_{scat,z}^{(s)}\left(x_{Rx}^{(s)}, y_{Rx}^{(s)}, f_i\right), \qquad s = 1, \dots, S, \qquad i = 1, \dots, F \qquad [2.17]$$

NOTE.– Such samples of the scattered electric field at multiple frequencies $E_{scat,z}^{(s)}\left(x_{Rx}^{(s)}, y_{Rx}^{(s)}, f_i\right)$ may be obtained directly from frequency-domain GPR systems or, more typically, by processing the data acquired by standard GPRs in time domain $E_{scat,z}^{(s)}\left(x_{Rx}^{(s)}, y_{Rx}^{(s)}, t\right), t \in [0, T_w]$ through a Fourier transformation (e.g. fast Fourier transform (FFT)) (Carlson and Crilly 2010).

However, the scattered field due to the buried targets is not so easy to retrieve in a practical situation, since it is impossible to remove the targets and have an exact measurement of the incident field. This problem is solved (in an approximate way) by estimating the scattered or incident field contributions by means of a suitable technique that processes the measured B-scan of the total field. As a first approximation, global averages or moving averages can be used to estimate the incident field from B-scans (since most incident field contributions remain the same across the various views). The retrieved estimation of the incident field is subsequently subtracted from measured signals to obtain the scattered contribution. This common method is known as *average removal*. Other more advanced examples, which have also been successfully validated with experimental data, include entropy-based approaches (Solimene et al. 2014), as well as adaptive filtering-based techniques (Schenone et al. 2022).

However, equation [2.16] is not yet ready to be exploited in multifrequency data inversion, because of the dependence on frequency of the contrast function $\chi(x, y, f)$. This quantity cannot be the unknown of the inverse problem if more than one frequency is jointly processed. To solve such an issue, a different definition of the dielectric properties in equation [2.16] is recommended. Let us consider the following frequency-independent quantity that describes the dielectric properties of the scenario under test:

$$\mathbf{x}(x, y) = \left[\frac{\varepsilon(x, y) - \varepsilon_b}{c_\varepsilon}, \qquad \frac{\sigma(x, y) - \sigma_b}{c_\sigma}\right]^T \qquad [2.18]$$

where superscript T denotes the transpose operator, and c_ε, c_σ are two normalization constants. In the subsoil outside the target, or in general, outside the investigation domain D, we have $\mathbf{x}(x, y) = \mathbf{0}$ for $(x, y) \notin D$ and $y < 0$.

If a transformation vector is defined as

$$\mathbf{T}(f) = \frac{1}{\varepsilon_b - j\frac{\sigma_b}{2\pi f}}[c_\varepsilon, \ c_\sigma]^T \qquad [2.19]$$

it turns out that the contrast function, at a given frequency f, can be expressed as

$$\chi(x, y, f) = \mathbf{x}(x, y) \cdot \mathbf{T}(f) \qquad [2.20]$$

The result is that equation [2.16] can be written in terms of $\mathbf{x}(x, y)$ as follows:

$$E_{scat,z}^{(s)}(x, y, f) = -k_b^2 \iint_D \mathbf{x}(x', y') \cdot \mathbf{T}(f) E_{tot,z}^{(s)}(x', y', f) g_{HS}(x, y, x', y', f) dx'dy' \qquad [2.21]$$

This equation, evaluated for points $(x, y) \in M$, can be exploited as the *data equation*, that is, to model the relationship between the measured data and the dielectric properties of D at frequency f:

$$E_{scat,z}^{(s)}(x, y, f) = -k_b^2 \iint_D \mathbf{x}(x', y') \cdot \mathbf{T}(f) E_{tot,z}^{(s)}(x', y', f) g_{HS}(x, y, x', y', f) dx'dy', (x, y) \in M \qquad [2.22]$$

Nevertheless, as previously noted, $E_{tot,z}^{(s)}$ depends on the buried target properties, that is, it is another unknown of the inverse problem and depends on $\mathbf{x}$. As a result, another equation is needed to solve the inverse problem. Evaluating [2.21] in the points $(x, y) \in D$ and using the definition of the scattered electric field, it is possible to obtain the *state equation* as

$$E_{tot,z}^{(s)}(x, y, f) = E_{inc,z}^{(s)}(x, y, f) - k_b^2 \iint_D \mathbf{x}(x', y') \cdot \mathbf{T}(f) E_{tot,z}^{(s)}(x', y', f) g_{HS}(x, y, x', y', f) dx'dy', (x, y) \in D \qquad [2.23]$$

Assuming $E_{inc,z}^{(s)}(x, y, f)$ is known, the joint use of [2.22] and [2.23] theoretically allows us to find both $E_{tot,z}^{(s)}$ and $\mathbf{x}$, that is, the dielectric properties of a cross-section of the buried targets.

To write the problem in a more compact form, let us define the following *data* and *state* integral operators:

$$\mathcal{G}_M^f a(x,y,f) = -k_b^2 \iint_D a(x',y',f) g_{HS}(x,y,x',y',f) dx'dy', (x,y) \in M \quad [2.24]$$

$$\mathcal{G}_D^f a(x,y,f) = -k_b^2 \iint_D a(x',y',f) g_{HS}(x,y,x',y',f) dx'dy', (x,y) \in D \quad [2.25]$$

If we apply such definitions, equations [2.22] and [2.23] become

$$E_{scat,z}^{(s)}(x,y,f) = \mathcal{G}_M^f[\mathbf{x}(x,y) \cdot \mathbf{T}(f)] E_{tot,z}^{(s)}(x,y,f), (x,y) \in M \quad [2.26]$$

$$E_{tot,z}^{(s)}(x,y,f) = E_{inc,z}^{(s)}(x,y,f) + \mathcal{G}_D^f[\mathbf{x}(x,y) \cdot \mathbf{T}(f)] E_{tot,z}^{(s)}(x,y,f), (x,y) \in D \quad [2.27]$$

Now, retrieving the total electric field from [2.27] as

$$E_{tot,z}^{(s)}(x,y,f) = \{I - \mathcal{G}_D^f[\mathbf{x}(x,y) \cdot \mathbf{T}(f)]\}^{-1} E_{inc,z}^{(s)}(x,y,f), (x,y) \in D, \quad [2.28]$$

I being the identity operator, and substituting this expression into [2.26], we find that

$$E_{scat,z}^{(s)}(x,y,f) = \mathcal{G}_M^f[\mathbf{x}(x,y) \cdot \mathbf{T}(f)]\{I - \mathcal{G}_D^f[\mathbf{x}(x,y) \cdot \mathbf{T}(f)]\}^{-1} E_{inc,z}^{(s)}(x,y,f), (x,y) \in M \quad [2.29]$$

This equation can be compactly written as

$$E_{scat,z}^{(s)}(x,y,f) = \mathcal{N}_f(\mathbf{x})(x,y), \qquad (x,y) \in M \quad [2.30]$$

where $\mathcal{N}_f$ is a nonlinear operator that describes the link between $E_{scat,z}^{(s)}$ and $\mathbf{x}$ at the frequency f.

At this point, the last step is to combine the available data for the various frequencies and views (see [2.17]) as a system of $S \cdot F$ nonlinear equations:

$$\begin{bmatrix} \mathcal{N}_1(\mathbf{x})\left(x_{Rx}^{(1)}, y_{Rx}^{(1)}\right) \\ \mathcal{N}_2(\mathbf{x})\left(x_{Rx}^{(2)}, y_{Rx}^{(2)}\right) \\ \vdots \\ \mathcal{N}_F(\mathbf{x})\left(x_{Rx}^{(S)}, y_{Rx}^{(S)}\right) \end{bmatrix} = \begin{bmatrix} E_{scat,z}^{(1)}\left(x_{Rx}^{(1)}, y_{Rx}^{(1)}, f_1\right) \\ E_{scat,z}^{(2)}\left(x_{Rx}^{(2)}, y_{Rx}^{(2)}, f_1\right) \\ \vdots \\ E_{scat,z}^{(S)}\left(x_{Rx}^{(S)}, y_{Rx}^{(S)}, f_F\right) \end{bmatrix} \quad [2.31]$$

Moreover, since the unknown $\mathbf{x}$ is real-valued, the whole problem can be translated to a real-valued one through the separation between real and imaginary parts as

$$\begin{bmatrix} \Re\left\{\mathcal{N}_1(\mathbf{x})\left(x_{Rx}^{(1)}, y_{Rx}^{(1)}\right)\right\} \\ \Im\left\{\mathcal{N}_1(\mathbf{x})\left(x_{Rx}^{(1)}, y_{Rx}^{(1)}\right)\right\} \\ \Re\left\{\mathcal{N}_2(\mathbf{x})\left(x_{Rx}^{(2)}, y_{Rx}^{(2)}\right)\right\} \\ \Im\left\{\mathcal{N}_2(\mathbf{x})\left(x_{Rx}^{(2)}, y_{Rx}^{(2)}\right)\right\} \\ \vdots \\ \Re\left\{\mathcal{N}_F(\mathbf{x})\left(x_{Rx}^{(S)}, y_{Rx}^{(S)}\right)\right\} \\ \Im\left\{\mathcal{N}_F(\mathbf{x})\left(x_{Rx}^{(S)}, y_{Rx}^{(S)}\right)\right\} \end{bmatrix} = \begin{bmatrix} \Re\left\{E_{scat,z}^{(1)}\left(x_{Rx}^{(1)}, y_{Rx}^{(1)}, f_1\right)\right\} \\ \Im\left\{E_{scat,z}^{(1)}\left(x_{Rx}^{(1)}, y_{Rx}^{(1)}, f_1\right)\right\} \\ \Re\left\{E_{scat,z}^{(2)}\left(x_{Rx}^{(2)}, y_{Rx}^{(2)}, f_1\right)\right\} \\ \Im\left\{E_{scat,z}^{(2)}\left(x_{Rx}^{(2)}, y_{Rx}^{(2)}, f_1\right)\right\} \\ \vdots \\ \Re\left\{E_{scat,z}^{(S)}\left(x_{Rx}^{(S)}, y_{Rx}^{(S)}, f_F\right)\right\} \\ \Im\left\{E_{scat,z}^{(S)}\left(x_{Rx}^{(S)}, y_{Rx}^{(S)}, f_F\right)\right\} \end{bmatrix} \quad [2.32]$$

If the following real-valued vector functions are defined,

$$\mathbf{N}^{MF}(\mathbf{x}) \triangleq \begin{bmatrix} \Re\left\{\mathcal{N}_1(\mathbf{x})\left(x_{Rx}^{(1)}, y_{Rx}^{(1)}\right)\right\} \\ \Im\left\{\mathcal{N}_1(\mathbf{x})\left(x_{Rx}^{(1)}, y_{Rx}^{(1)}\right)\right\} \\ \Re\left\{\mathcal{N}_2(\mathbf{x})\left(x_{Rx}^{(2)}, y_{Rx}^{(2)}\right)\right\} \\ \Im\left\{\mathcal{N}_2(\mathbf{x})\left(x_{Rx}^{(2)}, y_{Rx}^{(2)}\right)\right\} \\ \vdots \\ \Re\left\{\mathcal{N}_F(\mathbf{x})\left(x_{Rx}^{(S)}, y_{Rx}^{(S)}\right)\right\} \\ \Im\left\{\mathcal{N}_F(\mathbf{x})\left(x_{Rx}^{(S)}, y_{Rx}^{(S)}\right)\right\} \end{bmatrix}; \quad \mathbf{y} \triangleq \begin{bmatrix} \Re\left\{E_{scat,z}^{(1)}\left(x_{Rx}^{(1)}, y_{Rx}^{(1)}, f_1\right)\right\} \\ \Im\left\{E_{scat,z}^{(1)}\left(x_{Rx}^{(1)}, y_{Rx}^{(1)}, f_1\right)\right\} \\ \Re\left\{E_{scat,z}^{(2)}\left(x_{Rx}^{(2)}, y_{Rx}^{(2)}, f_1\right)\right\} \\ \Im\left\{E_{scat,z}^{(2)}\left(x_{Rx}^{(2)}, y_{Rx}^{(2)}, f_1\right)\right\} \\ \vdots \\ \Re\left\{E_{scat,z}^{(S)}\left(x_{Rx}^{(S)}, y_{Rx}^{(S)}, f_F\right)\right\} \\ \Im\left\{E_{scat,z}^{(S)}\left(x_{Rx}^{(S)}, y_{Rx}^{(S)}, f_F\right)\right\} \end{bmatrix} \quad [2.33]$$

the notation can be simplified as

$$\mathbf{N}^{MF}(\mathbf{x}) = \mathbf{y} \quad [2.34]$$

where $\mathbf{x} \in X$, $\mathbf{y} \in Y$, and $\mathbf{N}^{MF}: X \rightarrow Y$, with X being the functional space of the unknown in the investigation domain D, and Y the functional space of the data in the measurement domain M. The inverse problem, which is ill-posed and clearly nonlinear, consists of finding a suitable solution $\mathbf{x}$ given a (noisy) version of data $\mathbf{y}$.

2.2. Quantitative inversion

In order to solve the inverse problem in a quantitative way by retrieving the unknown $\mathbf{x}(x,y),(x,y) \in D$ from equation [2.34], several methods have been described in the scientific literature (Pastorino and Randazzo 2018). Here, we take the example of a recently proposed nonlinear deterministic technique based on the inexact-Newton method.

Two nested iterative cycles constitute the main structure of the considered algorithm, whose simplified flow chart is illustrated in Figure 2.2. Basically, the external loop performs an iterative linearization of [2.34] and estimates the problem solution $\mathbf{x}$ by means of a set of successive approximations $\mathbf{x}_k$, k being the iteration number. The internal loop is devoted to the regularized solution of the linearized equation, and in this case is composed by a truncated Landweber-like procedure formulated in Lebesgue spaces with non-constant exponents $L^{p(\cdot)}$ (Estatico et al. 2018; Fedeli et al. 2020). In more detail, the whole algorithm works as follows:

1) initialization of the method with an initial guess $\mathbf{x_0} = \mathbf{0} \in X$ (i.e. an investigation domain with homogeneous dielectric properties equal to those of soil) and an exponent value $p_0(x,y) = p_{min}$;

2) linearization of equation [2.34] through Newton's first-order expansion based on the Fréchet derivative of the operator $\mathbf{N}^{MF}$ around the current solution $\mathbf{x}_k$, denoted as $\mathbf{N}_{\mathbf{x}_k}^{MF\prime}$. The obtained linear equation is

$$\mathbf{N}_{\mathbf{x}_k}^{MF\prime}\boldsymbol{\delta}\mathbf{x}_k = \mathbf{y} - \mathbf{N}^{MF}(\mathbf{x}_k) \qquad [2.35]$$

where $\boldsymbol{\delta}\mathbf{x}_k \in X$, and $\mathbf{N}_{\mathbf{x}_k}^{MF\prime}: X \to Y$, and it is solved in the unknown $\boldsymbol{\delta}\mathbf{x}_k$ by using the truncated Landweber-like approach in $L^{p(\cdot)}$ spaces outlined below. An approximation $\widetilde{\boldsymbol{\delta}\mathbf{x}}_k$ of the unknown is found;

3) update of the current solution as

$$\mathbf{x}_{k+1}(x,y) = \mathbf{x}_k(x,y) + \widetilde{\boldsymbol{\delta}\mathbf{x}}_k(x,y) \qquad [2.36]$$

and the exponent function:

$$p_{k+1}(x,y) = p_{min} + (p_{max} - p_{min})\frac{|\mathbf{x}_{k+1}(x,y)|}{\min_{(x,y)\in D}|\mathbf{x}_{k+1}(x,y)|} \qquad [2.37]$$

4) repetition of steps (2)–(4) with $k \leftarrow k+1$ until a stopping rule is fulfilled (maximum number of external iterations K_{IN}, or minimum variation of data residual between subsequent steps).

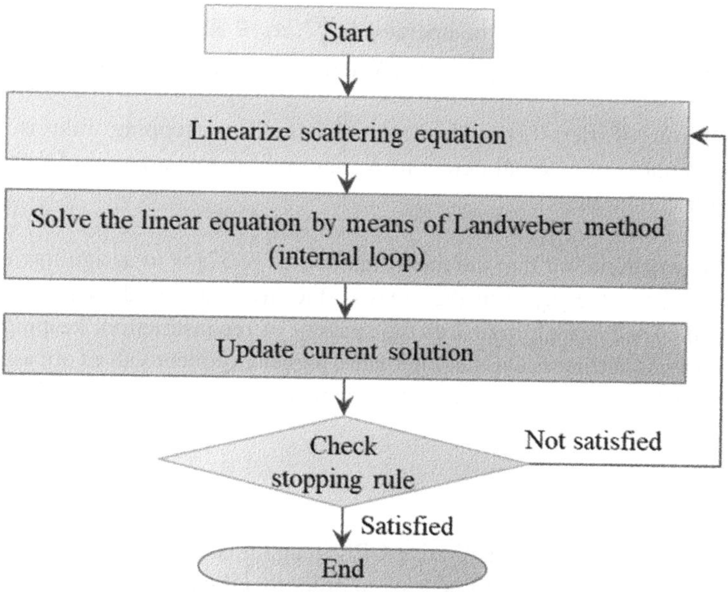

Figure 2.2. *Simplified flow chart of the quantitative inexact-Newton inversion method for GPR imaging*

In order to find the regularized solution of [2.35], the truncated Landweber-like method is applied, where X (space of unknown) is a variable-exponent Lebesgue space $L^{p_k(x,y)}$, while Y (space of data) is an L^{p_m} space with constant exponent (p_m is chosen to be equal to the average value of $p_k(x,y)$, $(x,y) \in D$). The objective of the inner solver is to minimize the cost functional $R_k: X \to \mathbb{R}$

$$R_k(\mathbf{x}) = \frac{1}{2}\left\|\mathbf{N}_{\mathbf{x}_k}^{MF'}\boldsymbol{\delta}\mathbf{x}_{k,n} - \mathbf{y}_k\right\|_Y^2 \qquad [2.38]$$

where $\mathbf{y}_k = \mathbf{y} - \mathbf{N}^{MF}(\mathbf{x}_k)$. The method, whose iterations are indicated with index n, has the following structure:

1) initialization of the solution as $\boldsymbol{\delta}\mathbf{x}_{k,0} = \mathbf{0}$;

2) iterative update of the solution as

$$\boldsymbol{\delta}\mathbf{x}_{k,n+1} = \mathbf{J}_{X^*}\left[\mathbf{J}_X\left(\boldsymbol{\delta}\mathbf{x}_{k,n}\right) - \alpha_k \mathbf{N}_{\mathbf{x}_k}^{MF'^*} \mathbf{J}_Y\left(\mathbf{N}_{\mathbf{x}_k}^{MF'} \boldsymbol{\delta}\mathbf{x}_{k,n} - \mathbf{y}_k\right)\right], \quad [2.39]$$

where $\mathbf{N}_{\mathbf{x}_k}^{MF'^*}$ is the adjoint of the operator $\mathbf{N}_{\mathbf{x}_k}^{MF'}$, $\alpha_k \in \mathbb{R}^+$is the step width and $\mathbf{J}_{X^*}$, $\mathbf{J}_X$, $\mathbf{J}_Y$ are the duality maps of spaces X^*, X and Y (Estatico et al. 2018);

3) repetition of step (2) with $n \leftarrow n+1$ until a stopping rule is fulfilled (maximum number of external iterations N_{LW}, or minimum variation of data residual between subsequent steps).

NOTE.– It is worth noting that the main function of [2.37] is to assign low exponent values (near the value p_{min}) to the areas of the investigation domain D where no targets are detected, which improves the sparsity of reconstruction, keeping it clean from background artifacts. On the other side, higher exponent values are assigned to the regions of D belonging to the targets (Estatico et al. 2018; Fedeli et al. 2020).

2.3. Hybrid techniques

Clearly, the problem of quantitative GPR imaging is very challenging, especially due to its nonlinearity and ill-posedness. In order to aid the inversion procedure, it is interesting to evaluate the possible combination with qualitative techniques, which are very common in GPR imaging and quite fast to apply (even though they cannot provide a full dielectric characterization of targets, but only some of their properties, like position and/or shape) (Persico 2014; Catapano et al. 2015). The result of such a combination takes the name of *hybrid* inversion techniques: their goal is to combine the benefits of both qualitative and quantitative methods.

Let us take as an example the well-known qualitative delay-and-sum method (Fear et al. 2002; Byrne et al. 2010). If a set of time-domain GPR measurements of the *z*-component of the scattered electric field are available as

$$E_{scat,z}^{(s)}\left(x_{Rx}^{(s)}, y_{Rx}^{(s)}, t\right), \qquad s = 1, \dots, S, \qquad t \in [0, T_W] \quad [2.40]$$

the method consists of calculating, for each point of the investigation domain D, the integral

$$\Gamma(x, y) = \int_0^{T_W} \left\{\sum_{s=1}^{S} E_{scat,z}^{(s)}\left(x_{Rx}^{(s)}, y_{Rx}^{(s)}, t - \tau_s(x, y)\right)\right\}^2 dt, \quad (x, y) \in D \quad [2.41]$$

In this expression,

$$\tau_s(x,y) = \frac{d_{Tx}^{(s)}(x,y) + d_{Rx}^{(s)}(x,y)}{v_b} \quad [2.42]$$

where v_b is the velocity of propagation of electromagnetic waves in the soil medium. If antennas are moved on a measurement line in contact with the air–soil interface, the terms $d_{Tx}^{(s)}(x,y)$ and $d_{Rx}^{(s)}(x,y)$ in [2.42] can be simply calculated as

$$d_{Tx}^{(s)}(x,y) = \sqrt{\left(x - x_{Tx}^{(s)}\right)^2 + \left(y - y_{Tx}^{(s)}\right)^2} \quad [2.43]$$

$$d_{Rx}^{(s)}(x,y) = \sqrt{\left(x - x_{Rx}^{(s)}\right)^2 + \left(y - y_{Rx}^{(s)}\right)^2} \quad [2.44]$$

The qualitative image obtained from the evaluation of [2.41], which is related to the reflectivity of targets in D, can also be normalized based on its maximum value as

$$\Gamma_n(x,y) = \frac{\Gamma(x,y)}{\max\limits_{(x,y)\in D} \Gamma(x,y)}, \qquad (x,y) \in D \quad [2.45]$$

Indeed, such a qualitative image can provide key information to a quantitative inversion method like the one described in section 2.2. For instance, the normalized qualitative map can be exploited to

– construct an initial map of the exponent function, assigning higher values of $p(x, y)$ to the points where the qualitative method detects the presence of the object (i.e. Γ_n is close to 1) and lower values outside of it (i.e. where $\Gamma_n \approx 0$);

– guide the updates of the unknown function to focus the quantitative reconstruction algorithm on the update of $\mathbf{x}(x,y)$ in points $(x,y) \in D$, where the qualitative procedure has found significant discontinuities with respect to the background properties.

Modifying the quantitative method in order to implement these two ideas, the hybridized inexact-Newton scheme now assumes the following structure:

1) Initialization of the method with an initial guess $\mathbf{x_0} = \mathbf{0} \in X$ and an exponent value derived from the qualitative image

$$p_0(x,y) = p_{min} + (p_{max} - p_{min})\Gamma_n(x,y) \quad [2.46]$$

2) Linearization of equation [2.34] through Newton's first-order expansion and regularized solution of the obtained equation through the truncated Landweber-like approach in $L^{p(\cdot)}$ spaces. An approximation $\widetilde{\boldsymbol{\delta}\mathbf{x}}_k$ of the unknown is found.

3) Guided update of the current solution by weighting $\widetilde{\boldsymbol{\delta}\mathbf{x}}_k$ by the qualitative indicator Γ_n,

$$\mathbf{x}_{k+1}(x,y) = \mathbf{x}_k(x,y) + \Gamma_n(x,y)\widetilde{\boldsymbol{\delta}\mathbf{x}}_k(x,y) \tag{2.47}$$

and update of the exponent function like in equation [2.37].

4) Repetition of steps (2)–(4) with $k \leftarrow k+1$ until a stopping rule is fulfilled (maximum number of external iterations K_{IN}, or minimum variation of data residual between subsequent steps).

2.4. Example of application in a simulated scenario

A numerical example is presented here to show and compare the potentialities of nonlinear and hybrid inversion techniques, with reference to the algorithms presented in the previous sections. Validation with real experimental data is presented in Schenone et al. (2022).

Let us consider the scenario in Figure 2.3, where a half-space configuration is assumed. The upper medium is air (modeled with the dielectric properties of vacuum, that is, with dielectric permittivity $\varepsilon_u = \varepsilon_0 \simeq 8.85 \times 10^{-12}$ F/m, and $\sigma_u = 0$ S/m). The lower medium is a medium-dry soil characterized by $\varepsilon_b = 5\varepsilon_0$ and $\sigma_b = 38 \times 10^{-3}$ S/m (Salucci et al. 2021). The interface between these two layers is the xz horizontal plane located at $y = 0$. A buried z-directed cylindrical target with circular cross-section of radius $r = 0.08$ m is centered at $(x_T, y_T) = (-0.25, -0.25)$ m. The target is characterized by the dielectric properties of vacuum.

The measurement system has a quasi-monostatic configuration with source (Tx) and one ideal probing point (Rx) sliding on a measurement line of length $l = 2$ m parallel to the x axis, at a height $h = 0.01$ m from the air–soil interface in the upper medium. On this line, measurements are taken with a fixed-offset approach, with Tx and Rx in $S = 100$ different positions (i.e. S views) with an offset $\Delta L = 0.2$ m. In particular, in the sth view ($s = 1, \dots, S$), the Tx is located at

$$\left(x_{Tx}^{(s)}, y_{Tx}^{(s)}\right) = \left(x_{start} + L\frac{s-1}{S-1} - \frac{\Delta L}{2}, h\right) \tag{2.48}$$

with $x_{start} = -1$ m, whereas the Rx is at

$$\left(x_{Rx}^{(s)}, y_{Rx}^{(s)}\right) = \left(x_{start} + L\frac{s-1}{S-1} + \frac{\Delta L}{2}, h\right) \quad [2.49]$$

Simulated data have been obtained by means of the open-source finite-difference time-domain (FD-TD) solver *gprMax* (Warren et al. 2016). In particular, the simulation domain has a size 3 m × 2 m, and it is discretized into $N_{fwd} = 300 \times 200$ square cells with 0.01 m side length (total cells: 60,000). A time window $T_w = 30$ ns has been simulated, with 1,273 FD-TD iterations (time step: 23.6 ps). In place of the Tx antenna, a line-current source has been considered, excited with a Ricker pulse with central frequency $f_c = 250$ MHz.

The resulting B-scans in the simulated GPR survey are reported in Figure 2.4. This figure shows the behavior of the z-component of incident, total and scattered electric fields in the time domain, as measured by the ideal Rx probe in the various points of the measurement domain M. Figure 2.4(a) highlights the reflection from the air-soil interface (i.e. the B-scan of the incident field), which also represents the most significant contribution to the measured total field (Figure 2.4(b)). As expected, the field scattered by the buried target is somehow hidden in Figure 2.4(b), being masked by the air–soil reflection phenomena. The true z-component of the electric field scattered by the target is graphed in Figure 2.4(c), which shows the classical hyperbolic pattern. However, as previously noted, this quantity is typically unknown in a practical case since an exact measurement of the incident field removing the target of interest cannot be done. The scattered field estimated with the mentioned preprocessing technique is shown in Figure 2.4(d). Despite the presence of a slight horizontal artifact, an acceptable estimation of the scattered field is achieved.

For the solution of the inverse scattering problem, a rectangular investigation domain D of size $h = 0.5$ m and $w = 0.8$ m on the xy plane has been considered, between points $\left(x_{D,min}, y_{D,min}\right) = (-0.4, -0.55)$ m and $\left(x_{D,max}, y_{D,max}\right) = (0.4, -0.05)$ m. The domain D has been discretized into a grid of $N_{inv} = 32 \times 20$ square cells with 0.025 m side length (total cells: 640).

First, let us present the results obtained with the pure quantitative method, applied to the scattered electric field estimated by the preprocessing technique and reported in Figure 2.4(d). In order to apply the inversion approach, the scattered electric field at $F = 4$ uniformly spaced frequencies (150, 233, 316 and 400 MHz) has been extracted from the time-domain data through the FFT. An additive white Gaussian noise with a mean value equal to zero and signal-to-noise ratio $SNR = 25$ dB has been also applied to the data. The inversion method has been executed with variable exponent parameters $p_{min} = 1.5$, $p_{max} = 2$ and $p_0(x,y) = p_{min}$, maximum numbers of iterations equal to $K_{IN} = 50$ and $N_{LW} = 50$ in outer inner loops, respectively, and minimum relative variation of the normalized residual between iterations $\Delta r = 10^{-4}$. The reconstructed distribution of the relative dielectric permittivity in the investigation domain D by means of the pure quantitative method is reported in Figure 2.5. As it can be noticed, the localization of the buried cylindrical target is good, but the contrast of the target is underestimated. Moreover, some artifacts appear on the top of the target, especially on the left side.

Now, let us consider the application of the described hybrid inversion method in the same simulated scenario. The result achieved by adopting the qualitative imaging technique based on the delay-and-sum method is shown in Figure 2.6, which evidences a correct target localization on the horizontal axis, even though it appears elongated and slightly shifted in the vertical direction. By exploiting this qualitative reconstruction, the hybrid imaging strategy leads to the result presented in Figure 2.7. The reconstructed distribution of the dielectric permittivity now appears significantly better than the one obtained by means of the quantitative method alone, reporting a very good characterization of the buried target: the position, shape, and dielectric properties of the cylinder are retrieved in a very accurate way, and the background is left without artifacts.

The results have also been analyzed through the calculation of relative reconstruction errors on the whole investigation domain (e_{inv}), on the target (e_{tar}) and on the background region (e_{bkg}) as follows:

$$e_{inv} = \frac{1}{N_{inv}} \sum_{n \in D} \frac{|\varepsilon_r(x_n, y_n) - \widehat{\varepsilon_r}(x_n, y_n)|}{|\varepsilon_b|} \qquad [2.50]$$

$$e_{tar} = \frac{1}{N_{tar}} \sum_{n \in D_{tar}} \frac{|\varepsilon_r(x_n, y_n) - \widehat{\varepsilon_r}(x_n, y_n)|}{|\varepsilon_b|} \qquad [2.51]$$

$$e_{bkg} = \frac{1}{N_{inv} - N_{tar}} \sum_{n \in D \setminus D_{tar}} \frac{|\varepsilon_r(x_n, y_n) - \widehat{\varepsilon_r}(x_n, y_n)|}{|\varepsilon_b|} \qquad [2.52]$$

where $\varepsilon_r(x_n, y_n)$ and $\widehat{\varepsilon_r}(x_n, y_n)$ are the actual and reconstructed relative dielectric permittivity in the nth cell of the investigation domain, and D_{tar} is the subset of D that encloses the cross-section of the target (N_{tar} being the number of cells of this region). These error metrics are reported, in the present case, in Table 2.1, where the pure qualitative method (top row) is compared with the hybrid approach (bottom row). It is evident that all the considered error quantities are lower when the hybrid method is applied, as visually confirmed by the reconstructed images.

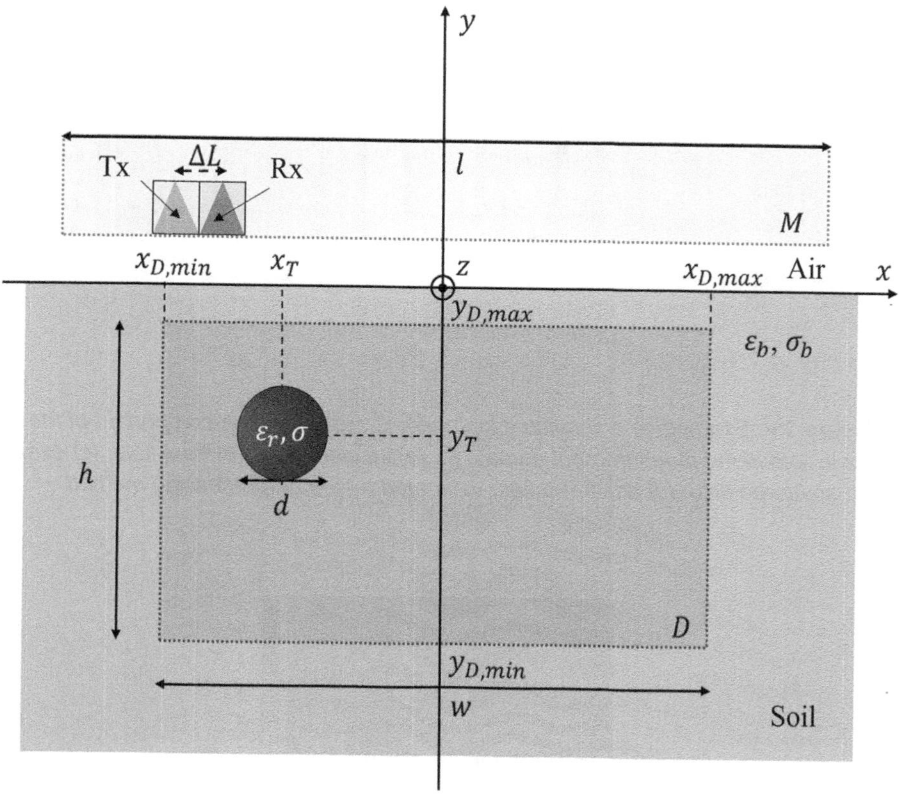

Figure 2.3. *Configuration of the numerical example, which includes a cylindrical buried target in a rectangular investigation domain*

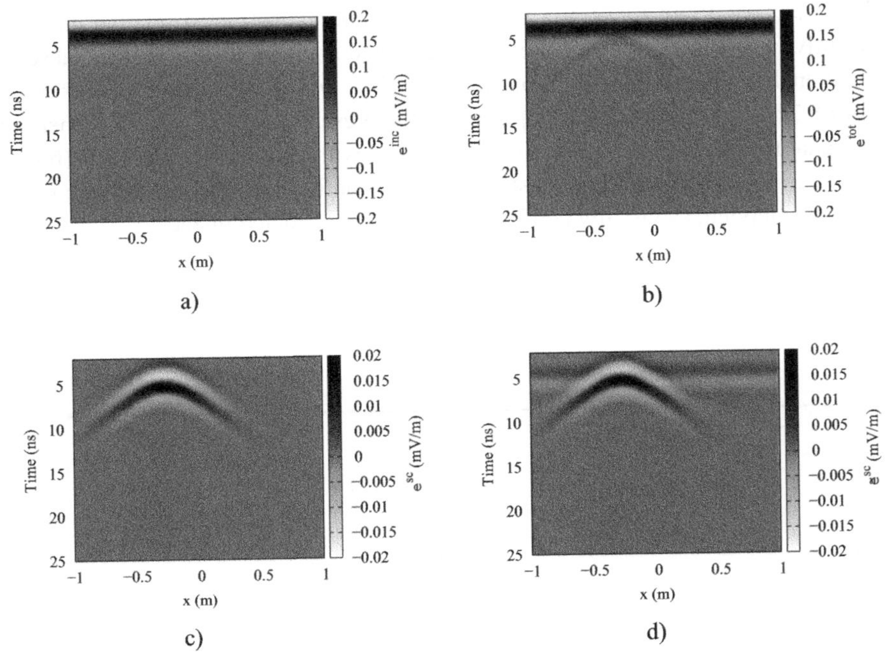

Figure 2.4. *Example in a simulated scenario: B-scans of the z-component of the electric field in the measurement domain M: (a) incident field; (b) total field; (c) exact scattered field; (d) scattered field estimated by the preprocessing method*

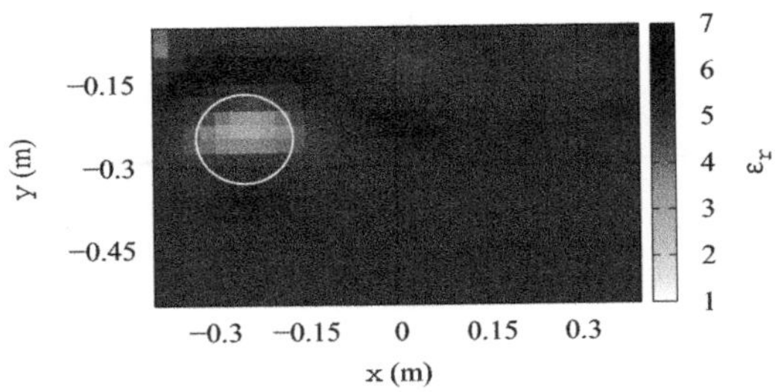

Figure 2.5. *Reconstructed distribution of the relative dielectric permittivity in the investigation domain D by means of the pure quantitative method*

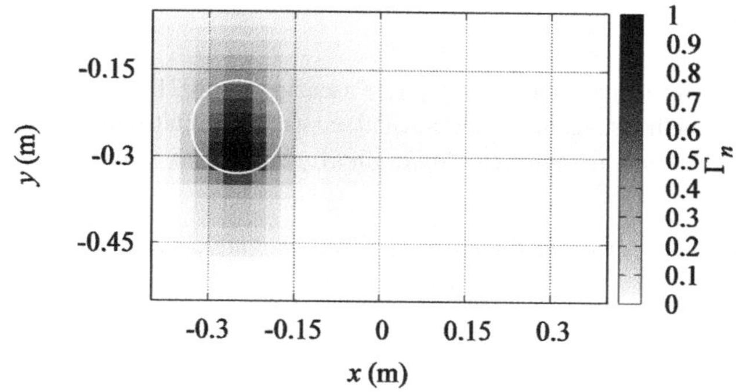

Figure 2.6. *Qualitative reconstruction of the target inside the investigation domain D by means of the delay-and-sum method*

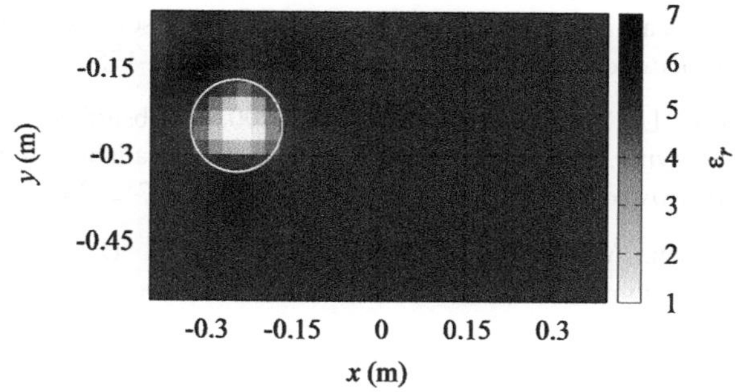

Figure 2.7. *Reconstructed distribution of the relative dielectric permittivity in the investigation domain D by means of the hybrid inversion method*

Method	Total error, e_{inv}	Target error, e_{tar}	Background error, e_{bkg}
Quantitative	0.0495	0.6062	0.0202
Hybrid	0.0292	0.4411	0.0075

Table 2.1. *Relative errors on the reconstruction of the whole investigation domain (e_{inv}), on the target (e_{tar}) and on the background region (e_{bkg})*

2.5. References

Ambrosanio, M., Bevacqua, M.T., Isernia, T., Pascazio, V. (2021). Performance analysis of tomographic methods against experimental contactless multistatic ground penetrating radar. *IEEE Journal of Selected Topics in Applied Earth Observations and Remote Sensing*, 14, 1171–1183.

Balanis, C.A. (2012). *Advanced Engineering Electromagnetics*, 2nd edition. John Wiley & Sons, Hoboken, NJ.

Baussard, A., Miller, E.L., Lesselier, D. (2004). Adaptive multiscale reconstruction of buried objects. *Inverse Problems*, 20(6), S1–S15.

Bevacqua, M.T. and Isernia, T. (2018). Boundary indicator for aspect limited sensing of hidden dielectric objects. *IEEE Geoscience and Remote Sensing Letters*, 15(6), 838–842.

Breard, A., Perrusson, G., Lesselier, D. (2008). Hybrid differential evolution and retrieval of buried spheres in subsoil. *IEEE Geoscience and Remote Sensing Letters*, 5(4), 788–792.

Bucci, O.M., Crocco, L., Isernia, T., Pascazio, V. (2000). Inverse scattering problems with multifrequency data: Reconstruction capabilities and solution strategies. *IEEE Transactions on Geoscience and Remote Sensing*, 38(4), 1749–1756.

Bucci, O.M., Crocco, L., Isernia, T., Pascazio, V. (2001). Subsurface inverse scattering problems: Quantifying, qualifying, and achieving the available information. *IEEE Transactions on Geoscience and Remote Sensing*, 39(11), 2527–2538.

Byrne, D., O'Halloran, M., Glavin, M., Jones, E. (2010). Data independent radar beamforming algorithms for breast cancer detection. *Progress in Electromagnetics Research*, 107, 331–348.

Carlson, A.B. and Crilly, P.B. (2010). *Communication Systems: An Introduction to Signals and Noise in Electrical Communication*, 5th edition. McGraw-Hill Higher Education, Boston, MA.

Catapano, I., Crocco, L., Isernia, T. (2008). Improved sampling methods for shape reconstruction of 3-D buried targets. *IEEE Transactions on Geoscience and Remote Sensing*, 46(10), 3265–3273.

Catapano, I., Randazzo, A., Slob, E., Solimene, R. (2015). GPR imaging via qualitative and quantitative approaches. In *Civil Engineering Applications of Ground Penetrating Radar*, Benedetto, A. and Pajewski, L. (eds). Springer International Publishing, Cham.

Chen, X., Wei, Z., Li, M., Rocca, P. (2020). A review of deep learning approaches for inverse scattering problems. *Progress in Electromagnetics Research*, 167, 67–81.

Daniels, D.J. (2004). *Ground Penetrating Radar*, 2nd edition. The Institution of Electrical Engineers, London.

De Coster, A. and Lambot, S. (2019). Full-wave removal of internal antenna effects and antenna–medium interactions for improved ground-penetrating radar imaging. *IEEE Transactions on Geoscience and Remote Sensing*, 57(1), 93–103.

Di Donato, L. and Crocco, L. (2015). Model-based quantitative cross-borehole GPR imaging via virtual experiments. *IEEE Transactions on Geoscience and Remote Sensing*, 53(8), 4178–4185.

Estatico, C., Fedeli, A., Pastorino, M., Randazzo, A. (2015a). A multifrequency inexact-Newton method in Lp Banach spaces for buried objects detection. *IEEE Transactions on Antennas and Propagation*, 63(9), 4198–4204.

Estatico, C., Fedeli, A., Pastorino, M., Randazzo, A. (2015b). Buried object detection by means of a Lp Banach-space inversion procedure. *Radio Science*, 50(1), 41–51.

Estatico, C., Fedeli, A., Pastorino, M., Randazzo, A. (2018). Quantitative microwave imaging method in Lebesgue spaces with nonconstant exponents. *IEEE Transactions on Antennas and Propagation*, 66(12), 7282–7294.

Estatico, C., Fedeli, A., Pastorino, M., Randazzo, A. (2019). Variable exponent Lebesgue space inversion for cross-borehole subsurface imaging. *URSI Radio Science Letters*, 1, 1–5.

Fear, E.C., Li, X., Hagness, S.C., Stuchly, M.A. (2002). Confocal microwave imaging for breast cancer detection: Localization of tumors in three dimensions. *IEEE Transactions on Biomedical Engineering*, 49(8), 812–822.

Fedeli, A. (2021). Microwave tomography with LSTM-based processing of the scattered field. *IEEE Open Journal of Antennas and Propagation*, 2, 213–223.

Fedeli, A., Pastorino, M., Randazzo, A. (2017). Advanced inversion techniques for ground penetrating radar. *Journal of Telecommunications and Information Technology*, 3, 37–42.

Fedeli, A., Estatico, C., Randazzo, A., Pastorino, M. (2020). Multifrequency microwave tomography in Lebesgue spaces with nonconstant exponents. *URSI Radio Science Letters*, 2, 1–4.

Gennarelli, G., Ludeno, G., Catapano, I., Soldovieri, F. (2020). Full 3-D imaging of vertical structures via ground-penetrating radar. *IEEE Transactions on Geoscience and Remote Sensing*, 58(12), 8857–8873.

Gurbuz, T.U., Aslanyurek, B., Karabulut, E.P., Akduman, I. (2014). An efficient nonlinear imaging approach for dielectric objects buried under a rough surface. *IEEE Transactions on Geoscience and Remote Sensing*, 52(5), 3013–3022.

Hajebi, M. and Hoorfar, A. (2021). A multiscale optimization technique for large-scale subsurface profiling. *IEEE Geoscience and Remote Sensing Letters*, 18(10), 1706–1710.

Hajebi, M., Tavakoli, A., Dehmollaian, M., Dehkhoda, P. (2018). An iterative modified diffraction tomography method for reconstruction of a high-contrast buried object. *IEEE Transactions on Geoscience and Remote Sensing*, 56(7), 4138–4148.

Jia, H., Takenaka, T., Tanaka, T. (2002). Time-domain inverse scattering method for cross-borehole radar imaging. *IEEE Transactions on Geoscience and Remote Sensing*, 40(7), 1640–1647.

Liu, B., Ren, Y., Liu, H., Xu, H., Wang, Z., Cohn, A.G., Jiang, P. (2021). GPRInvNet: Deep learning-based ground-penetrating radar data inversion for tunnel linings. *IEEE Transactions on Geoscience and Remote Sensing*, 59(10), 8305–8325.

Pastorino, M. (2010). *Microwave Imaging*. John Wiley & Sons, Hoboken, NJ.

Pastorino, M. and Randazzo, A. (2018). *Microwave Imaging Methods and Applications*. Artech House, Boston, MA.

Persico, R. (2014). *Introduction to Ground Penetrating Radar: Inverse Scattering and Data Processing*. John Wiley & Sons, Hoboken, NJ.

Salucci, M., Poli, L., Anselmi, N., Massa, A. (2017). Multifrequency particle swarm optimization for enhanced multiresolution GPR microwave imaging. *IEEE Transactions on Geoscience and Remote Sensing*, 55(3), 1305–1317.

Salucci, M., Estatico, C., Fedeli, A., Oliveri, G., Pastorino, M., Povoli, S., Randazzo, A., Rocca, P. (2021). 2-D TM GPR imaging through a multiscaling multifrequency approach in Lp spaces. *IEEE Transactions on Geoscience and Remote Sensing*, 59(12), 10011–10021.

Schenone, V., Fedeli, A., Estatico, C., Pastorino, M., Randazzo, A. (2022). Experimental assessment of a novel hybrid scheme for quantitative GPR imaging. *IEEE Geoscience and Remote Sensing Letters*, 19, 1–5.

Solimene, R., Cuccaro, A., Dell'Aversano, A., Catapano, I., Soldovieri, F. (2014). Ground clutter removal in GPR surveys. *IEEE Journal of Selected Topics in Applied Earth Observations and Remote Sensing*, 7(3), 792–798.

Stinson, D.C. (1976). *Intermediate Mathematics of Electromagnetics*. Prentice-Hall, Englewood Cliffs, NJ.

Sun, S., Kooij, B.J., Yarovoy, A.G. (2017). Linearized 3-D electromagnetic contrast source inversion and its applications to half-space configurations. *IEEE Transactions on Geoscience and Remote Sensing*, 55(6), 3475–3487.

Wang, J., Liu, H., Jiang, P., Wang, Z., Sui, Q., Zhang, F. (2022). GPRI2Net: A deep-neural-network-based ground penetrating radar data inversion and object identification framework for consecutive and long survey lines. *IEEE Transactions on Geoscience and Remote Sensing*, 60, 1–20.

Warren, C., Giannopoulos, A., Giannakis, I. (2016). gprMax: Open source software to simulate electromagnetic wave propagation for ground penetrating radar. *Computer Physics Communications*, 209, 163–170.

Wei, Z. and Chen, X. (2019). Deep-learning schemes for full-wave nonlinear inverse scattering problems. *IEEE Transactions on Geoscience and Remote Sensing*, 57(4), 1849–1860.

3

Sensor Deployment in Subsurface GPR Imaging

Maria Antonia MAISTO[1], Angela DELL'AVERSANO[1], Antonio CUCCARO[2] and Raffaele SOLIMENE[1]

[1]*Dipartimento di Ingegneria, Università della Campania Luigi Vanvitelli, Caserta, Italy*

[2]*Department of Computer, Modeling, Electronic and Systems Engineering, University of Calabria, Rende, Italy*

3.1. Introduction

The problem of imaging buried targets (subsurface imaging) finds application in a very large number of research fields. Civil engineering diagnostics (Pastorino 1998; Soldovieri et al. 2006; Hugenschmidt and Kalogeropoulos 2009; Randazzo and Estatico 2012), archaeological and geophysical prospecting (Conyers 1997; Piro et al. 2003; Orlando and Soldovieri 2008; Catapano et al. 2012) and cultural heritage monitoring (Masini et al. 2010; Masini and Soldovieri 2011) are only a few examples. Besides these classical fields of application, new fields such as through-wall imaging (Ahmad et al. 2005; Baranoski 2008; Solimene et al. 2009), contaminant detection (Bevan et al. 2003; Porsani et al. 2004), tunnel and underground facility detection (Meschino et al. 2013) and planet exploration (Watters et al. 2006) are gaining increasing interest.

In order to succeed in subsurface imaging, a number of requirements, which span from hardware to signal processing, must be complied with. For instance, it is often

For a color version of all the figures in this chapter, see www.iste.co.uk/serhir/radar.zip.

Ground Penetrating Radar,
coordinated by Mohammed SERHIR and Dominique LESSELIER.

required that imaging systems be of relatively small dimension (for portability), exhibit ultra-wide-band behavior and low levels of the direct coupling between TX/RX antennas. Moreover, as the targets are usually buried within extremely complicated (in-homogeneous) scenes, with media that can be lossy and dispersive, a number of "preliminary tasks" must be carried out before imaging. For example, clutter suppression procedures must be run in order to extract scattered field data (Solimene et al. 2014a). Also medium estimation is mandatory in order to obtain properly focused reconstructions (Gorriti and Slob 2005; Minet et al. 2010; Di Matteo et al. 2013; Solimene et al. 2014b). Finally, antenna deconvolution is generally needed (Daniels 2004; Lambot et al. 2004; Jadoon et al. 2011; Patriarca et al. 2013), the latter being particularly difficult when antennas are deployed in close proximity of the air–soil interface (Lambot and André 2014).

In this chapter, we mainly focus on the imaging procedure.

Subsurface imaging entails solving an electromagnetic inverse scattering problem (Pastorino 2010). As it is well known, such problems are extremely difficult to solve due to their nonlinearity (Chew 1999) and ill-posedness (Bertero and Boccacci, 1998). A great deal of research activities have been developed and documented in the literature to address nonlinear inverse scattering problems (Caorsi et al. 1994, 1995; Lobel et al. 1997; Franchois and Pichot 1997; Dourthe et al. 2000; Liu et al. 2002; Benedetti et al. 2006; Salvade et al. 2009; Meles et al. 2010; Ali and Moghaddam 2010; Pastorino 2010; Patriarca et al. 2011; Crocco and Soldovieri 2011). In this framework, the reconstruction/imaging is usually recast as an optimization problem where the minimum of a suitable cost function accounting for the misfit between the scattered field data and the model data must be found. Newton-like iterative methods (Roger 1981; Cui et al. 2001; Estatico et al. 2005a; Rubek et al. 2007; Estatico et al. 2012; Ostadrahimi et al. 2013) and gradient-based solution procedures (Harada et al. 1995; Lobel et al. 1996; Isernia et al. 1997; Van den Berg and Abubakar 2001) are generally the approaches upon which deterministic nonlinear inversions are based. As it is well known, this kind of inversion method can suffer from convergence problems and can experience limited reliability due to the occurrence of false solutions (Isernia et al. 2001). This stimulated researchers to look for strategies aimed at overcoming/mitigating the drawbacks mentioned above. For example, on the one hand, approximate and novel nonlinear scattering models, such as quadratic ones, have been developed to reduce the nonlinearity (Leone et al. 2003; Isernia et al. 2004; Estatico et al. 2005b; Catapano et al. 2006). On the other hand, it is possible to gradually increase the number of unknowns as minimization procedure evolves (Baussard et al. 2004; Catapano et al. 2004; Donelli et al. 2006; Oliveri et al. 2012) to more efficiently exploit the ratio between independent data/number of unknowns; in fact, the latter affects the curvature of the functional range and hence the occurrence of local minima. Also stochastic minimization schemes have been largely used (Rahmat-Samii and Michielssen 1999; Pastorino et al. 2000; Pastorino 2007; Rekanos 2008; Randazzo 2012). These methods are in principle able to overcome the problem of local minima, but their numerical complexity is often greater when compared to deterministic approaches. Finally, despite advancement in numerical modeling and in computer science, effective use of nonlinear inversion methods is still limited to a relatively small spatial region.

The problem is drastically simplified if linearized scattering models are exploited. Linearization can be achieved by invoking the Born or Rytov approximations (Chew 1999; Pastorino 2010) in the case of penetrable scatterers or the Kirchhoff one for metallic objects (Pierri et al. 2006). The resulting inverse problem is usually addressed as radar imaging and the corresponding methods can be coarsely grouped in "migration" methods and "inverse-filtering" procedures (Soldovieri and Solimene 2010). Migration algorithms (Bleistein 2001) were first developed within the framework of seismic prospecting. For a lossless medium, they exploit the time "reversibility" of the wave equation in order to enjoy an FFT-based implementation. Related imaging algorithms are the back-propagation (Hansen and Johansen 2000), the time-reversal (Fink 1992) and the omega-kappa algorithm, the latter usually in SAR imaging (Cafforio et al. 1991). When the scenario becomes lossy, FFT can no longer be used and migration implementation can be conveniently cast by approximating the inverse operator in terms of the adjoint of the scattering operator (Soldovieri and Solimene 2010). Inverse filtering casts the imaging as a linear inverse problem where reconstructions are obtained by inverting the linearized scattering operator for some object functions that account for the support and the electromagnetic features of the scatterers (Soldovieri and Solimene 2010).

Linear inversion imaging is computationally effective as compared to nonlinear methods and hence allows us to deal with scattering scenarios, which can be very large (in terms of wavelength) and do not suffer from reliability problems. However, ill-posedness still remains an issue. To this end, linear inversions must always be paired with some regularization scheme in order to establish a trade-off between accuracy and stability against noise (Bertero and Boccacci, 1998). In particular, while migration "implicitly" regularizes the problem, inverse filtering requires explicit use of a regularization scheme, which is obtained by enforcing some constraint on the solution, dictated by a priori information, or by adopting iterative solving procedures where regularization is achieved by controlling the number of iterations. However, we must remark that, since the class of scatterers for which linear models work is extremely limited, such imaging procedures generally do not allow us to obtain quantitative reconstructions. Therefore, detection, localization and rough information about the scatterers' shape is what we can expect to retrieve through linear inversion methods (Slaney et al. 1984; Pike and Sabatier 2002; Pastorino 2010). Fortunately, in many applicative contexts this is exactly what we are interested in.

The computational advantages of the linear model based inversion schemes are also shared by the so-called "qualitative" methods (Cakoni 2006). Such methods directly aim at retrieving only the scatterers' supports and do not require simplified models of the scattering phenomenon. In particular, they identify the spatial support of the scatterers by adopting certain indicator functions that assume very different values in norm depending whether they are evaluated within or outside the scatterer supports. The linear sampling (Colton et al. 2000) and the factorization (Kirsch 1999) methods as well as MUSIC-based approaches (Marengo et al. 2007) belong to this class of inversion algorithms. They require a multistatic/multi-view/single-frequency configuration. It has been noted, however, that their performances are reduced when aspect-limited configurations are adopted (Brandfass et al. 2001; Liseno and Pierri 2004).

In this chapter, we essentially consider subsurface radar imaging problems. In particular, we are interested in determining a strategy for sensor arrangement.

3.2. Linearized scattering model

A 2D scattering problem is considered. Invariance is assumed along the y-axis and $\mathbf{r} = (x, z) \in \mathbb{R}^2$ is the 2D position vector. The background medium is assumed to be nonmagnetic (i.e. its magnetic permeability is everywhere the one of free-space μ_0) and nondispersive so that its dielectric permittivity is denoted as $\epsilon_b = \epsilon_b(\mathbf{r})$.

Two types of background medium will be considered. The first type refers to the homogeneous free-space medium with $\epsilon_b = \epsilon_0$ everywhere. The second type, instead, concerns a two-layered medium with the *flat* separation interface located at $z = 0$. In this case, the upper half-space is assumed to be the free-space, hence, $\epsilon_b = \epsilon_u = \epsilon_0$ for $z > 0$. The lower half-space is electromagnetically denser so that $\epsilon_b = \epsilon_l$ for $z > 0$ and $\epsilon_l > \epsilon_u$. Of course, when $\epsilon_l = \epsilon_u = \epsilon_0$, the free-space scenario is restored. The half-space medium is a commonly employed simplified description of scattering scenarios of interest in GPR imaging. Accordingly, the primary field source as well as the sensors that collect the scattered field are located on the air side (upper half-space), whereas the target to be imaged is located (*buried*) in the lower half-space. Note that here, the free-space case is considered as well in order to establish a benchmark against which to appreciate the role played by the lower half-space permittivity.

Assume that the probing field is radiated by a 2D electric dipole directed along the y-axis, that is $\mathbf{J}(\omega, \mathbf{r}) = S(\omega)\delta(\mathbf{r} - \mathbf{r}_s)\hat{y}$. Here, ω is the angular frequency, $S(\omega)$ the radiated pulse time Fourier spectrum and $\mathbf{r}_s \in \Gamma_s$ the source position, with Γ_s being a curve of the $x - z$ plane with $z > 0$. Accordingly, the problem is also scalar since the electric field has only the y component, that is, $\underline{E}(\omega, \mathbf{r}) = E(\omega, \mathbf{r})\hat{y}$. Furthermore, assume that there is a target in the lower half-space, which is described in terms of its dielectric permittivity $\epsilon_t(\mathbf{r})$, which is supported over a compact not necessarily connected set D_T included within the investigation domain ID, that is $D_T \subset ID$. Indeed, ID is contained in the lower half-space and represents the spatial region where the target is a priori presumed to reside. Under the described framework, the scattering phenomenon is governed (in the frequency domain) by the following Helmholtz equation

$$\nabla_{\mathbf{r}}^2 E(\omega,\mathbf{r},\mathbf{r}_s) + k_b^2(\mathbf{r})E(\omega,\mathbf{r},\mathbf{r}_s) = -k_b^2(\mathbf{r})\chi(\mathbf{r})E(\omega,\mathbf{r},\mathbf{r}_s) + j\omega\mu_0 S(\omega)\delta(\mathbf{r}-\mathbf{r}_s) \quad [3.1]$$

where $k_b(\mathbf{r}) = \omega\sqrt{\epsilon_b(\mathbf{r})\mu_0}$ is the background medium wavenumber, $\mathbf{r}$ is the field point and $\chi(\mathbf{r})$ describes the targets in terms of the so-called contrast function, which in turn is defined as $\chi(\mathbf{r}) = \epsilon_t(\mathbf{r})/\epsilon_b(\mathbf{r}) - 1$. Clearly, χ is just supported over D_T.

It is noted that for homogeneous background medium, the wavenumber is everywhere the one of the free-space, that is $k_b = k_0$, whereas for half-space medium $k_b = k_0$ if $z > 0$ and $k_b = k_l = \omega\sqrt{\epsilon_l\mu_0} = k_0\sqrt{\epsilon_l}$ if $z < 0$.

The field $E(\omega,\mathbf{r},\mathbf{r}_s)$ is the so-called total field that can be decomposed as $E_i(\omega,\mathbf{r},\mathbf{r}_s) + E_s(\omega,\mathbf{r},\mathbf{r}_s)$, with $E_i(\omega,\mathbf{r},\mathbf{r}_s)$ being the incident field, that is the field which exists when no target is present in ID, and $E_s(\omega,\mathbf{r},\mathbf{r}_s)$ is the scattered field. For the case in hand, $E_i(\omega,\mathbf{r},\mathbf{r}_s) = -j\omega\mu_0 S(\omega)G(\omega,\mathbf{r},\mathbf{r}_s)$, where $G(\omega,\mathbf{r},\mathbf{r}_s)$ is the background medium Green function that in turn solves the following differential equation:

$$\nabla_{\mathbf{r}}^2 G(\omega,\mathbf{r},\mathbf{r}_s) + k_b^2(\mathbf{r})G(\omega,\mathbf{r},\mathbf{r}_s) = -\delta(\mathbf{r}-\mathbf{r}_s) \quad [3.2]$$

The method of the Green function can be exploited to get an expression for the scattered field and transform [3.1] into an integral equation. This is the well-known Lippmann–Swinger equation. In particular, on assuming that the scattered field is collected over $\mathbf{r}_o \in \Gamma_o$, with the observation curve Γ_o residing entirely in the upper half-space, it yields

$$E_s(\omega,\mathbf{r}_o,\mathbf{r}_s) = -k_b^2 \int_{ID} G(\omega,\mathbf{r}_o,\mathbf{r})\chi(\mathbf{r})E(\omega,\mathbf{r},\mathbf{r}_s)d\mathbf{r} \quad [3.3]$$

The inverse scattering problem consists of solving [3.3] for χ once E_s has been collected over a certain observation curve Γ_o and for a frequency band $\omega \in \Omega$ and/or for different source positions $\mathbf{r}_s \in \Gamma_s$ or both. The problem is, however, nonlinear since E depends on χ through [3.1]. Adopting a linearized scattering model drastically simplifies the problem. For the case of penetrable targets, one option is to invoke the Born approximation. This way the total field E inside the target (and hence inside the integral term of [3.3]) is replaced by E_i, so that [3.3] is approximated as

$$E_s(\omega,\mathbf{r}_o,\mathbf{r}_s) = jk_b^2\omega\mu_o \int_{ID} G(\omega,\mathbf{r}_o,\mathbf{r})\chi(\mathbf{r})G(\omega,\mathbf{r},\mathbf{r}_s)d\mathbf{r} \quad [3.4]$$

From a physical point of view, Born approximation entails neglecting multiple scatterings occurring inside the target. This requires the target to be a *weak* scatterer, that is it should only slightly perturb the background medium. To this end, the contrast function needs to meet rather restrictive conditions that hardly hold true in most interesting scattering scenarios. Nonetheless, linearized scattering models are widely employed in radar imaging. This is because often they are the only convenient option, for example, when the spatial region to be imaged is large in terms of wavelength. Moreover, even when targets are highly contrasted, linear inversions act as *qualitative* methods. Indeed, linear inversions allow for target detection and localization and to retrieve information about the target's size. This statement can been given a rigorous characterization by embedding the inverse problem within the theory of pseudodifferential operators and by exploiting their property to preserve the singularities of the function they transform.

Turning back to [3.4], this equation defines the linear operator

$$\mathcal{A} : \chi(\mathbf{r}) \in X = L^2(ID) \to E_s(\omega, \mathbf{r}_o, \mathbf{r}_s) \in Y = L^2(\Omega \times \Gamma_o \times \Gamma_s) \qquad [3.5]$$

It is noted that the unknown and the field spaces have been equipped with the usual L^2 Hilbert structure. Different choices for X can be made if a priori information about the unknown are available (Solimene et al. 2018a). Also, the field space Y is chosen as large enough to account for noise and uncertainties that always corrupt the field measurements.

Equation [3.5] subtends the most general configuration set up since the field space consists of fields that are collected by varying the source and observation positions as well as the working frequency. This configuration is addressed in literature as multistatic/multi-view/multifrequency. For the considered 2D geometry, two-dimensional data are enough. To this end, multi-static/multi-view/single-frequency configurations are common in conjunction with nonlinear inverse methods. However, the multi-bistatic/multifrequency or the multi-monostatic/multifrequency are by far the most common in GPR imaging. More in detail, in the multi-bistatic case, for each source position $\mathbf{r}_s$, the scattered field is collected only at single offset position $\mathbf{r}_o = \mathbf{r}_s + \Delta$ while $\mathbf{r}_s$ changes over the source curve. When the offset is null, that is, $\Delta = 0$, $\mathbf{r}_s = \mathbf{r}_o$ and the multi-monostatic case is obtained. Different configurations entail a different radar system complexity and possibly different achievable performances during the imaging stage. Herein, we just limit to consider the multi-monostatic configuration. However, similar results can be obtained for the other cases. Eventually, the considered scattering model reads as

$$\mathcal{A} : \chi(\mathbf{r}) \in X = L^2(ID) \to E_s(\omega, \mathbf{r}_o, \mathbf{r}_s) \in Y = L^2(\Omega \times \Gamma) \qquad [3.6]$$

with

$$E_s(\omega, \mathbf{r}_o) = (\mathcal{A}\chi)(\omega, \mathbf{r}_o) = jk_b^2\omega\mu_o \int_{ID} G^2(\omega, \mathbf{r}_o, \mathbf{r})\chi(\mathbf{r})d\mathbf{r} \qquad [3.7]$$

where $\mathbf{r}_s = \mathbf{r}_o \in \Gamma$ has been considered and the symmetry of the Green function, that is $G(\omega, \mathbf{r}_o, \mathbf{r}) = G(\omega, \mathbf{r}, \mathbf{r}_o)$, is exploited.

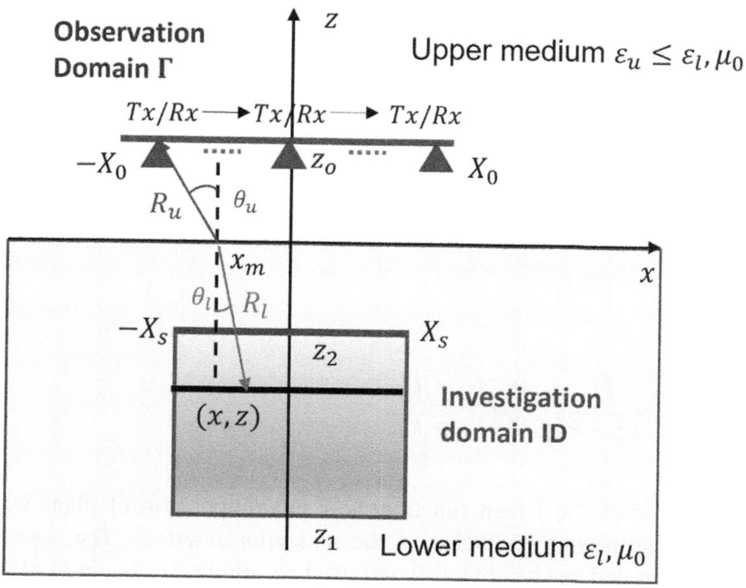

Figure 3.1. *Geometry of the problem*

To obtain an explicit expression of [3.7], we need to specify the measurement curve, the investigation domain and give the Green function expression. As to the measurement curve, it is assumed that the $\Gamma = [-X_o, X_o]$, that is a segment along the x-axis, hence parallel to the separation interface, located at the height $z_o > 0$. The investigation domain, instead, is the rectangle $ID = [-X_s, X_s] \times [z_1, z_2]$, with $z_1 < z_2 < 0$. A pictorial view of the considered scattering geometry is sketched in Figure 3.1. Also, denote as $\Omega_k = [k_{0min}, k_{0max}]$ the free-space wavenumber band corresponding to the frequency band Ω. It is convenient to employ the Weyl expansion for the Green function. Accordingly, for the two-layered background medium and $z_o > 0, z < 0$, this yields

$$G(k_0, \mathbf{r}_o, \mathbf{r}) = \frac{j}{2\pi} \int_{-\infty}^{\infty} \hat{G}(k_x, z_o, z)e^{-jk_x(x_o - x)}dk_x \qquad [3.8]$$

where $\hat{G}$ is the Fourier transform of the Green function (with respect to x, k_x being the corresponding spectral variable) given by

$$\hat{G}(k_x, z_o, z_s) = \frac{1}{2}\tau(k_x)e^{jk_{zl}z}e^{-jk_{zu}z_o}$$

with

$$\tau(k_x) = \frac{2}{k_{zl} + k_{zu}}$$

and

$$k_{zl} = \begin{cases} \sqrt{k_l^2 - k_x^2} & \text{if} \quad k_x \leq k_l \\ -j\sqrt{k_x^2 - k_l^2} & \text{if} \quad k_x > k_l \end{cases}$$

and

$$k_{zu} = \begin{cases} \sqrt{k_0^2 - k_x^2} & \text{if} \quad k_x \leq k_0 \\ -j\sqrt{k_x^2 - k_0^2} & \text{if} \quad k_x > k_0 \end{cases}$$

Basically, [3.8] gives the Green function as a superimposition of plane-waves with τ being the transmission coefficient at the separation interface. The case of a homogeneous background can be obviously retrieved by setting $k_l = k_0$ in [3.8].

It is assumed that $z_2 > \lambda_l$, with λ_l being the wavelength in the lower half-space. This entails that we are excluding targets very close to the interface. Accordingly, the Green function can be uniformly approximated by employing asymptotic arguments in Fortuny-Guasch (2002)

$$G(k_0, \mathbf{r}_o, \mathbf{r}) = \sqrt{h(k_0, \mathbf{r}_o, \mathbf{r})}e^{-jk_0\phi(\mathbf{r}_o, \mathbf{r})} \qquad [3.9]$$

where $\sqrt{h(k_0, \mathbf{r}_o, \mathbf{r})}$ takes into account the relevant amplitude factors, and $\phi(\mathbf{r}_o, \mathbf{r}) = R_u + \sqrt{\epsilon_l}R_l$ is the phase term that takes into account the propagation path. In particular, $R_u = \sqrt{(x_o - x_m)^2 + z_o}$ is the propagation path in the upper half-space, $R_l = \sqrt{(x_m - x)^2 + z^2}$ in the lower half-space and x_m is the refraction point at the interface given by the Snell's law

$$\frac{x_o - x_m}{\sqrt{(x_o - x_m)^2 + z_o^2}} = \sqrt{\epsilon_l}\frac{x_m - x}{\sqrt{(x_m - x)^2 + z_o^2}} \qquad [3.10]$$

Therefore, the linearized scattering operator is rewritten as

$$E_s(k_0, \mathbf{r}_o) = (\mathcal{A}\chi)(k_0, \mathbf{r}_o) = jk_b^2 k_0 \zeta_0 \int_{ID} h(\omega, \mathbf{r}_o, \mathbf{r}) e^{-j2k_0\phi(\mathbf{r}_o, \mathbf{r})} \chi(\mathbf{r}) d\mathbf{r} \quad [3.11]$$

with $\zeta_0 = \sqrt{\mu_0/\epsilon_0}$ being the free-space impedance.

3.3. Inversion

Radar imaging entails addressing the inverse problem related to solution of equation [3.11] for the χ function. Say

$$\mathcal{B} : E_s \in Y \rightarrow \tilde{\chi} \in X \quad [3.12]$$

a reconstruction operator, that starting from data returns an estimation of the unknown contrast

$$\tilde{\chi}(\mathbf{r}) = (\mathcal{B}E_s)(\mathbf{r}) \quad [3.13]$$

where X and Y are the unknown and the data spaces that depend on the considered configuration as detailed above.

In the case of a point like target, the contrast is $\chi = \delta(\mathbf{r} - \mathbf{r}')$ and [3.13] returns the so-called point-spread function (psf) $psf(\mathbf{r}, \mathbf{r}') = \mathcal{B}\mathcal{A}\delta(\mathbf{r} - \mathbf{r}')$. In general, because of the linear formulation, the psf provides the link between the unknown and its reconstructed version as

$$\tilde{\chi}(\mathbf{r}) = \int_{ID} psf(\mathbf{r}, \mathbf{r}')\chi(\mathbf{r}')d\mathbf{r}' \quad [3.14]$$

Equation [3.14] highlights that the contrast function, during the reconstruction process, undergoes a filtering which is dictated by the psf features. The latter in general depends on the parameters of the configuration (Maisto et al. 2019a, 2019b, 2020a, 2021a) and on the adopted inversion (what is $\mathcal{B}$) scheme. Ideally, we would like the psf to be a Dirac delta. This would entail a perfect reconstruction. In practice, this is hardly possible because of the unavoidable presence of noise/uncertainties on data and the mathematical features of the linearized scattering operator $\mathcal{A}$. More in detail, in order to obtain "meaningful" reconstructions, we are forced to establish a compromise between accuracy and stability. Accuracy is related to the resolution at which the unknown can be reconstructed and it is directly linked to the psf characteristics, and in particular to the width of its main beam (Maisto et al. 2019a, 2019b, 2020a, 2021a) The stability is instead related to how the reconstruction operator "propagates" the noise from data to the reconstructions. This statement can be given a rigorous characterization: radar imaging entails dealing with a *linear ill-posed* inverse problem. According to Hadamard, a problem is *well-posed* if the solution exists, it is unique and depends continuously on data. If at least one of the previous conditions is lacking, the problem is ill-posed.

It is convenient to dwell a bit more on this crucial aspect of radar imaging.

Firstly, we remark that, for the considered configuration, it can be shown that $\mathcal{A}$ is injective. Hence, if the solution exists, it is unique. This is not necessarily the case. Indeed, (linearly) non-scattering targets can be easily found, for example, in the single-frequency case. In any case, when the operator to be inverted is non-injective, we can refer to the generalized solution (the one of minimum norm).

The other two requirements can be conveniently investigated by introducing the singular spectrum (singular value decomposition (SVD)) of the scattering operator (Solimene et al. 2013a, 2013b). In particular, it can be easily verified that $\mathcal{A}$ is a Hilbert–Schmidt operator defined over a separable Hilbert space. Hence, it is also compact and its singular spectrum can be employed. Denote as $\{u_n, v_n, \sigma_n\}_{n=0}^{\infty}$ the singular system of $\mathcal{A}$ defined as

$$\begin{aligned} \mathcal{A}^{\dagger}\mathcal{A}u_n &= \sigma_n^2 u_n \\ \mathcal{A}\mathcal{A}^{\dagger}v_n &= \sigma_n^2 v_n \\ \mathcal{A}u_n &= \sigma_n v_n \\ \mathcal{A}^{\dagger}v_n &= \sigma_n u_n \end{aligned} \tag{3.15}$$

with $\mathcal{A}^{\dagger}$ being the adjoint operator of $\mathcal{A}$, that is

$$\begin{aligned} &\mathcal{A}^{\dagger} : f \in Y \to (\mathcal{A}^{\dagger}f)(\mathbf{r}) = \\ &-jk_b^2 k_0 \zeta_0 \int_{\Omega_k \times \Gamma} G^{2*}(k_0, \mathbf{r}_o, \mathbf{r}) f(k_0, \mathbf{r}_o) dk_0 d\mathbf{r}_o \in X \end{aligned} \tag{3.16}$$

or by using the expression of the linearized scattering operator in [3.11]

$$\begin{aligned} &\mathcal{A}^{\dagger} : f \in Y \to (\mathcal{A}^{\dagger}f)(\mathbf{r}) = \\ &-jk_b^2 k_0 \zeta_0 \int_{\Omega_k \times \Gamma} h^*(k_0, \mathbf{r}_o, \mathbf{r}) e^{j2k_0\phi(\mathbf{r}_o,\mathbf{r})} f(k_0, \mathbf{r}_o) dk_0 d\mathbf{r}_o \in X \end{aligned} \tag{3.17}$$

Denote now as $R(\mathcal{A})$ and $R(\mathcal{A}^{\dagger})$ the ranges of $\mathcal{A}$ and $\mathcal{A}^{\dagger}$, respectively, and as $N(\mathcal{A})$ and $N(\mathcal{A}^{\dagger})$ the corresponding null spaces. In particular, $\overline{R(\mathcal{A})} = N(\mathcal{A}^{\dagger})^{\perp}$ and

$\overline{R(\mathcal{A}^{\dagger})} = N(\mathcal{A})^{\perp}$. For the case at hand, as mentioned above, $N(\mathcal{A})$ is empty. The sets $\{u_n\}_{n=0}^{\infty}$ and $\{v_n\}_{n=0}^{\infty}$ are orthonormal and span $X = \overline{R(\mathcal{A}^{\dagger})}$ and $\overline{R(\mathcal{A})}$. Hence, according to the spectral representation theorem

$$E_s = \sum_n < E_s, v_n >_Y v_n \qquad [3.18]$$

$$\chi = \sum_n < \chi, u_n >_X u_n \qquad [3.19]$$

with, $< \cdot, \cdot >_Y$ and $< \cdot, \cdot >_X$ being the scalar products in the data and unknown spaces, and, because of [3.15],

$$< E_s, v_n >_Y = \sigma_n < \chi, u_n >_X \qquad [3.20]$$

Therefore, the solution can be formally given as

$$\tilde{\chi} = \sum_n \frac{< E_s, v_n >_Y}{\sigma_n} u_n \qquad [3.21]$$

This is equivalent to the reconstruction operator having the following expression:

$$(\mathcal{B}E_s)(\mathbf{r}) = \int_{\Omega_k \times \Gamma} \sum_n \frac{u_n(\mathbf{r}) v_n(k_0, \mathbf{r}_o)}{\sigma_n} E_s(k_0, \mathbf{r}_o)) dk_0 d\mathbf{r}_o \qquad [3.22]$$

which leads to the nice expression for the point-spread function

$$psf(\mathbf{r}, \mathbf{r}') = \sum_n u_n(\mathbf{r}) u_n^*(\mathbf{r}') \qquad [3.23]$$

Ill-posedness occurs when data are noisy. In this case, in [3.21], E_s is replaced by $\tilde{E}_s = E_s + \mathcal{N}$, with $\mathcal{N}$ being the noise. In particular, since the singular values $\sigma_n \to 0$ as $n \to 0$ (because $\mathcal{A}$ is compact), it is seen that the inverse operator $\mathcal{B}$ is not continuous and the solution exists as long as the Picard's condition is satisfied, that is, $\tilde{E}_s \in R(\mathcal{A}) \cup N(\mathcal{A}^{\dagger})$. In order to counteract the role of noise and extend the reconstruction operator to the whole Y, a regularized inverse must be used. In terms of SVD expansion, the regularized reconstruction operator is basically expressed as

$$(\mathcal{B}_r E_s)(\mathbf{r}) = \int_{\Omega_k \times \Gamma} \sum_n \frac{W_n(\mathcal{N}) u_n(\mathbf{r}) v_n(k_0, \mathbf{r}_o)}{\sigma_n} E_s(k_0, \mathbf{r}_o)) dk_0 d\mathbf{r}_o \qquad [3.24]$$

which corresponds to the point-spread function

$$psf_r(\mathbf{r},\mathbf{r}') = \sum_n W_n(\mathcal{N}) u_n(\mathbf{r}) u_n^*(\mathbf{r}') \tag{3.25}$$

$W_n(\mathcal{N})$ is a windowing function that depends on the noise and the strategy adopted to achieve regularization. Details concerning regularization can be found in many textbooks (see, for example, Bertero (1989)). What matters here is that it is such a windowing function that is responsible for the trade-off between resolution and stability. What is more, the features of $\mathcal{A}$, reflected in its singular value behavior, play a major role while establishing this trade-off.

3.3.1. *Migration*

In order to perform the reconstruction, the linearized scattering operator can be inverted by suitably employing the adjoint operator. Indeed, these types of inversion schemes are very common in radar imaging, where depending on the particular field of research, they are addressed as migration, back-propagation, time-reversal and so on Cassereau and Fink (1992); Gilmore et al. (2006) and (Zhuge et al. 2010). Basically, their focusing ability is a consequence of the Schwarz's inequality. Therefore, those methods basically act as a matched filter. Moreover, they generally are effective from the computational point of view and easy implementation, which can often be accomplished via FFT. Also, migration inversions implicitly give regularized inversion. However, unlike SVD-based inversion, the regularization cannot be matched to the noise level. While this avoids the (generally open) problem of optimally setting the regularization parameter, on the other side performance cannot be improved even when noise is reduced.

Maybe, the simplest scheme in this class is time reversal. By this scheme, reconstruction is obtained as

$$\tilde{\chi}(\mathbf{r}) = [(\mathcal{A}^\dagger)^* E_S^*](\mathbf{r}) \tag{3.26}$$

From the physical point of view [3.26] means that the collected field, after conjugation (time reversal), is re-propagated in the medium. Details on the physical justification and mathematical derivation can be found in a number of papers. Here, we only mention that this scheme is equivalent to the Porter–Bojarski equation for field reconstruction (Solimene et al. 2014c). Consequently, the achievable resolution is limited. The latter statement can be readily shown by employing the SVD expansion. Accordingly,

$$\tilde{\chi} = \sum_n \sigma_n < E_S^*, v_n^* > u_n^* \tag{3.27}$$

Since σ_n tend to zero, it is clear that [3.27] returns stable inversions. However, it is not a regularized inverse since, even in absence of noise, does not lead to the generalized solution. This is evident by looking at the corresponding point-spread function, that is

$$psf(\mathbf{r}, \mathbf{r}') = \sum_n \sigma_n^2 u_n^*(\mathbf{r}) u_n(\mathbf{r}') \qquad [3.28]$$

which clearly is a smoothed version of [3.23].

More generally, reconstruction through a migration scheme is a weighted adjoint inversion, by which data are first properly filtered and then back-propagated by the adjoint operator. Here, we consider some more details since they are useful in the subsequent section concerning the sampling strategy.

The reconstruction returned by a migration scheme can be formally written as

$$\tilde{\chi}(\mathbf{r}) = -jk_b^2 k_0 \zeta_0 \int_{\Omega_k \times \Gamma} W(k_0, \mathbf{r}_o, \mathbf{r}) h^*(k_0, \mathbf{r}_o, \mathbf{r}) e^{j2k_0\phi(\mathbf{r}_o, \mathbf{r})} \times E_s(k_0, \mathbf{r}_o) dk_0 d\mathbf{r}_o \qquad [3.29]$$

where $W(k_0, \mathbf{r}_o, \mathbf{r})$ is the filtering function. The point-spread function is then given by

$$psf_M(\mathbf{r}, \mathbf{r}') = -(k_b^2 k_0 \zeta_0)^2 \times \int_{\Omega_k \times \Gamma} W(k_0, \mathbf{r}_o, \mathbf{r}) h^*(k_0, \mathbf{r}_o, \mathbf{r}) h(k_0, \mathbf{r}_o, \mathbf{r}') e^{j2k_0[\phi(\mathbf{r}_o, \mathbf{r}) - \phi(\mathbf{r}_o, \mathbf{r}')]} dk_0 d\mathbf{r}_o \qquad [3.30]$$

By approximating the phase term as $\phi(\mathbf{r}_o, \mathbf{r}) - \phi(\mathbf{r}_o, \mathbf{r}') = \nabla_{\mathbf{r}} \phi(\mathbf{r}_o, \mathbf{r}) \mid_{\mathbf{r}=\mathbf{r}'} \cdot (\mathbf{r} - \mathbf{r}')$, setting $\mathbf{w} = 2k_0 \nabla_{\mathbf{r}} \phi(\mathbf{r}_o, \mathbf{r}) \mid_{\mathbf{r}=\mathbf{r}'}$, [3.30] is rewritten as

$$psf_M(\mathbf{r}, \mathbf{r}') = \int_{\Omega_k \times \Gamma} W(k_0, \mathbf{r}_o, \mathbf{r}) H(k_0, \mathbf{r}_o, \mathbf{r}) e^{j\mathbf{w} \cdot (\mathbf{r} - \mathbf{r}')} dk_0 d\mathbf{r}_o \qquad [3.31]$$

with $H(k_0, \mathbf{r}_o, \mathbf{r})$ collecting the amplitude terms in [3.30]. Now, assuming that $\mathbf{w} : (k_0, \mathbf{r}_o) \to 2k_0 \nabla_{\mathbf{r}} \phi(\mathbf{r}_o, \mathbf{r}) \mid_{\mathbf{r}=\mathbf{r}'}$ is invertible (this indeed may depend on the considered configuration and on the background medium features) and by defying $\Omega_{\mathbf{w}}(\mathbf{r}) = \{\nabla_{\mathbf{r}} \phi(\mathbf{r}_o, \mathbf{r}) \mid_{\mathbf{r}=\mathbf{r}'}, (k_0, \mathbf{r}_o) \in \Omega_k \times \Gamma\}$, [3.31] assumes the nice expression

$$psf_M(\mathbf{r}, \mathbf{r}') = \int_{\Omega_{\mathbf{w}(\mathbf{r})}} W(\mathbf{w}, \mathbf{r}) H(\mathbf{w}, \mathbf{r}) \left| \frac{\partial(k_0, \mathbf{r}_o)}{\partial \mathbf{w}} \right| e^{j\mathbf{w} \cdot (\mathbf{r} - \mathbf{r}')} d\mathbf{w} \qquad [3.32]$$

Equation [3.32] highlights the reconstruction kernel (i.e. the psf) as a pseudodifferential operator which due to its microlocal features justifies migration as a reconstruction scheme modulo smooth errors, that is, it succeeds in the reconstruction of the contrast function singularities. The weighting function $W(\mathbf{w}, \mathbf{r})$ plays the same role as the analogous in [3.25]. In particular, it is chosen to compensate the propagation amplitude term as well as the transformation Jacobian matrix so that [3.32] becomes

$$psf_M(\mathbf{r}, \mathbf{r}') = \int_{\Omega_{\mathbf{w}(\mathbf{r})}} e^{j\mathbf{w}\cdot(\mathbf{r}-\mathbf{r}')} d\mathbf{w} \qquad [3.33]$$

This way the psf is actually a filtered version of a Dirac delta with the spatial frequency band just restricted to $\Omega_{\mathbf{w}}(\mathbf{r})$. That is how regularization is achieved. Note that differently from [3.25], the weighting function is generally set by not accounting for the level of noise.

As mentioned at the beginning of this section, many adjoint-based inversion schemes have been developed in literature. All of them attempt to obtain [3.33]. Different schemes in this realm mainly entail a difference in the numerical complexity required for their implementation, and only weak difference on the achievable performance.

3.3.2. *A simple pedagogical example*

In this section, we find it instructive to pause and address a very simple radar imaging problem.

The background medium is assumed to be the free-space, the contrast just supported over a one-dimensional domain, that is $\chi(\mathbf{r}) = \chi(x)\delta(z)$ with $-X_s \le x \le X_s$, and the scattered field collected in far-field at a single frequency. Employing the far-field approximation of the Green function and by properly normalizing the scattered field, [3.7] can be simplified as

$$\mathcal{A}_{ff} : \chi \in L^2[-X_s, X_s] \rightarrow E_s(u) = \int_{-X_s}^{X_s} e^{-j2ux}\chi(x)dx \in L^2[\Omega_u] \qquad [3.34]$$

where $E_s(u)$ is the normalized scattered field, $u = k_0 \sin\theta$, $\theta_{min} \le \theta \le \theta_{max}$ is the observation angular interval and $\Omega_u = [k_0 \sin\theta_{min}, k_0 \sin\theta_{max}]$. As a further simplification, we assume $\theta_{min} = -\theta_{max}$ so that Ω_u is symmetric. It turns out that the scattering model in this simple case is given by a compactly supported Fourier transformation, which can be rewritten in terms of projection operators as

$$\mathcal{A}_{ff} = P_{\Omega_u}\mathcal{F}P_{[-X_s,X_s]} \qquad [3.35]$$

with $\mathcal{F}$ denoting the Fourier transformation and $P_{[-X_s,X_s]}$ the spatial limiting projector over $[-X_s, X_s]$.

For the case at hand, migration means performing the inversion by $\mathcal{A}_{ff}^{\dagger}$. Hence, the corresponding point spread function is given by

$$psf_M(x,x') = \mathcal{A}_{ff}^{\dagger}\mathcal{A}_{ff}\delta(x-x') = \frac{\sin[2k_0 \sin\theta_{max}(x-x')]}{x-x'} \qquad [3.36]$$

from which the filtering introduced by the migration scheme is evident. In fact, the achievable resolution (Rayleigh's limit) is given by $\pi/(2k_0 \sin\theta_{max})$.

Turn now to consider inversion in terms of the SVD. For the case at hand, we need to consider the eigenvalue problem

$$\mathcal{A}_{ff}^{\dagger}\mathcal{A}_{ff}u_n = P_{[-X_s,X_s]}B_{\Omega_u}P_{[-X_s,X_s]}u_n = \sigma_n^2 u_n \qquad [3.37]$$

with B_{Ω_u} being the band-limiting (over Ω_u) projector.

It can be easily verified that the kernel function of the operator $\mathcal{A}_{ff}^{\dagger}\mathcal{A}_{ff}$ is given by psf_M function. This is a very well-studied kernel. In particular, the singular system can be given in terms of the so-called prolate spheroidal wavefunctions (Slepian and Pollak 1961), which have a number of nice properties. For our discussion, the most important one concerns the singular values that exhibit a step-like behavior. Indeed, the singular values are nearly constant until a critical index, beyond which they exponentially decay to zero. The fast decay to zero characterizes compact operators with analytical kernels. However, the step-like behavior is typical of band-limited kernels as in [3.36]. In these cases, the kernel is an entire function of exponential type. The critical index is naturally linked to the parameter of the configuration through the so-called spatial-bandwidth, that is $N_c = 2(2X_s k_0 \sin\theta_{max})/\pi$. Note that N_c is often addressed as the number of degrees of freedom of the problem (Toraldo Di Francia 1969) and (Piestun and Miller 2000). For the case at hand, it also coincides to the number of samples taken at the Nyquist rate that fall within the scatterer's domain, $[-X_s, X_s]$.

The mentioned particular singular value behavior marks the related inverse problem as being severely ill-posed. Basically this means that noisy data should be handled with a lot of care. In particular, by referring to [3.21], unless unfeasible very high signal to noise ratio is available, we are forced to recover an essential finite dimensional unknown version spanned by the first N_c singular functions. In these cases, therefore, inversion returns practically the same performance as migration. Note that for the latter the filtering is explicit and embodied within the shape of Ω_{w}.

For a more general set up, the singular values do not precisely exhibit a step-like behavior. However, the oscillatory nature of the kernel still entails (indeed for investigation and/or measurement domains not smaller than certain sizes) an abrupt decay of the singular values beyond a critical index, which is dependent on the configuration parameters. This explains why migration methods have been so popular in radar imaging.

Changing perspective, the singular value step-like behavior, due by the oscillatory nature of the kernel, equivalently entails that the range of the linearized scattering operator is essentially finite dimensional (i.e. the NDf is finite). This is the very basic reason that enables us to address the question of determining how to deploy the sensors. Indeed, in practical cases data can be collected only over a finite set of spatial positions. Hence, it is natural to inquire how many measurements to employ and where to collect them. Here, these questions are answered by selecting the spatial measurements as those that allow us to approximate the singular values preceding the abrupt decay (Maisto et al. 2019c, 2020c, 2021c, 2021d).

3.3.3. *The case of unbounded observation domain*

In this section, we shall still consider a simplified one-dimensional problem, but remove the working hypothesis concerning the far-field and the homogeneity of the background medium. This will give further examples to show the essentially finite dimensional nature of the range of the scattering operator and, what is more, provide results necessary for the sensor deployment strategy.

Consider the following one-dimensional version of [3.7]

$$E_s(x_o) = (\mathcal{A}\chi)(x_o) = jk_b^2 k_0 \zeta_0 \int_{-X_s}^{X_s} G^2(x_o, x)\chi(x)dx \qquad [3.38]$$

The observation curve Γ can be the whole real line or a bounded segment. Using the Green function Weyl expansion, [3.3] can be synthetically rewritten as

$$\mathcal{A} = P_\Gamma \mathcal{F}^{-1} |\widehat{G^2}|^2 \mathcal{F} P_{[-X_s, X_s]} \qquad [3.39]$$

with P_Γ being the identity operator when the observation domain is unbounded. In this case, we denote the scattering operator as $\mathcal{A}_{ub}$. Assuming the evanescent contribution is negligible, [3.39] is approximated as (see Solimene et al. (2018b) for more details concerning the role of evanescent waves)

$$\mathcal{A} = P_\Gamma \mathcal{F}^{-1} P_{[-2k_0, 2k_0]} |\widehat{G^2}|^2 \mathcal{F} P_{[-X_s, X_s]} \qquad [3.40]$$

and

$$\mathcal{A}_{ub} = \mathcal{F}^{-1} P_{[-2k_0, 2k_0]} |\widehat{G^2}|^2 \mathcal{F} P_{[-X_s, X_s]} \quad [3.41]$$

Also, we introduce the operator

$$\mathcal{A}_{max} = \mathcal{F}^{-1} P_{[-2k_0, 2k_0]} G_M \mathcal{F} P_{[-X, X]} \quad [3.42]$$

with $G_M = \max_{[-2k_0, 2k_0]}\{|\widehat{G^2}|^2\}$

It is observed that

$$\sigma_n(\mathcal{A}) \leq \sigma_n(\mathcal{A}_{ub}) \leq \sigma_n(\mathcal{A}_M) \quad [3.43]$$

Moreover,

$$\mathcal{A}_M^\dagger \mathcal{A}_M = G_M P_{[-X,X]} \mathcal{B}_{[-2k_0, 2k_0]} P_{[-X,X]} \quad [3.44]$$

Accordingly, as remarked in the previous section, the $\sigma(\mathcal{A}_M)$ have a step-like behavior. In particular, here, $N_c = 4Xk_0/\pi$. Similar results can be obtained for the case of bounded Γ (Solimene et al. 2015). However, [3.43] is enough to show that in general the range of the scattering operator is essentially of finite dimension.

3.4. Sensor deployment strategy

Regardless of the inversion method we may want to employ, the scattering problem must be first translated into its finite dimensional counterpart through a suitable discretization process. To this end, it is necessary to choose a finite dimensional unknown representation as well as to collect data over a finite set of measurements. If a priori information about the unknown are available, the base functions can be accordingly tailored. Depending on the features of the unknown, this can significantly reduce the dimension of the unknown space. In general, however, the unknown space is very large and its dimension is much greater than that of the data space. Therefore, the related inverse problem is underdetermined and can be cast as a semi-discrete problem. In detail, [3.6] becomes

$$A : \chi(\mathbf{r}) \in X = L^2(ID) \rightarrow \mathbf{E}_s \in C^L \quad [3.45]$$

where $\mathbf{E}_s$ is the scattered field vector resulting from a vectorization of the data matrix collected over the points $\mathbf{r}_{on} \in \Gamma$ for the frequencies $k_{0m} \in \Omega_k$, with $n = 1, 2, \cdots, N$, $m = 1, 2, \cdots, M$ and $L = NM$.

The choice of how many measurements and where to collect them is clearly crucial for the inverse problem. Indeed, the measurement configuration must be designed in order to trade-off different aspects of the problem. On the one hand, to keep the number of measurements as low as possible favorably affects the data acquisition time and the computational and the storage burden. On the other hand, the number and the way data are sampled impacts on the eigenspectrum of the matrix model arising from discretization and hence on the achievable performance and the resilience against noise and uncertainties (Bertero 1989).

The setting of the measurement configuration can be cast as a sensor selection problem (Reeves and Heck 1995). This can be roughly stated as follows: given an $L \times D$ matrix model to be inverted, where L is the number of measurements and D the number of unknowns, select $L' < L$ so that the achievable performance is optimized. The selection problem presents a combinatorial complexity and hence in practice cannot be addressed by an exhaustive exploration across all of the possible measurement point arrangements. To overcome this drawback, a number of methods based on convex optimization, greedy methods and heuristics have been developed in order to optimize the selection against some error metrics, such as the confidence ellipsoid, the mean square error (Joshi and Boyd 2009) and the frame potential (Ranieri et al. 2014; Jiang et al. 2016), which are all linked to the singular values of the problem and try to shape their behavior (Capozzoli et al. 2017). All of these methods are of general purpose but require us to run iterative procedures. Also, they usually require us to a priori fix the size of the unknown space and sometimes the number of measurements to be retained as well.

The selection of the measurement points can be more specifically addressed by employing the properties of the scattered field. Indeed, since the field is assumed to be in the set of square integral functions (indeed it is even more regular), it can surely be represented in terms of its samples by employing the interpolation scheme dictated by the Shannon sampling theorem. This is because the set of band-limited functions is dense in the set of finite energy functions. Therefore, the representation error can be controlled by carefully choosing a sufficiently large band. In this case, no iterative procedures are required and the number of measurements to be employed in the reconstructions is the sampling points (at the Nyquist rate) that fall within the measurement aperture and the adopted frequency band. The point is to properly set such a band in order to avoid collecting too much data. In this framework, a common approach is based in the Green function plane-wave expansion [3.8]. Indeed, once evanescent waves are neglected, the field has spatial band $[-2k_0, 2k_0]$ (note that this band estimation takes into account that the Green function is squared because of the assumed monostatic configuration), which leads to a spatial sampling step equal to

$\pi/2k_0 = \lambda/4$, with λ being the wavelength. Stationary phase arguments are employed in some previous studies (Soumekh 1992, 1999; Lopez-Sanchez and Fortuny-Guasch 2000; Fortuny-Guasch and Lopez-Sanchez 2001); the resulting spatial sampling step depends on the extent of the spatial region to be imaged and is generally greater than $\lambda/4$. This is remarkable since many data points can be saved.

Nonetheless, the number of measurements returned by the method in Soumekh (1992, 1999); Lopez-Sanchez and Fortuny-Guasch (2000); Fortuny-Guasch and Lopez-Sanchez (2001) is still high with respect to the NDF of the problem. Indeed, as we have discussed above, the range of the scattering operator is essentially finite dimensional. Therefore, it is reasonable to expect that the number of measurements be of the same order as the NDF.

Accordingly, we have developed a sampling strategy that takes into account this distinctive feature of the inverse problem. This way, the advantages offered by the sampling approach are paired with a relevant reduction of the required measurement points, which can be much below those returned by previous approaches as in the literature (Soumekh 1992; Lopez-Sanchez and Fortuny-Guasch 2000). In detail, the sensors' positions are determined so that the kernel of $A^{\dagger}A$ approximates the one of $\mathcal{A}^{\dagger}\mathcal{A}$. Since these kernels are mainly affected by the singular functions corresponding to the highest (nearly constant) part of the singular spectrum (see [3.28]), this entails that the semi-discrete operator singular spectrum (which is by definition finite rank) should approximate the most important part of the singular spectrum of $\mathcal{A}$, which is of major importance in the inversions (Maisto et al. 2019c, 2020c, 2021c, 2021d).

The proposed approach is inspired by the rewriting of the kernel function in terms of the spectral variable $\mathbf{w}$ as in [3.32], though the first-order approximation is not assumed. Since this transformation warps the original observation variables into the new one $\mathbf{w}$, this approach is termed as the *warping* method. The details of the method are provided in the next sections. Here, we will summarize its main features as follows:

– The spatially varying filtering introduced by near-field propagator (usually in subsurface radar imaging) (see [3.32] where this is evident from the dependence on $\mathbf{r}$ of the band $\Omega_{\mathbf{w}}$) is taken into account.

– The sampling points are non-uniformly arranged across the observation domain. In particular, here, for the sake of simplicity, we consider optimizing only the spatial measurements point, whereas frequencies are uniformly sampled according to some standard approach.

– The sampling law takes explicitly into account the dielectric permittivity of the half-spaces which model the subsurface scattering scenario.

– The number of required data points (depending on the parameters of the configuration) is much lower than the ones commonly employed and returned by the literature results.

3.4.1. *Warping method*

As discussed above, the sensors' deployment strategy is devised to approximate the eigenvalue problem

$$\mathcal{A}^{\dagger}\mathcal{A}u_n = \sigma_n^2 u_n \quad [3.46]$$

We need the explicit expression of the kernel function of $\mathcal{A}^{\dagger}\mathcal{A}$. In particular, by considering the scattering configuration depicted in Figure 3.1 and exploiting the operators' representation in [3.11] and [3.17], such a kernel function is given by

$$\begin{aligned} kern(\mathbf{r},\mathbf{r}') = & \int_{k_{0min}}^{k_{0max}} dk_0 \times \\ & \int_{-X_o}^{X_o} dx_o H(k_0, x_o, \mathbf{r}, \mathbf{r}') e^{j2k_0[\phi(x_o,\mathbf{r})-\phi(x_o,\mathbf{r}')]} dk_0 dx_o \end{aligned} \quad [3.47]$$

where, as above, $H(k_0, x_o, \mathbf{r}) = -(k_b^2 k_0 \zeta_0)^2 h^*(k_0, x_o, \mathbf{r}) h(k_0, x_o, \mathbf{r}')$. Note that [3.47] still considers the ideal cases where data are continuously collected over the frequency band and the measurement domain. Of course, practical scenarios require us to discretize the problem by properly sampling the spatial and frequency variables x_o and k_0.

In the following we are mainly concerned with the determination of the spatial sampling. This allows us to simplify the matter and, after all, the number of spatial sensors have a major role on the system cost and acquisition time. It must be noted that the case of full sampling, that is, both in the spatial and the frequency domains, can be obtained as well. This only requires us to deal with minor technical problems in order to generalize the following derivation. However, it can be shown that the resulting sensors' deployment within $\Omega_k \times \Gamma$ is of less practical use since it requires spatial positions change with frequencies (see Maisto et al. (2021d) for more details).

Therefore, the sampling of the wavenumber band is achieved by employing standard arguments based on the range extent of the area to be imaged. Accordingly, wavenumber frequencies are uniformly sampled with a step $\Delta k_0 = \pi/\sqrt{\epsilon_l}|z_1 - z_2|$. The resulting wavenumber discrete set is denoted as $k_{0m} = k_{0min} + (m-1)\Delta k_0$. Hence, [3.47] is approximated as

$$\begin{aligned} kern(\mathbf{r},\mathbf{r}') \approx & \Delta k_0 \times \\ & \sum_m \int_{-X_o}^{X_o} dx_o H(k_{0m}, x_o, \mathbf{r}, \mathbf{r}') e^{j2k_{0m}[\phi(x_o,\mathbf{r})-\phi(x_o,\mathbf{r}')]} dx_o \end{aligned} \quad [3.48]$$

Now, in order to devise the spatial sampling, we focus on the spatial integral appearing in the kernel function expression [3.48], which for convenience is rewritten as follows:

$$kern_{k_{0m}}(\mathbf{r},\mathbf{r}') = \int_{-X_o}^{X_o} dx_o H(k_{0m}, x_o, \mathbf{r}, \mathbf{r}') e^{j2k_{0m}[\phi(x_o,\mathbf{r})-\phi(x_o,\mathbf{r}')]} dx_o \quad [3.49]$$

where the frequency k_{0m} is considered as a parameter. In order to establish a sampling approximation that works for each frequency and for each position within the investigation domain, in [3.50], we set $k_0 = k_{0max}$ and $z = z_2$. In particular, $z = z_2$ is the most critical for discrete kernel function approximation since the scattered field arising from targets that are closer to the observation domain has a larger spatial bandwidth and hence requires more samples. Accordingly, [3.50] is particularized as

$$kern_{k_{0max}}(\mathbf{x}, \mathbf{x}') = \int_{-X_o}^{X_o} dx_o H(k_{0max}, x_o, x, x') e^{j2k_{0m}[\phi(x_o,x)-\phi(x_o,x')]} dx_o \qquad [3.50]$$

where the dependence on z has been understood since it was set at the constant value z_2.

The next step is to rewrite $kern_{k_{0max}}$ so that Fourier-based arguments can be used to gain insights on the spatial sampling (Maisto et al. 2020c, 2021d, 2020b). To this end, we first introduce a warping transformation

$$\eta : x \in [-X_s, X_s] \rightarrow \eta(x) \in [\eta(-X_s), \eta(X_s)] \qquad [3.51]$$

that basically stretches the interval $[-X_s, X_s]$ into $[\eta(-X_s), \eta(X_s)]$. At this juncture, such a function does not need to be defined yet. What is important is that $\eta(x)$ is monotonic with $\frac{d\eta}{dx} > 0$. Then, by employing the first-order integral form of the Taylor remainder, the phase term is rewritten as

$$2k_{0max}[\phi(x_o, x) - \phi(x_o, x')) = k_{0max}(\eta_2 - \eta_1)w(\eta_2, \eta_1, x_o) \qquad [3.52]$$

with

$$w(\eta_2, \eta_1, x_o) = 2\int_0^1 \frac{\partial\phi(x_o, x(\eta))}{\partial\eta}\bigg|_{\eta=\eta_1+\nu(\eta_2-\eta_1)} d\nu \qquad [3.53]$$

and $\eta_2 = \eta(x)$ and $\eta_1 = \eta(x')$. Note that [3.53] collapses in the approximation employed in [3.31] when arrested at the first order. The function $w(\eta_2, \eta_1, x_o)$ is continuous and monotonic decreasing with respect to x_o (hence invertible) $\forall\eta_2, \eta_1$ (see next section). This allows us to replace the integration in x_o with the integration in w. In particular, on denoting

$$\Omega_w(\eta_2, \eta_1, X_0) = \frac{w(\eta_2, \eta_1, -X_0) - w(\eta_2, \eta_1, X_0)}{2} \qquad [3.54]$$

and

$$w_{avg}(\eta_2, \eta_1, X_0) = \frac{w(\eta_2, \eta_1, -X_0) + w(\eta_2, \eta_1, X_0)}{2} \qquad [3.55]$$

and by setting $w = \bar{w} + w_{avg}$, expression [3.50] can be rewritten as

$$kern_{k_{0max}}(\eta_2, \eta_1) = e^{jk_{0max}w_{avg}[\eta_2 - \eta_1]} \times \int_{-\Omega w}^{\Omega w} K(\bar{w}, \eta_2, \eta_1, k_{0max}) e^{jk_{0max}\bar{w}[\eta_2 - \eta_1]} d\bar{w} \qquad [3.56]$$

where $K(\bar{w}, \eta_2, \eta_1, k_{0max}) = -H(x_o(\bar{w}), \eta_2, \eta_1, k_{0max}) \frac{dx_o}{d\bar{w}}$. Note that, for the sake of notation simplicity, we omitted to explicitly report the dependence on η_2 and η_1 of Ω_w and w_{avg}. However, this dependence is indicated in [3.54] and [3.55] and it highlights the spatially varying filtering due to the near-field propagator.

The transformation $\eta(x)$ is conveniently set as (Maisto et al. 2020a)

$$\eta(x) = \phi(-X_0, x) - \phi(X_0, x) \qquad [3.57]$$

This way, the kernel function bandwidth Ω_w is made constant and equal to 1 (see next section for the details) so that the spatially varying behavior mentioned above results now encoded into the nonlinear link between the variables x and η, which clearly implies a spatially varying resolution. Also, by noting that the leading order contribution in [3.56] occurs for $\eta_2 - \eta_1 = 0$ (Cheney and Bonneau 2004), the amplitude factor is approximated as $K(\bar{w}, \eta_2, \eta_1, k_0) \approx K(\bar{w}, \eta_1, \eta_1, k_0) = K(\bar{w}, \eta_1, k_0)$. Accordingly, [3.56] becomes

$$kern_{k_{0max}}(\eta_2, \eta_1) \simeq e^{jk_{0max}w_{avg}[\eta_2 - \eta_1]} \int_{-1}^{1} K(\bar{w}, \eta_1, k_{0max}) \times e^{jk_{0max}\bar{w}[\eta_2 - \eta_1]} d\bar{w} \qquad [3.58]$$

Now, [3.58] resembles a Fourier transformation (indeed, it exhibits a pseudodifferential structure) that can be approximated by assuming the collection of data over a set of spatial positions so that $\bar{w}$ results are uniformly sampled. Let us say

$\Delta\bar{w} = \Delta w$ is the corresponding sampling step and $\bar{w}_m = m\Delta w$ is the corresponding sampling points, then

$$kern_{k_{0max}}(\eta_2, \eta_1) \simeq e^{jk_{0max}w_{avg}[\eta_2-\eta_1]}\Delta w \times \sum_m K(\bar{w}_m, \eta', k_{0max})e^{jk_{0max}m\Delta w[\eta_2-\eta_1]} \quad [3.59]$$

Hence, standard Fourier arguments, as the one, for example, invoked for avoiding grating lobes in array antenna theory, can be exploited to choose the sampling step for $\bar{w}$. From [3.59], it is evident that $kern_{k_{0max}}$ in terms of $k_{0max}\Delta w[\eta_2 - \eta_1]$ is periodic with period 2π. Hence, Δw should be chosen so that $kern_{k_{0max}}$ runs over one single period when η_2 and η_1 vary in $[\eta(-X_s), \eta(X_s)]$. These are the same arguments invoked for avoiding greeting lobes in array antenna theory. Note that the most critical situation occurs for $\eta_2 = \eta(X_s)$ and $\eta_1 = \eta(-X_s) = -\eta(X_s)$. Hence,

$$2\eta(X_s)k_{0max}\Delta w < 2\pi \quad [3.60]$$

By highlighting $\bar{w}_m$ in terms of the corresponding sampling points in x_o, that is x_{om}, [3.60] is rewritten as

$$2\eta(X_s)k_{0max}[w(\eta(X_s), -\eta(X_s), x_{om}) - w(\eta(X_s), -\eta(X_s), x_{om+1})] < 2\pi \quad [3.61]$$

with $x_{om} < x_{om+1}$ being two consecutive sampling points in x_o. Now, by using [3.53] to express $w(\eta(X_s), -\eta(X_s), x_{om})$ and $w(\eta(X_s), -\eta(X_s), x_{om+1})$, it yields

$$k_{0max}[[\phi(x_{om}, X_s) - \phi(x_{om}, -X_s) - \phi(x_{om+1}, X_s) + \phi(x_{om+1}, -X_s)]] < \pi \quad [3.62]$$

so that, eventually, the equation for finding the sampling points x_{om} is

$$\phi(x_{om}, -X) - \phi(x_{om}, X) = \frac{n\pi}{\alpha_{OS}k_{0max}} \quad [3.63]$$

where α_{OS} is an oversampling factor. In general, an oversampling factor could be required so that the kernel function side-lobes do not start to increase for positions, which are close to the transverse boundaries (i.e. $\eta_2 = \pm\eta(X_s)$ and $\eta_1 = \mp\eta(X_s)$). This is particularly true for small relative frequency bandwidths. For large bandwidths, having set the sampling step in correspondence to k_{0max} guarantees a certain degree of oversampling for most of the frequencies. In the following, we consider a slight oversampling factor $\alpha_{OS} = 1.1$. Accordingly, the required number of spatial samples (to be deployed non-uniformly, though symmetrically) turns out to be

$$N_w = \frac{2\Omega w}{\Delta w} = \frac{2k_{0max}\alpha_{OS}\eta(X)}{\pi}$$

that, for a homogeneous scenario, becomes

$$N_w = \frac{4\alpha_{OS}}{\lambda_{min}}[R(-X_0, X, z_2) - R(X_0, X, z_2)] \qquad [3.64]$$

and for the half-space case

$$N_w = \frac{4\alpha_{OS}}{\lambda_{min}}[R_u(-X_0, X, z_2) - R_u(X_0, X, z_2)$$

$$+\sqrt{\epsilon_l}R_l(-X_0, X, z_2) - \sqrt{\epsilon_l}R_l(X_0, X, z_2)] \qquad [3.65]$$

For comparison purposes, we end this section by reporting some common literature results. In particular, we consider the estimation on the number of spatial measurements (to be collected uniformly) reported in Soumekh (1992); Lopez-Sanchez and Fortuny-Guasch (2000). In particular, those studies refer to a free-space configuration. Therefore, considering a homogeneous scenario (or $z_o = 0$ so that the sampling point estimation can be pursued as for a homogeneous background medium equal to the lower half-space), in Soumekh (1992); Lopez-Sanchez and Fortuny-Guasch (2000) two sampling criteria are suggested. One, which besides the configuration parameters (frequency, ID size, its distance from the observation line), also depends on the measurement aperture, and another one from which, after *signal compression*, such a dependence is removed. The latter one requires fewer spatial samples, that is

$$N^c_{x_o} = \frac{8X_0X_s}{\lambda_{min}z_2} \qquad [3.66]$$

and is considered as a benchmark in the following. It can be easily shown that, for fixed X_0 and X_s, $N_w \le N^c_{x_o}$ with equality that tends to be reached when $|z_2|$ increases; for fixed $|z_2|$ and increasing X_0, N_w grows more slowly than $N^c_{x_o}$. This comparison is graphically illustrated in Figure 3.2. It is clearly seen that $N^c_{x_o}$ is generally much larger than N_w and that this is enhanced as n increases. If T is the time required to collect data at a fixed spatial position, then the time saved by the new sampling scheme grows with the linear law $(N^c_{X_O} - N_w)T$. Hence, it can be concluded that the proposed spatial sampling scheme proves to be more effective for near-zone configurations.

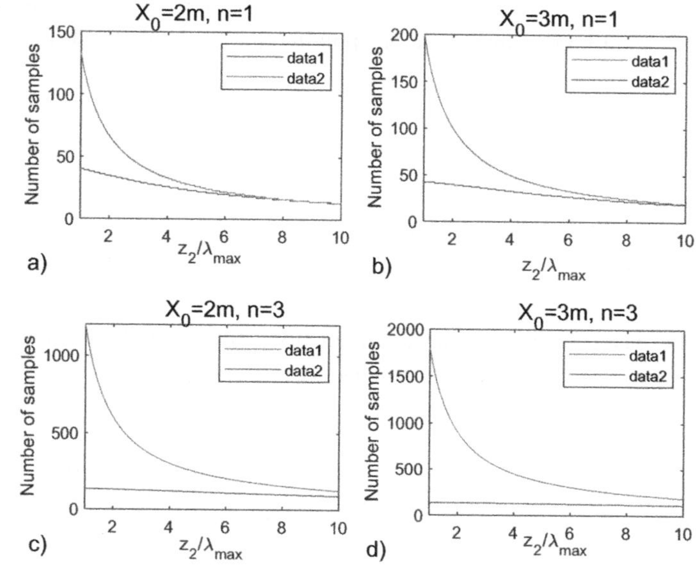

Figure 3.2. *Comparison between N_w and $N^c_{X_0}$ as a function of $|z_2|$ for two different values of X_0 and n. The other parameters are $z_o = 0$, $X_s = 1.5m$, $f_{min} = 500MHz$ and $f_{max} = 1GHz$*

3.4.1.1. *Supplement to section 3.4.1*

This section is devoted to provide some details that for the sake of readability have been previously omitted.

3.4.1.1.1. Derivation of equation [3.52]

Consider the phase term in [3.50], that is

$$2[\phi(x_o, x) - \phi(x_o, x')] = 2[\phi(x_o, x(\eta_2)) - \phi(x_o, x(\eta_1))] \quad [3.67]$$

Then, by using the fundamental theorem of calculus (FTC), [3.67] can be rewritten as

$$2[\phi(x_o, x) - \phi(x_o, x')] = 2\int_0^1 \frac{\partial}{\partial\nu}\phi(x_o, x(\eta))|_{\eta=\eta_1+\nu(\eta_2-\eta_1)} d\nu \quad [3.68]$$

which is basically the integral form of the first-order Taylor series remainder. Now, equations [3.52] and [3.53] simply arise on applying the chain rule for differentiation, that is

$$2[\phi(x_o, x) - \phi(x_o, x')] = 2(\eta_2 - \eta_1)\times$$

$$\int_0^1 \frac{\partial}{\partial \eta}\phi(x_o, x(\eta))|_{\eta=\eta_1+\nu(\eta_2-\eta_1)} d\nu \qquad [3.69]$$

3.4.1.1.2. Invertibility of w with respect to x_o

This can be easily shown by direct calculation. To this end, we first need to express $\frac{\partial}{\partial \eta}\phi(x_o, x(\eta))$. By recalling that $\phi(x_o, x(\eta)) = R_u[x_o, x_m(x_o, x(\eta))] + \sqrt{\epsilon_l}R_l[x_m(x_o, x(\eta)), x)]$, it is obtained that

$$\frac{\partial}{\partial \eta}\phi(x_o, x(\eta)) = \frac{\partial R_u}{\partial x_m}\frac{\partial x_m}{\partial x}\frac{dx}{d\eta} + \sqrt{\epsilon_l}\left[\frac{\partial R_l}{\partial x_m}\frac{\partial x_m}{\partial x}\frac{dx}{d\eta} + \frac{\partial R_l}{\partial x}\frac{dx}{d\eta}\right] \qquad [3.70]$$

which can be more conveniently rewritten as

$$\frac{\partial}{\partial \eta}\phi(x_o, x(\eta)) = -(\hat{x}\cdot\hat{R}_u + \sqrt{\epsilon_l}\hat{x}\cdot\hat{R}_l)\frac{\partial x_m}{\partial x}\frac{dx}{d\eta} + \sqrt{\epsilon_l}\hat{x}\cdot\hat{R}_l\frac{dx}{d\eta} \qquad [3.71]$$

where $\hat{x}$, $\hat{R}_u$ and $\hat{R}_l$ are unitary direction vectors (see Figure 3.1) and $\cdot$ denotes the scalar product. In particular, for $x_o < x$, $\hat{R}_u = (-\sin\theta_u, \cos\theta_u)$ and $\hat{R}_l = (\sin\theta_l, -\cos\theta_l)$, whereas for $x_o > x$, $\hat{R}_u = (\sin\theta_u, \cos\theta_u)$ and $\hat{R}_l = (-\sin\theta_l, -\cos\theta_l)$. Hence, because of Snell's law $\sin\theta_u = \sqrt{\epsilon_l}\sin\theta_l$, [3.71] is simplified as

$$\frac{\partial}{\partial \eta}\phi(x_o, x(\eta)) = \sqrt{\epsilon_l}\hat{x}\cdot\hat{R}_l\frac{dx}{d\eta} \qquad [3.72]$$

Now, in order to show that $w(\eta_2, \eta_1, x_o)$ is monotonically decreasing, it is sufficient to check that $w(\eta_2, \eta_1, x_{o2}) - w(\eta_2, \eta_1, x_{o1}) < 0$ when $x_{o2} > x_{o1}$ and $\forall \eta_1, \eta_2$. According to [3.72] and the definition [3.52], such a difference can be expressed as

$$2\sqrt{\epsilon_l}\int_0^1 \hat{x}\cdot(\hat{R}_l(x_{o2}, x(\eta)) - \hat{R}_l(x_{o1}, x(\eta))\frac{dx}{d\eta}|_{\eta=\eta_1+\nu(\eta_2-\eta_1)} d\nu \qquad [3.73]$$

It can be easily verified that when $x_{o2} > x_{o1}$, $\hat{x}\cdot(\hat{R}_l(x_{o2}, x(\eta)) - \hat{R}_l(x_{o1}, x(\eta))$ is a real negative function of η. Hence, by recalling that (by construction) $dx/d\eta > 0$, then the integral [3.73] is always negative. This proves the monotonic behavior of w with respect to x_o.

3.4.1.1.3. Invertibility of η with respect to x

With reference to the transformation introduced in [3.57], to show that $\eta(x)$ is an increasing monotonic function we have to prove that $d\eta/dx > 0$. In detail,

$$\frac{d\eta}{dx}(x) = \frac{\partial\phi(-X_0, x)}{\partial x} - \frac{\partial\phi(X_0, x)}{\partial x} \qquad [3.74]$$

which, upon employing [3.72], can be rewritten as

$$\frac{d\eta}{dx}(x) = \sqrt{\epsilon_l}(\hat{x} \cdot \hat{R}_l(-X_0, x) - \hat{x} \cdot \hat{R}_l(X_0, x)) \qquad [3.75]$$

from which, by employing the same geometrical arguments exploited above, we can conclude that $d\eta/dx > 0$.

3.4.1.1.4. Expression of the bandwidth Ωw

According to [3.54], the bandwidth can be expressed as

$$\Omega w(\eta_2, \eta_1, X_0) = \\ \int_0^1 \frac{\partial}{\partial\eta}[\phi(-X_0, x(\eta)) - \phi(X_0, x(\eta))]|_{\eta=\eta_1+\nu(\eta_2-\eta_1)} d\nu \qquad [3.76]$$

with

$$\frac{\partial}{\partial\eta}[\phi(-X_0, x(\eta)) - \phi(X_0, x(\eta))] = \\ \frac{\partial}{\partial x}[\phi(-X_0, x(\eta)) - \phi(X_0, x(\eta))]\frac{dx}{d\eta} = 1 \;\; \forall\eta \qquad [3.77]$$

where $\eta(x)$ defined in [3.57] has been employed. Accordingly, [3.76] says that $\Omega w = 1 \; \forall\eta_2, \eta_1$.

3.4.1.2. *Numerical results*

In order to check the sampling scheme introduced above, a few numerical examples are shown. We focus on the comparison between the "exact" $kern(\mathbf{r}, \mathbf{r}')$ and the one obtained by employing the sampling scheme in [3.63]. We call the latter $kern_e(\mathbf{r}, \mathbf{r}')$. In detail, by exact $kern(\mathbf{r}, \mathbf{r}')$ we mean [3.47] numerically implemented by employing a very fine and uniform grid of points within the data space Σ. The kernel function $kern_e(\mathbf{r}, \mathbf{r}')$ is still obtained from [3.47] but by taking into account the proposed sampling strategy, which suggests deploying the spatial points in a non-uniformly way according to [3.63]. As a final remark before addressing the numerical examples, we advise the reader that in the following, the approximate Green function introduced in section 3.2 is not exploited. Indeed, the Green function is computed by using its plane-wave expansion (Weyl expansion).

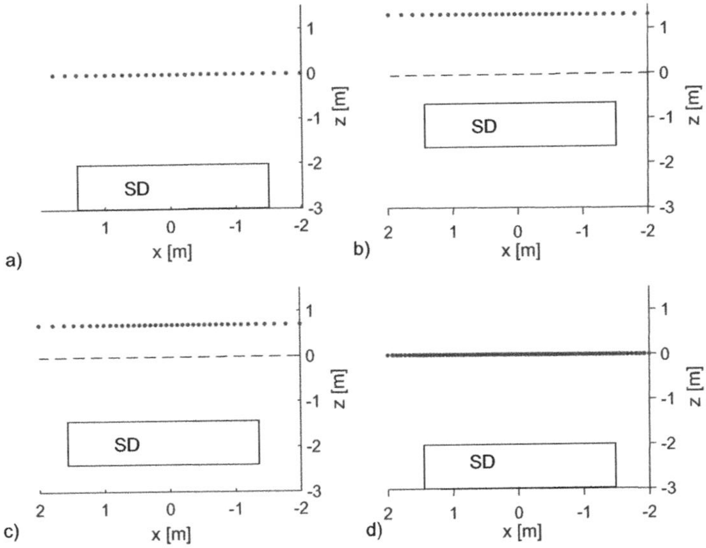

Figure 3.3. *Schematic representation of the geometrical parameters of the considered scattering configurations and sampling point locations for* $f_{min} = 500MHz$, $f_{max} = 1GHz$ $X_s = 1.5m$ *and* $X_0 = 2m$. *(a) Scattering scenario* 1 *with* $z_o = 0$, $z_2 = -2m$, $z_1 = -3m$; *the corresponding number of samples are* $N_w = 29$ *and* $N^c_{x_o} = 40$ *(b and c) Scattering scenario* 2: *the dashed line presents where the interface is located and* $n = 3$. *In particular, in (b)* $z_o = 1.3m$, $z_{min} = -0.7m$, $z_{max} = -1.7m$ *and the sampling points are* $N_w = 34$, *whereas for (c)* $z_o = 0.7m$, $z_{min} = -1.3m$, $z_{max} = -2.3m$ *and* $N_w = 38$. *Finally, (d) scattering scenario* 3 *is presented with* $n = 3$, $z_o = 0$, $z_2 = -2m$, $z_1 = -3m$, $N_w = 87$ *and* $N^c_{x_o} = 120$. *Note that in this case the interface has not been highlighted since it coincides with the measurement line*

In order to quantify the approximation error between $kern(\mathbf{r}, \mathbf{r}')$ and $kern_e(\mathbf{r}, \mathbf{r}')$, the correlation coefficient is defined as

$$C(\boldsymbol{r}') = \frac{\int_{SD} kern_e(\boldsymbol{r}, \boldsymbol{r}') kern^*(\boldsymbol{r}, \boldsymbol{r}') d\boldsymbol{r}}{\|kern_e\|(\boldsymbol{r}')\|kern\|(\boldsymbol{r}')} \quad [3.78]$$

where $\|kern_e\|(\boldsymbol{r}')$ and $\|kern\|(\boldsymbol{r}')$ are the norms computed with respect to $\boldsymbol{r}$ and hence they remain function of $\boldsymbol{r}'$.

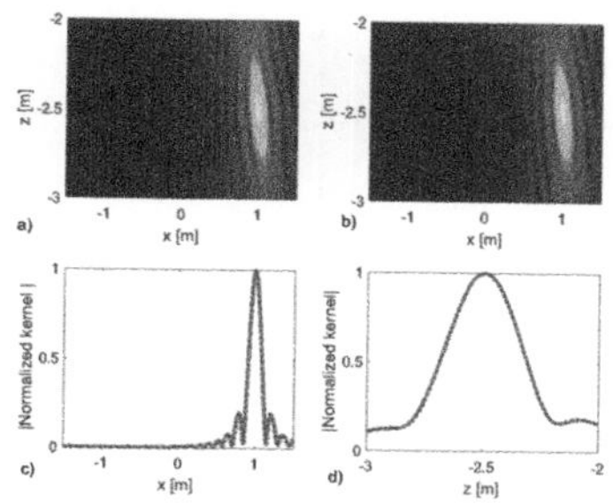

(a) Corresponds to (a) in Figure 3.3, $|C(\boldsymbol{r}')| = 0.99$.

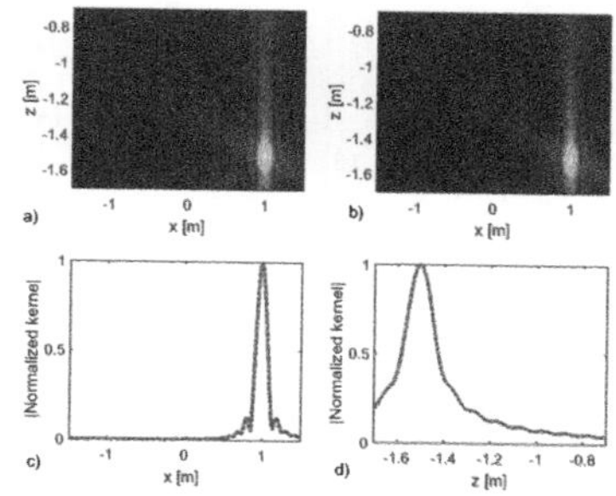

(b) Corresponds to (b) in Figure 3.3, $|C(\boldsymbol{r}')| = 0.99$.

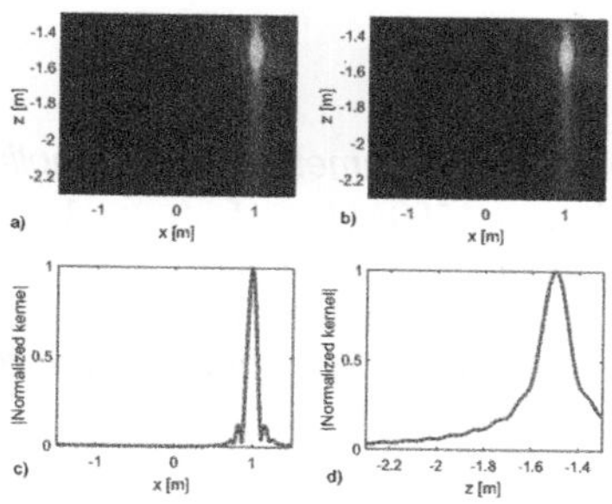

(c) Corresponds to (c) in Figure 3.3, $|C(\boldsymbol{r}')| = 0.99$.

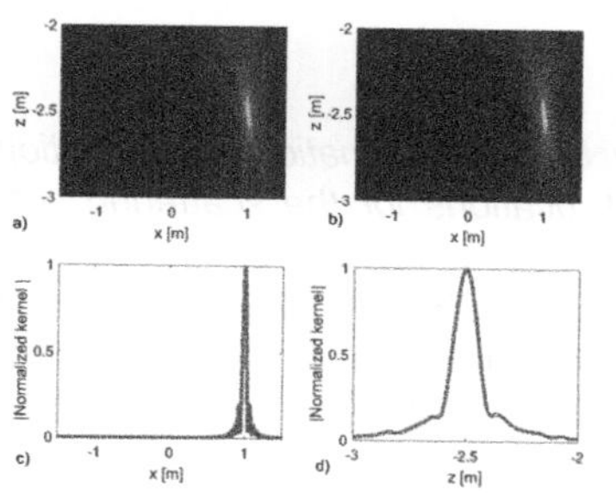

(d) Corresponds to (d) in Figure 3.3, $|C(\boldsymbol{r}')| = 0.99$.

Figure 3.4. *Illustrating the normalized kernel function amplitudes. Each panel refers to the analogous reported in Figure 3.3 and shows the kernel function and its cut-views. The blue solid lines refer to* $kern(\mathbf{r}, \mathbf{r}')$, *and the red lines refer to* $kern_e(\mathbf{r}, \mathbf{r}')$

The numerical examples deal basically with three scattering scenarios: the free-space case (scenario 1), the half-space case with measurements taken away from the separation interface (scenario 2) and the half-space case with measurements just collected over the separation interface (scenario 3). In particular, the free-space case is addressed in order to have a reference for the subsurface configuration. Also, it allows an easy vis-a-vis comparison for the estimation of the required number of samples reported in Soumekh (1992), which has been indeed worked out for such a case. Of course, $N^c_{x_o}$ can be generalized to the half-space case. In particular, for the case, the scattered field is observed just over the separation interface, and the number of samples is computed by using the wavenumber of the lower half-space as the scattering phenomenon was taking place in a homogeneous medium. However, for the case of the standoff configuration, we did not pursue the generalization of $N^c_{x_o}$.

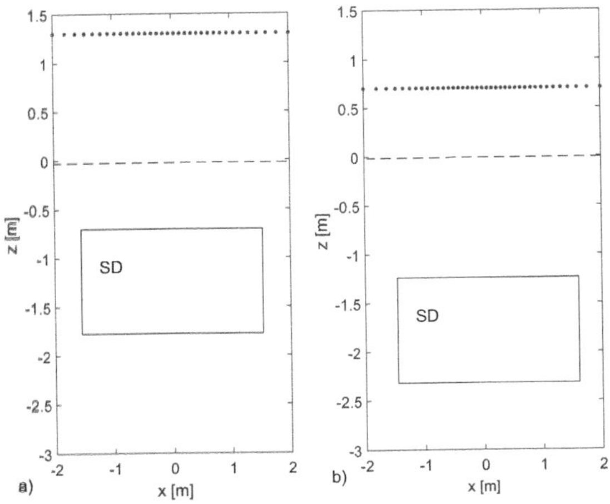

Figure 3.5. *Schematic representation of the geometrical parameters and sampling point locations for the scattering scenario* 2 *for* $f_{min} = 500MHz$, $f_{max} = 1GHz$ $X_s = 1.5m$ *and* $X_0 = 2m$ *but* $n = 4$. *In particular, in (a)* $z_o = 1.3m$, $z_{min} = -0.7m$, $z_{max} = -1.7m$ *and the sampling points are* $N_w = 34$, *whereas for (b)* $z_o = 0.7m$, $z_{min} = -1.3m$, $z_{max} = -2.3m$ *and* $N_w = 39$. *The dashed line shows where the interface is located*

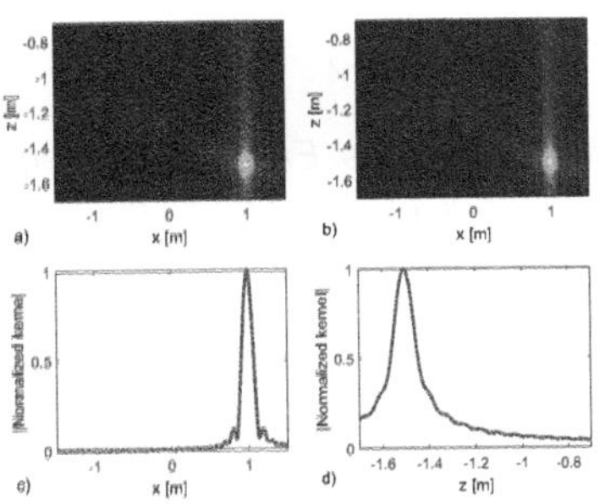

(a) Corresponds to (a) in Figure 3.5, $|C(\boldsymbol{r}')| = 0.99$.

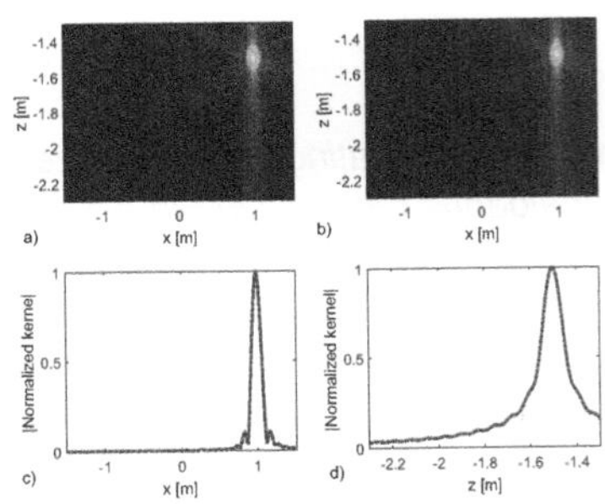

(b) Corresponds to (b) in Figure 3.5, $|C(\boldsymbol{r}')| = 0.99$.

Figure 3.6. *Illustrating the normalized kernel function amplitudes. Each panel refers to the analogous reported in Figure 3.3 and shows the kernel function and its cut-views. The blue solid lines refer to* $kern(\mathbf{r}, \mathbf{r}')$, *and the red lines refer to* $kern_e(\mathbf{r}, \mathbf{r}')$

The considered examples are reported in Figure 3.3, therein, the number and the spatial measurement deployment across the measurement line $[-X_0, X_0] = [-2, 2]m$ are shown. The parameters of the scattering scenarios are provided in the figure caption. Note that in order to highlight the role played by the background medium in all the cases, the sizes of OD and SD as well as $|z_o - z_2|$ are kept constant. In detail, panels (b) and (c) address the case of scenario 2 but with a different standoff distance from the interface, whereas panels (a) and (d) refer to scattering scenarios 1 and 3, respectively. From such a figure, they appreciated the expected non-uniform arrangement of sensors and the reduction in the number of spatial samples N_w compared to $N_{x_o}^c$ (see panels (a) and (d) for which the comparison has been pursued). Moreover, N_w increases while moving from case (a) to (d). This indeed could be expected since this is somehow like passing from free-space to a denser medium corresponding to the lower half-space. In Figure 3.4, the comparison between $kern(\mathbf{r}, \mathbf{r}')$ and $kern_e(\mathbf{r}, \mathbf{r}')$ corresponding to the cases depicted in Figure 3.3 is shown. As can be seen, for all scenarios, the two curves practically overlap. The correlation coefficient is reported in each sub-panel of Figure 3.4 and, as can be seen, it is always very high (i.e. > 0.9). Accordingly, the proposed sensor arrangement works very well for estimating the kernel function and, as a consequence, also the most "significant" eigenvalues of $\mathcal{A}^\dagger\mathcal{A}$.

While moving toward the case relative to scenario 3, the main beam of the kernel function tends to become narrower. This is consistent with the higher number of samples that are required to represent such a function. Since such a kernel function is also the point-spread function obtained by [3.28], this also entails an improvement of the achievable resolution (Maisto et al. 2020a).

In order to establish the role of the dielectric permittivity on the sampling points and the resolution, some examples are shown with a different refractive index. In particular, for the case of scenario 3, it is readily seen from equations [3.64] and [3.65] that the number of samples grows linearly with n. Therefore, we just consider scenario 2 with a refractive index $n = 4$. The corresponding results are shown in Figures 3.5 and 3.6. In particular, the first figure shows the sampling point locations, whereas Figure 3.6 reports the kernel functions. From 3.5, it is evident that the increase in n does not change significantly N_w and the distribution of the sampling points (as opposed to case relative to scenario 3). On contrary, as expected, the number of required frequencies increases from 4 to 14. By considering the resolution, the increase in n does not affect significantly the transverse resolution that remains practically unchanged, instead, depth resolution improves (see Figure 3.6). Accordingly, this example suggests that for stand-off distance only the depth resolution and the number of required frequencies are affected by the half-space dielectric. Of course, this effect *quantitatively* depends on the geometrical parameters of the configuration (i.e., OD, SD and $|z_o - z_{in}|$). However, again, the proposed sampling scheme works very well in estimating the kernel function. Accordingly, the proposed theoretical framework allows us to foresee what is going on when the configuration parameters have been set.

3.5. References

Ahmad, F., Amin, M.G., Kassam, S.A. (2005). Synthetic aperture beamformer for imaging through a dielectric wall. *IEEE Transactions on Aerospace and Electronic Systems*, 41, 271–283.

Ali, M.A. and Moghaddam, M. (2010). 3D nonlinear super-resolution microwave inversion technique using time-domain data. *IEEE Transactions on Antennas and Propagation*, 58, 2327–2336.

Baranoski, E.J. (2008). Through-wall imaging: Historical perspective and future directions. *Journal of the Franklin Institute*, 345, 556–569.

Baussard, A., Miller, E.L., Prémel, D. (2004). Adaptive B-spline scheme for solving an inverse scattering problem. *Inverse Problems*, 20, 347–365.

Benedetti, M., Donelli, M., Martini, A., Pastorino, M., Rosani, A., Massa, A. (2006). An innovative microwave-imaging technique for nondestructive evaluation: Applications to civil structures monitoring and biological bodies inspection. *IEEE Transactions on Instrumentation and Measurement*, 55, 1878–1884.

Bertero, M. (1989). Linear inverse and ill-posed problems. *Advances in Electronics and Electron Physics*, 75, 539–545.

Bertero, M. and Boccacci, P. (1998). *Introduction to Inverse Problems in Imaging*. IOP Publishing, Bristol.

Bevan, M.J., Endres, A.L., Rudolph, D.L., Parkin, G. (2003). The non-invasive characterization of pumping-induced dewatering using ground penetrating radar. *Journal of Hydrology*, 281, 55–69.

Bleistein, N. (2001). *Mathematics of Multidimensional Seismic Imaging, Migration, and Inversion*. Springer, New York.

Brandfass, M., Lanterman, A.D., Warnick, K.F. (2001). A comparison of the Colton-Kirsch inverse scattering methods with linearized tomographic inverse scattering. *Inverse Problems*, 17, 1797–1816.

Cafforio, C., Prati, C., Rocca, F. (1991). SAR data focusing using seismic migration techniques. *IEEE Transactions on Aerospace and Electronic Systems*, 27, 194–207.

Cakoni, F. (2006). *Qualitative Methods in Inverse Scattering Theory: An Introduction*. Springer, Berlin/New York.

Caorsi, S., Gragnani, G.L., Medicina, S., Pastorino, M., Zunino, G. (1994). Microwave imaging based on a Markov random field model. *IEEE Transactions on Antennas and Propagation*, 42, 293–303.

Caorsi, S., Ciaramella, S., Gragnani, G.L., Pastorino, M. (1995). On the use of regularization techniques in numerical inverse-scattering solutions for microwave imaging applications. *IEEE Transactions on Microwave Theory and Techniques*, 43, 632–640.

Capozzoli, A., Curcio, C., Liseno, A. (2017). Singular value optimization in inverse electromagnetic scattering. *IEEE Antennas and Wireless Propagation Letters*, 16, 10941097.

Cassereau, D. and Fink, M. (1992). Time-reversal of ultrasonic fields. Part III: Theory of the closed time-reversal cavity. *IEEE Transactions on Ultrasonics, Ferroelectronics and Frequency Control*, 39(5), 579–592.

Catapano, I., Crocco, L., Isernia, T. (2004). A simple two-dimensional inversion technique for imaging homogeneous targets in stratified media. *Radio Science*, 39(1), 1–14.

Catapano, I., Crocco, L., D' Urso, M., Isernia, T. (2006). A novel effective model for solving 3-D nonlinear inverse scattering problems in lossy scenarios. *IEEE Geoscience and Remote Sensing Letters*, 3, 302–306.

Catapano, I., Crocco, L., Napoli, R.D., Soldovieri, F., Brancaccio, A., Pesando, F., Aiello, A. (2012). Microwave tomography enhanced GPR surveys in Centaur's Domus, Regio VI of Pompeii, Italy. *Journals of Geophysics and Engineering*, 9, S92–S99.

Cheney, M. and Bonneau, R.J. (2004). Imaging that exploits multipath scattering from point scatterers. *Inverse Problems*, 20, 1691–1711.

Chew, W.C. (1999). *Waves and Fields in Inhomogeneous Media*. IEEE Press, New York.

Colton, D., Giebermann, K., Monk, P. (2000). A regularized sampling method for solving three-dimensional inverse scattering problems. *SIAM Journal on Scientific Computing*, 21, 2316–2330.

Conyers, L.B. (1997). *Ground-penetrating Radar: An Introduction for Archaeologists*. AltaMira Press, Walnut Creek.

Crocco, L. and Soldovieri, F. (2011). Nonlinear inversion algorithms. In *Subsurface Sensing*, Turk, A.S., Hocaoglu, K.A., Vertiy, A.A. (eds). John Wiley & Sons, New York.

Cui, T.J., Chew, W.C., Aydiner, A.A., Chen, S. (2001). Inverse scattering of two-dimensional dielectric objects buried in a lossy earth using the distorted Born iterative method. *IEEE Transactions on Geoscience and Remote Sensing*, 39, 339–346.

Daniels, D.J. (2004). *Ground Penetrating Radar*, 2nd edition. Institution of Electrical Engineers, London.

Di Matteo, A., Pettinelli, E., Slob, E. (2013). Early-time GPR signal attributes to estimate soil dielectric permittivity: A theoretical study. *IEEE Transactions on Geoscience and Remote Sensing*, 51, 1643–1654.

Donelli, M., Franceschini, G., Martini, A., Massa, A. (2006). An integrated multiscaling strategy based on a particle swarm algorithm for inverse scattering problems. *IEEE Transactions on Geoscience and Remote Sensing*, 44, 298–312.

Dourthe, C., Pichot, C., Dauvignac, J.Y., Cariou, J. (2000). Inversion algorithm and measurement system for microwave tomography of buried object. *Radio Science*, 35, 1097–1108.

Estatico, C., Bozza, G., Massa, A., Pastorino, M., Randazzo, A. (2005a). A two-step iterative inexact-Newton method for electromagnetic imaging of dielectric structures from real data. *Inverse Problems*, 21, S81–S94.

Estatico, C., Pastorino, M., Randazzo, A. (2005b). An inexact-Newton method for short-range microwave imaging within the second-order Born approximation. *IEEE Transactions on Geoscience and Remote Sensing*, 43, 2593–2605.

Estatico, C., Pastorino, M., Randazzo, A. (2012). A novel microwave imaging approach based on regularization in Lp Banach spaces. *IEEE Transactions on Antennas and Propagation*, 60, 3373–3381.

Fink, M. (1992). Time reversal of ultrasonic fields I. Basic principles. *IEEE Transactions on Ultrasonics, Ferroelectrics and Frequency Control*, 39, 555–566.

Fortuny-Guasch, J. (2002). A novel 3-D subsurface radar imaging technique. *IEEE Transactions on Geoscience and Remote Sensing*, 40(2), 443–452.

Fortuny-Guasch, J. and Lopez-Sanchez, J.M. (2001). Extension of the 3-D range migration algorithm to cylindrical and spherical scanning geometries. *IEEE Transactions on Antennas and Propagation*, 49(10), 1434–1444.

Franchois, A. and Pichot, C. (1997). Microwave imaging-complex permittivity reconstruction with a Levenberg-Marquardt method. *IEEE Transactions on Antennas and Propagation*, 45, 203–215.

Gilmore, G., Jeffry, I., Lo Vetri, J. (2006). Derivation and comparison of SAR and frequency-wavenumber migration within a common inverse scalar wave problem formulation. *IEEE Transactions on Geoscience and Remote Sensing*, 44(16), 1454–1461.

Gorriti, A.G. and Slob, E.C. (2005). A new tool for accurate S-parameters measurements and permittivity reconstruction. *IEEE Transactions on Geoscience and Remote Sensing*, 43, 1727–1735.

Hansen, T.B. and Johansen, P.M. (2000). Inversion scheme for ground penetrating radar that takes into account the planar air-soil interface. *IEEE Transactions on Geoscience and Remote Sensing*, 38, 496506.

Harada, H., Wall, D.J.N., Takenaka, T., Tanaka, M. (1995). Conjugate gradient method applied to inverse scattering problem. *IEEE Transactions on Antennas and Propagation*, 43, 784–792.

Hugenschmidt, J. and Kalogeropoulos, A. (2009). The inspection of retaining walls using GPR. *Journal of Applied Geophysics*, 67(4), 335–344.

Isernia, T., Pascazio, V., Pierri, R. (1997). A nonlinear estimation method in tomographic imaging. *IEEE Transactions on Geoscience and Remote Sensing*, 35, 910–923.

Isernia, T., Pascazio, V., Pierri, R. (2001). On the local minima in a tomographic imaging technique. *IEEE Transactions on Geoscience and Remote Sensing*, 39, 1596–1607.

Isernia, T., Crocco, L., D'Urso, M. (2004). New tools and series for forward and inverse scattering problems in lossy media. *IEEE Geoscience and Remote Sensing Letters*, 1, 327–331.

Jadoon, K.Z., Lambot, S., Slob, E.C., Vereecken, H. (2011). Analysis of horn antenna transfer functions and phase-center position for modeling off-ground GPR. *IEEE Transactions on Geoscience and Remote Sensing*, 49, 1649–1662.

Jiang, C., Soh, Y., Li, H. (2016). Sensor placement by maximal projection on minimum eigenspace for linear inverse problems. *IEEE Transactions on Signal Processing*, 64(21), 5595–5610.

Joshi, S. and Boyd, S. (2009). Sensor selection via convex optimization. *IEEE Transactions on Signal Processing*, 57(2), 451–462.

Kirsch, A. (1999). Factorization of the far-field operator for the inhomogeneous medium case and an application in inverse scattering theory. *Inverse Problems*, 15, 413–429.

Lambot, S. and André, F. (2014). Full-wave modeling of near-field radar data for planar layered media reconstruction. *IEEE Transactions on Geoscience and Remote Sensing*, 52, 2295–2303.

Lambot, S., Slob, E.C., van den Bosch, I., Stockbroeckx, B., Vanclooster, M. (2004). Modeling of ground-penetrating radar for accurate characterization of subsurface electric properties. *IEEE Transactions on Geoscience and Remote Sensing*, 42, 2555–2568.

Leone, G., Persico, R., Solimene, R. (2003). A quadratic model for electromagnetic subsurface prospecting. *AEU – International Journal of Electronics and Communications*, 57, 33–46.

Liseno, A. and Pierri, R. (2004). Shape reconstruction by the spectral data of the far-field operator: Analysis and performances. *IEEE Transactions on Antennas and Propagation*, 52, 899–903.

Liu, Q.H., Zhang, Z.Q., Wang, T.T., Bryan, J.A., Ybarra, G.A., Nolte, L.W., Joines, W.T. (2002). Active microwave imaging I. 2-D forward and inverse scattering methods. *IEEE Transactions on Microwave Theory and Techniques*, 50, 123–133.

Lobel, P., Kleinman, R.E., Pichot, C., Blanc-Feraud, L., Barlaud, M. (1996). Conjugate-gradient method for solving inverse scattering with experimental data. *IEEE Transactions on Antennas and Propagation*, 38, 48.

Lobel, P., Blanc-Féraud, L., Pichot, C., Barlaud, M. (1997). A new regularization scheme for inverse scattering. *Inverse Problems*, 13, 403.

Lopez-Sanchez, J.M. and Fortuny-Guasch, J. (2000). 3-D radar imaging using range migration techniques. *IEEE Transactions on Antennas and Propagation*, 48(5), 728–737.

Maisto, M.A., Solimene, R., Pierri, R. (2019a). Depth resolution in strip current reconstructions in near non-reactive zone. *Journal of the Optical Society of America*, 36(6), 975–982.

Maisto, M.A., Solimene, R., Pierri, R. (2019b). Resolution limits in inverse source problem for strip currents not in Fresnel zone. *Journal of the Optical Society of America*, 36(5), 826–833.

Maisto, M.A., Solimene, R., Pierri, R. (2019c). Sampling approach for singular system computation of a radiation operator. *Journal of the Optical Society of America*, 36(3), 353–361.

Maisto, M.A., Solimene, R., Pierri, R. (2020a). Transverse resolution in microwave imaging for strip objects buried in a half-space medium. *Progress in Electromagnetics Research M*, 88, 145–157.

Maisto, M.A., Pierri, R., Solimene, R. (2020b). Near-field warping sampling scheme for broad-side antenna characterization. *Electronics*, 9, 1047.

Maisto, M.A., Pierri, R., Solimene, R. (2020c). Spatial sampling in monostatic radar imaging. *IEEE Geoscience and Remote Sensing Letters*, 19, 1–5.

Maisto, M.A., Pierri, R., Solimene, R. (2021a). Near-field transverse resolution in planar source reconstructions. *IEEE Transactions on Antennas and Propagation*, 69(8), 4836–4845.

Maisto, M.A., Pierri, R., Solimene, R. (2021b). Sensor arrangement in monostatic subsurface radar imaging. *IEEE Open Journal of Antennas and Propagation*, 2, 3–13.

Maisto, M.A., Leone, G., Brancaccio, A., Solimene, R. (2021c). Efficient planar near-field measurements for radiation pattern evaluation by a warping strategy. *IEEE Access*, 9, 62255–62265.

Maisto, M.A., Masoodi, M., Leone, G., Solimene, R., Pierri, R. (2021d). Scattered far-field sampling in multi-static multi-frequency configuration. *Sensors*, 21, 4724.

Marengo, E.A., Gruber, F.K., Simonetti, F. (2007). Time-reversal MUSIC imaging of extended targets. *IEEE Transactions on Image Processing*, 16, 1967–1984.

Masini, N. and Soldovieri, F. (2011). Integrated non-invasive sensing techniques and geophysical methods for the study and conservation of architectural, archaeological and artistic heritage. *Journal of Geophysical Engineering*, 8(3), E01.

Masini, N., Persico, R., Rizzo, E. (2010). Some examples of GPR prospecting for monitoring of the monumental heritage. *Journal of Geophysical Engineering*, 7, 190–199.

Meles, G.A., Van der Kruk, J., Greenhalgh, S.A., Ernst, J.R., Maurer, H., Green, A.G. (2010). A new vector waveform inversion algorithm for simultaneous updating of conductivity and permittivity parameters from combination crosshole/borehole-to-surface GPR data. *IEEE Transactions on Geoscience and Remote Sensing*, 48, 3391–3407.

Meschino, S., Pajewski, L., Pastorino, M., Randazzo, A., Schettini, G. (2013). Detection of subsurface metallic utilities by means of a sap technique: Comparing MUSIC- and SVM-based approaches. *Journal of Applied Geophysics*, 97, 60–68.

Minet, J., Lambot, S., Slob, E.C., Vanclooster, M. (2010). Soil surface water content estimation by full-waveform GPR signal inversion in the presence of thin layers. *IEEE Transactions on Geoscience and Remote Sensing*, 48, 1138–1150.

Oliveri, G., Lizzi, L., Pastorino, M., Massa, A. (2012). A nested multi-scaling inexact-Newton iterative approach for microwave imaging. *IEEE Transactions on Antennas and Propagation*, 60, 971–983.

Orlando, L. and Soldovieri, F. (2008). Two different approaches for georadar data processing: A case study in archaeological prospecting. *Journal of Applied Geophysics*, 64, 1–13.

Ostadrahimi, M., Mojabi, P., Zakaria, A., LoVetri, J., Shafai, L. (2013). Enhancement of Gauss–Newton inversion method for biological tissue imaging. *IEEE Transactions on Microwave Theory and Techniques*, 61, 3424–3434.

Pastorino, M. (1998). Short-range microwave inverse scattering techniques for image reconstruction and applications. *IEEE Transactions on Instrumentation and Measurement*, 27(3), 1419–1427.

Pastorino, M. (2007). Stochastic optimization methods applied to microwave imaging: A review. *IEEE Transactions on Antennas and Propagation*, 55, 538–548.

Pastorino, M. (2010). *Microwave Imaging*. John Wiley & Sons, New York.

Pastorino, M., Massa, A., Caorsi, S. (2000). A microwave inverse scattering technique for image reconstruction based on a genetic algorithm. *IEEE Transactions on Instrumentation and Measurement*, 49, 573–578.

Patriarca, C., Lambot, S., Mahmoudzadeh, M.R., Minet, J., Slob, E. (2011). Reconstruction of sub-wavelength fractures and physical properties of masonry media using full-waveform inversion of proximal penetrating radar. *Journal of Applied Geophysics*, 74, 26–37.

Patriarca, C., Miorali, M., Slob, E., Lambot, S. (2013). Uncertainty quantification in off-ground monostatic ground penetrating radar. *IEEE Transactions on Antennas and Propagation*, 61, 3334–3344.

Pierri, R., Liseno, A., Solimene, R., Soldovieri, F. (2006). Beyond physical optics SVD shape reconstruction of metallic cylinders. *IEEE Transactions on Antennas and Propagation*, 54, 655–665.

Piestun, R. and Miller, D.A.B. (2000). Electromagnetic degrees of freedom of an optical system. *Journal of the Optical Society of America A*, 17, 892–902.

Pike, E.R. and Sabatier, P.C. (2002). *Scattering: Scattering and Inverse Scattering in Pure and Applied Science*. Academic Press, San Diego.

Piro, S., Goodman, D., Nishimura, Y. (2003). The study and characterization of Emperor Traiano's Villa (Altopiani di Arcinazzo, Roma) using high-resolution integrated geophysical surveys. *Archaeological Prospecting*, 10, 1–25.

Porsani, J.L., Filho, W.M., Elis, V.R., Shimeles, F., Dourado, J.C., Moura, H.P. (2004). The use of GPR and VES in delineating a contamination plume in a landfill site: A case study in SE Brazil. *Journal of Applied Geophysics*, 55, 199–209.

Rahmat-Samii, Y. and Michielssen, E. (1999). *Electromagnetic Optimization by Genetic Algorithms*. John Wiley & Sons, New York.

Randazzo, A. (2012). Swarm optimization methods in microwave imaging. *International Journal of Microwave Science and Technology*, 1–12, 491713.

Randazzo, A. and Estatico, C. (2012). A regularisation scheme for electromagnetic inverse problems: Application to crack detection in civil structures. *Nondestructive Testing and Evaluation*, 27(3), 189–197.

Ranieri, J., Chebira, A., Vetterli, M. (2014). Near-optimal sensor placement for linear inverse problems. *IEEE Transactions on Signal Processing*, 62(5), 1135–1146.

Reeves, S.J. and Heck, L.P. (1995). Selection of observations in signal reconstruction. *IEEE Transactions on Signal Processing*, 43(3), 788–791.

Rekanos, I.T. (2008). Shape reconstruction of a perfectly conducting scatterer using differential evolution and particle swarm optimization. *IEEE Transactions on Geoscience and Remote Sensing*, 46, 1967–1974.

Roger, A. (1981). Newton-Kantorovitch algorithm applied to an electromagnetic inverse problem. *IEEE Transactions on Antennas and Propagation*, 29, 232–238.

Rubek, T., Meaney, P.M., Meincke, P., Paulsen, K.D. (2007). Nonlinear microwave imaging for breast-cancer screening using Gauss–Newton's method and the CGLS inversion algorithm. *IEEE Transactions on Antennas and Propagation*, 55, 2320–2331.

Salvade, A., Pastorino, M., Monleone, R., Bozza, G., Randazzo, A. (2009). A new microwave axial tomograph for the inspection of dielectric materials. *IEEE Transactions on Instrumentation and Measurement*, 58, 2072–2079.

Slaney, M., Kak, A.C., Larsen, L.E. (1984). Limitations of imaging with first-order diffraction tomography. *IEEE Transactions on Microwave Theory and Techniques*, 32, 860–874.

Slepian, D. and Pollak, H.O. (1961). Prolate spheroidal wave functions, Fourier analysis and uncertainty I. *Bell System Technical Journal*, 40(1), 43–63.

Soldovieri, F. and Solimene, R. (2010). Ground penetrating radar subsurface imaging of buried objects. *Radar Technology*, 105–126.

Soldovieri, F., Persico, R., Utsi, E., Utsi, V. (2006). The application of inverse scattering techniques with ground penetrating radar to the problem of rebar location in concrete. *NDT & E International*, 39, 602–607.

Solimene, R., Soldovieri, F., Prisco, G., Pierri, R. (2009). Three-dimensional through-wall imaging under ambiguous wall parameters. *IEEE Transactions on Geoscience and Remote Sensing*, 47, 1310–1317.

Solimene, R., Maisto, M.A., Pierri, R. (2013a). Role of diversity on the singular values of linear scattering operators: The case of strip objects. *Journal of the Optical Society of America*, 30(11), 2266–2272.

Solimene, R., Maisto, M.A., Romeo, G., Pierri, R. (2013b). On the singular spectrum of the radiation operator for multiple and extended observation domains. *International Journal of Antennas and Propagation*, 585238.

Solimene, R., Catapano, I., Cuccaro, A., Dell'Aversano, A., Soldovieri, F. (2014a). Ground clutter removal in GPR surveys. *IEEE Journal of Selected Topics in Applied Earth Observations and Remote Sensing*, 7, 792–798.

Solimene, R., D'Alterio, A., Gennarelli, G., Soldovieri, F. (2014b). Estimation of soil permittivity in presence of antenna-soil interactions. *IEEE Journal of Selected Topics in Applied Earth Observations and Remote Sensing*, 7, 805–812.

Solimene, R., Catapano, I., Gennarelli, G., Cuccaro, A., Dell'Aversano, A., Soldovieri, F. (2014c). A unified mathematical overview of SAR imaging algorithms and some unconventional applications. *IEEE Signal Processing Magazine*, 31(4), 90–98.

Solimene, R., Mola, C., Gennarelli, G., Soldovieri, F. (2015). On the singular spectrum of radiation operators in the non-reactive zone: The case of strip sources. *Journal of Optics*, 17(2), 025605.

Solimene, R., Maisto, M.A., Pierri, R. (2018a). Information content in inverse source with symmetry and support priors. *Progress in Electromagnetics Research C*, 80, 39–54.

Solimene, R., Maisto, M.A., Pierri, R. (2018b). Inverse source in the near field: The case of a strip current. *Journal of the Optical Society of America*, 35(5), 755–763.

Soumekh, M. (1992). A system model and inversion for synthetic aperture radar imaging. *IEEE Transactions on Image Processing*, 1(1), 64–76.

Soumekh, M. (1999). Depth-focused interior echo imaging. *IEEE Transactions on Image Processing*, 8(11), 1608–1618.

Toraldo Di Francia, G. (1969). Degrees of freedom of an image. *Journal of the Optical Society of America A*, 59, 799–804.

Van den Berg, P.M. and Abubakar, A. (2001). Contrast source inversion method: State of art. *Progress in Electromagnetics Research*, 34, 189–218.

Watters, T.R., Leuschen, C.J., Plaut, J.J., Picardi, G., Safaeinili, A., Clifford, S.M., Farrell, W.M., Ivanov, A.B., Phillips, R.J., Stofan, E.R. (2006). MARSIS radar sounder evidence of buried basins in the northern lowlands of Mars. *Nature*, 444, 905–908.

Zhuge, X., Yarovoy, A.G., Savelyev, T., Ligthart, L. (2010). Modified Kirchhoff migration for UWB MIMO array-based radar imaging. *IEEE Transactions on Geoscience and Remote Sensing*, 48(6), 2692–2703.

4

Ground Penetrating Radar Tomography for Cultural Heritage

Ilaria CATAPANO, Giovanni LUDENO, Gianluca GENNARELLI, Giuseppe ESPOSITO, Lorenzo CROCCO and Francesco SOLDOVIERI

National Research Council of Italy, Institute for Electromagnetic Sensing of the Environment, Naples, Italy

4.1. Introduction

Nowadays, non-invasive sensing technologies working at different spatial scales represent a recognized tool to support archaeological research, as well as sustainable management of cultural heritage assets. These technologies offer significant advantages in planning stratigraphic assays, while reducing costs and the time spent on excavations, and supporting holistic approaches where cultural heritage management, protection and fruition aspects appear under a unified context (Masini and Soldovieri 2011; Lasaponara and Masini 2013; Tejedor et al. 2022). In this frame, ground penetrating radar (GPR) is an assessed and widespread technology (Vaughan 1986; Goodman 1994; Conyers 2013; Lai et al. 2018), because of its capability to provide high-resolution images of the interior of investigated domains (underground, manmade structures and so on), with a spatial resolution that can vary from centimeters to few meters depending on the application's needs.

For a color version of all the figures in this chapter, see www.iste.co.uk/serhir/radar.zip.

Ground Penetrating Radar,
coordinated by Mohammed SERHIR and Dominique LESSELIER.

Despite the large number of successful surveys available in the literature (see, for example, Trinks et al. (2010) to mention one relevant case study), one of the main issues faced when using GPRs is to achieve easily interpretable and objective images, that are not subject to the user's expertise. This is particularly significant in the cultural heritage field, wherein the outcomes of the GPR survey are exploited to make decisions that have an economic impact or may impact the integrity of archaeological finds. Accordingly, on the one hand, data processing software is an integral part of any system, and, on the other hand, continuous efforts have been made toward the development of approaches capable of enhancing imaging capabilities (Annan 2005; Daniels 2005). In detail, research activities constantly address the improvement of GPR data interpretability and visualization (Economou et al. 2017), as well as the development of automatic data processing strategies and target recognition tools (Küçükdemirci and Sarris 2022).

The approaches that are used the most among those proposed in the open literature range from simple filtering procedures, aimed at removing artifacts and noise to enhance the useful content of the collected raw data, to advanced imaging strategies. The latter aim to provide a focused image, giving clear information about the position and geometrical features of the targets so that they can be distinguished in migration strategies (Stolt 1978; Feng and Sato 2004) and microwave tomography (MWT) approaches (Persico 2014; Catapano et al. 2019).

MWT is a family of data processing approaches, exploiting physics-based models of electromagnetic scattering, which exhibit considerable operational flexibility. MWT allows the data collected to be managed under the standard single view–single receiver radar systems, as well as data supplied by new generation multiple input–multiple output (MIMO) architectures. Moreover, MWT allows the imaging to be faced in canonical two-dimensional homogeneous reference scenarios, while also accounting for the actual three-dimensional nature of the signal-material interaction and signal propagation in inhomogeneous media. On the other hand, while MWT allows, in principle, a quantitative reconstruction of the electromagnetic properties of the inspected scenario (Pastorino 2010), the need for reliable and computationally efficient imaging approaches favors the use of simplified (linear) electromagnetic models. These linear MWT approaches are more robust against uncertainties in the features of the scenario and allow us to achieve the localization and a rough estimation of the geometrical features of the detected objects, as demonstrated in several GPR applications (Soldovieri et al. 2007; Persico 2014; Catapano et al. 2019).

This chapter focuses on linear MWT and provides an overview of some cultural heritage GPR surveys in which such an approach has been used to process the collected data. Specifically, the chapter is organized as follows. In the next section, the basic concepts underlying MWT enhanced GPR surveys or, briefly, GPRT, are recalled, showing the flexibility of MWT in handling the imaging task under different measurement configurations. In section 4.2, some analytical tools to study the performances achievable by MWT processing under the different measurement configurations are recalled. Section 4.3 analyzes the GPRT computational burden and discusses a strategy referred to as shifting and zoom (SZ) (Persico et al. 2017) to reduce it. To demonstrate the effectiveness of GPRT surveys in cultural heritage, examples regarding surveys performed at famous archeological sites are presented in section 4.4. Finally, the potential of new generation GPR systems mounted on unmanned aerial vehicles (UAVs) and UAV-based GPRT for large-scale archaeological prospection is discussed.

4.2. GPRT general concepts

The GPRT data processing chain, as shown in Figure 4.1, involves two main steps: pre-processing and data inversion. The pre-processing step implements standard time domain (TD) procedures to remove the direct coupling between the antennas, reduce noise and improve the targets' response. The data inversion faces the imaging task as the solution of a linear inverse scattering problem (Pastorino 2010) in the frequency domain.

TD data processing begins with the *start time correction*. After which, all the gathered waveforms are processed by means of *data filtering* strategies, in order to reduce low- and high-frequency noise components and remove/mitigate undesired signal components. The *linear gain* procedure compensates the amplitude decay due to the signal propagation. This step also involves the setting of the average dielectric permittivity of the host medium, whose value is required to perform the data inversion (*wave speed selection*). This task is faced by means of the hyperbola matching procedure (Goodman 1994), which looks for the best fitting between a synthetic hyperbola (depending on the estimated dielectric permittivity) with the collected diffraction patterns.

Once TD data processing is completed, the Fourier transform is exploited to move in the frequency domain. In this step, the actual data spectrum is fixed as the frequency range, where the normalized average spectrum of the gathered waveforms, represented over a logarithmic scale, is not lower than 10 dB, its maximum value.

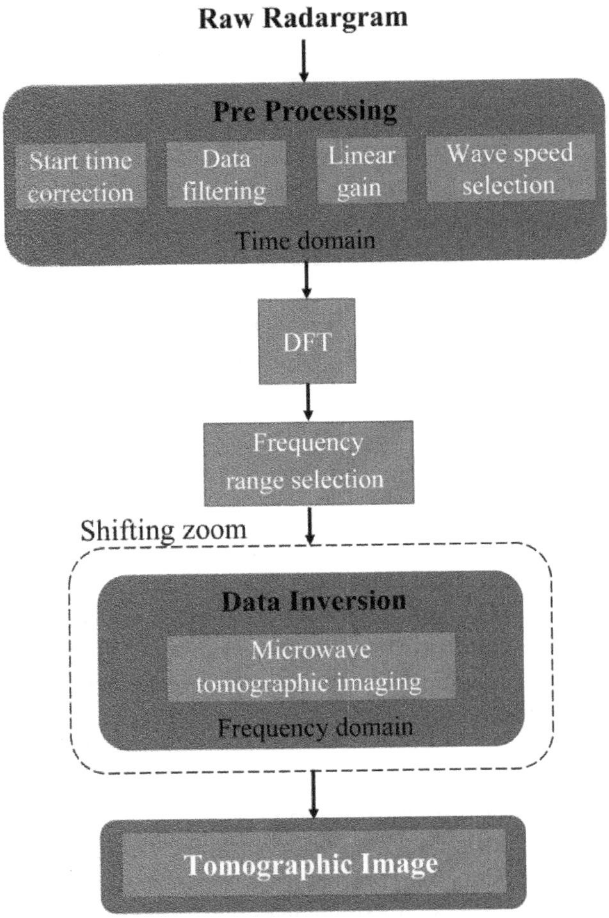

Figure 4.1. *Flow chart of the data processing strategy*

The data inversion step consists of the application of the MWT approach. The latter faces the imaging as the solution of an inverse scattering problem and exploits the Born approximation (Chew 1995) to define the mathematical model describing the relationship between data and unknowns. Although 3D MWT approaches that are able to account for the vectorial nature of signal propagation have been proposed (Catapano et al. 2009; Galli et al. 2013; Catapano et al. 2014; Gennarelli et al. 2020), the data inversion step is herein described with respect to the canonical 2D homogeneous case and transverse magnetic polarization of the probing signal. Specifically, the reference scenario considered in the following is depicted in Figure 4.2.

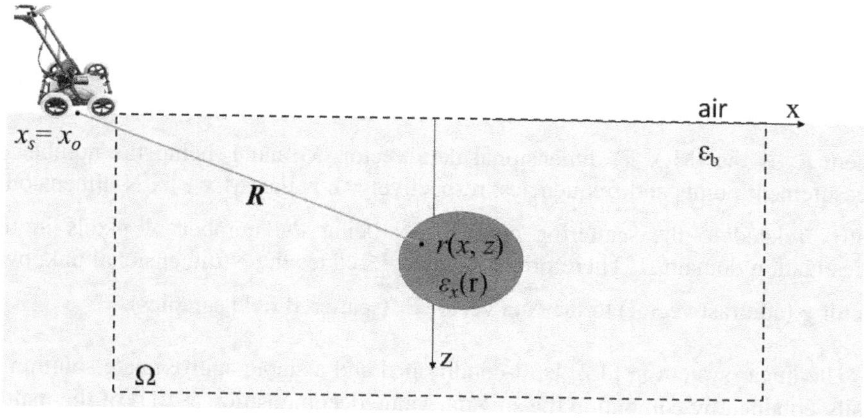

Figure 4.2. *GPRT reference scenario*

The GPR collects data along a single profile and the surveyed medium is modeled as a homogeneous lossless half-space, with relative dielectric permittivity ε_b. The targets are cylinders with an arbitrary cross-section located in the investigated domain Ω and they are looked for as electromagnetic anomalies with relative dielectric permittivity $\varepsilon_x(r)$, r being the generic point in Ω. The unknown quantity to be reconstructed is the contrast function $\chi(r) = \frac{\varepsilon_x(r)}{\varepsilon_b} - 1$, which represents the relative difference between the permittivity of the targets and the permittivity of the surveyed medium.

By assuming and dropping $e^{j\omega t}$ as time dependence, the scattering phenomenon is described, at each angular frequency ω in the selected range $[\omega_{min}, \omega_{max}]$, by means of the integral equation (Chew 1995; Balanis 2012):

$$E_s(x_s, x_o, \omega) = k_b^2 \iint_\Omega G(x_o, r, \omega) E_{inc}(x_s, r, \omega) \chi(r) dr \qquad [4.1]$$

In equation [4.1], E_s denotes the scattered field measured at x_o when the probing source is at x_s, that is, the field sensed by the GPR receiving antenna (after the data filtering step), k_b is the wave number into the background medium, G is the known Green's function that refers to the scenario at hand (Chew 1995) and E_{inc} is the incident field, that is, the field radiated by the GPR transmitting antenna into the probed region Ω in the absence of any target. The latter is modeled as the field radiated by a filamentary line source directed along the invariance axis of the targets, and fed by a unitary electric current.

The integral relationship in equation [4.2] is discretized using the method of moments (Balanis 2012) and rewritten as the matrix problem:

$$\mathbf{E}_s = \underline{\underline{\mathbf{L}}}\boldsymbol{\chi} \qquad [4.2]$$

where $\mathbf{E}_s$ is the (M x F) dimensional data vector, M and F being the number of measurement points and frequencies, respectively; $\underline{\underline{\mathbf{L}}}$ is the (M x F) x N dimensional matrix related to the scattering operator, N being the number of pixels in the investigation domain Ω. Therefore, the matrix $\underline{\underline{\mathbf{L}}}$ relates the N-dimensional unknown vector $\boldsymbol{\chi}$ (contrast vector) to the data vector $\mathbf{E}_s$ (scattered field samples).

The linear system in [4.2] is ill-conditioned and a stable approximate solution is easily obtained by computing the singular value decomposition (SVD) of the matrix $\underline{\underline{\mathbf{L}}}$ (Bertero 1989). By exploiting the truncated SVD (TSVD) as a regularization scheme, an approximated expression of the unknown contrast values in Ω is given by:

$$\tilde{\boldsymbol{\chi}} = \sum_{n=1}^{T} \frac{1}{\sigma_n} \langle \mathbf{E}_{s,}, \mathbf{u}_{n,} \rangle \mathbf{v}_{n,} \qquad [4.3]$$

In [4.3], σ_n is the *n*th singular value of the matrix $\underline{\underline{\mathbf{L}}}$, while $\mathbf{v}_n$ and $\mathbf{u}_n$ are its *n*th left and right singular vectors, and <·,·> denotes the scalar product in the data space. The threshold $T \leq \min\{(M \text{ x } F), N\}$ is set in such a way as to guarantee a good trade-off between the accuracy and stability of the results based on the noise level of data (Bertero 1989; Soldovieri et al. 2007).

The outcome of the processing chain, specifically of the MWT approach, is a spatial map, named tomographic image, where the significant values of the reconstructed contrast function identify the estimated location of targets and give information about their geometry.

The processing chain in Figure 4.1 has been successfully adopted to detect and localize targets hidden in media without a null tangent loss. Moreover, equation [4.1] provides a general formulation of the scattering phenomenon; indeed, changing the relevant key ingredients, that is, the expressions of the incident field and Green's function, it allows us to deal with arbitrary complex electromagnetic scenarios. Moreover, it is worth pointing out that equation [4.1] requires knowledge about the positions of the transmitting and receiving antennas (i.e. x_s and x_o in the case at hand), but does not involve any constraint on their spatial distribution. Therefore, it is suitable to manage data collected by canonical GPR systems, which exploit a single pair of transmitting and receiving antennas at a negligible distance in terms of the probing wavelength, and an odometer assuring evenly spaced measurement points, as well as data gathered by innovative devices, such as MIMO GPR and vehicle mounted systems. Of course, the expression of the kernel of the integral equation, and thus the scattering matrix $\underline{\underline{\mathbf{L}}}$, depends on the measurement configuration, as well as on the scattering model adopted to describe the signal propagation and its interaction with the investigated scenario, and it has to be determined accordingly.

4.3. GPRT performance assessment

Whatever the model adopted to express the kernel of the integral relationship in equation [4.1], the singular values of the matrix $\underline{\underline{\mathbf{L}}}$ in equation [4.2] go to zero as the index n grows up (Bertero 1989). Therefore, the inversion process is unstable and a regularized solution, that is, an estimate of the contrast vector $\boldsymbol{\chi}$, is achieved by truncating the summation to the regularization parameter T. This implies that higher order singular vectors associated with faster spatial variations of $\boldsymbol{\chi}$ are filtered and only a smoothed version can be reconstructed (Persico 2014).

The filtering effects introduced by the regularized inversion of the matrix $\underline{\underline{\mathbf{L}}}$ can be investigated by means of the spectral content (Persico 2014), which qualifies the retrievable spatial harmonics of $\boldsymbol{\chi}$ and is defined as:

$$SC(k_x, k_z) = \sum_{n=1}^{T} |\hat{\mathbf{v}}_n(k_x, k_z)|^2 \qquad [4.4]$$

where k_x, k_z are the spectral variables corresponding to x, z and $\hat{\mathbf{v}}_n(k_x, k_z)$ are the 2D Fourier transforms of the singular vectors $\mathbf{v}_n(x, z)$.

From a quantitative perspective, the MWT spatial resolution can be estimated by evaluating the point spread function (PSF) in the spatial domain, that is, the regularized reconstruction of a point-like target, whose TSVD-based expression is written as:

$$\mathrm{PSF}(x, z) = \sum_{n=1}^{T} \mathbf{v}_n^*(x_0, z_0)\mathbf{v}_n(x_0, z_0) \qquad [4.5]$$

where (x_0, z_0) is the position of the point target and * denotes the conjugation operation.

The spectral content and the PSF are powerful mathematical tools that are useful when predicting the GPRT reconstruction capabilities and understanding how the measurement configuration affects them (Persico 2014). They allow tracing guidelines devoted to a suitable design of the measurement configuration in terms of the extent of a GPR profile and working frequency range (Pierri et al. 1999), and establish how the availability of a limited amount of data affects the imaging capabilities (Pierri and Soldovieri 1998). The latter aspect is particularly important when data collected along multiple profiles are accounted for (Gennarelli et al. 2017; Catapano et al. 2021b). The spectral content and the PSF are also mathematical tools to compare the imaging capabilities related to different scattering models and establish, for the specific case study at hand, the model assuring the best trade-off between computational burden and imaging performance in terms of achievable resolution (Catapano et al. 2021a).

The analysis of the spectral content of the scattering operator and of the PSF within a GPRT approach has been performed in a number of previous works (for example, see, Persico et al. 2005; Gennarelli et al. 2015; Gennarelli et al. 2020; Negishi et al. 2020). In particular, the imaging performance of classical ground-based multibistatic GPR configuration and the complex multi-view/multistatic configuration have been compared in Persico et al. (2005) in the case of a 2D scenario. Such an analysis was extended to a 3D homogenous scenario in Gennarelli et al. (2015). Recently, multimonostatic and multi-view/multistatic GPR configurations have been compared in relation to the imaging of circular structures such as columns or pillars (in Negishi et al. (2020)), which attract considerable interest in cultural heritage monitoring applications. As shown in Negishi et al. (2020), where signal propagation into an inhomogeneous medium is considered, the multi-view/multistatic/multifrequency configuration has a resolution comparable to the classical multibistatic/multifrequency one, but it turns out to be more robust when the data are corrupted by the noise.

4.4. Effective implementation of GPRT

As described in section 4.2, linear MWT requires the computation of the matrix $\underline{\underline{\mathbf{L}}}$ and of its SVD to solve, by means of the TSVD regularization scheme, the imaging problem. However, the calculation of these quantities may be computationally intensive and requires significant computing resources, especially when large domains (in terms of the probing wavelength) are investigated. In detail, let K be the overall amount of data and N be the number of pixels in Ω, then the computational cost for a GPRT image is given as:

$$\sim O(K^{2N}) \qquad [4.6]$$

To overcome this significant drawback, the SZ concept (Persico et al. 2017) is a key tool that can speed up the computational time.

The SZ approach consists of processing data on partially overlapping intervals and in combination with the achieved tomographic images in such a way as to get an overall focused image. Figure 4.3 shows a schematic representation of the SZ procedure, whose main steps are as follows:

– the overall measurement profile Γ and the whole inspected domain Ω are divided into N partially overlapping measurement and imaging subdomains Γ_i and Ω_i, respectively, with $i = 1,2,\cdots,P$;

– for each subdomain Ω_i, the tomographic reconstruction $\tilde{\chi}_i$ is obtained by using the TSVD inversion scheme indicated in equation [4.2];

– the tomographic image of the overall surveyed area Ω is obtained by combining the P reconstructions $\tilde{\chi}_i$ achieved for each sub-domain Ω_i.

It is worth noting that the partial overlapping of the sub-domains allows us to avoid the occurrence of artifacts in each investigated sub-domain due to the presence of the peripheral and external targets close to it.

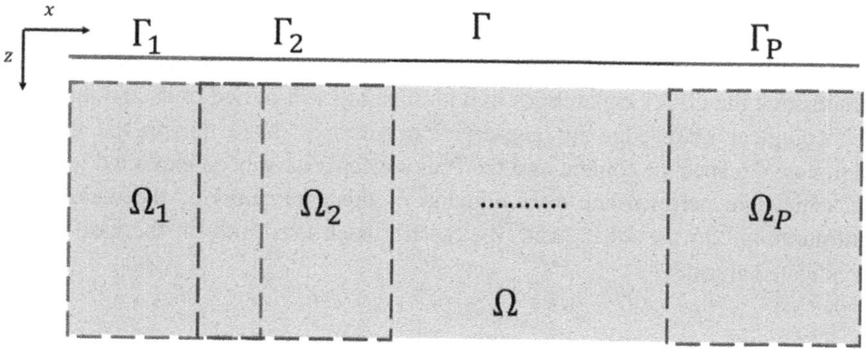

Figure 4.3. *Shift and Zoom concept*

In order to appreciate the computational advantage introduced by the SZ procedure, let us consider the common case of evenly spaced measurement points along the profile Γ. In this case, the relative distances between the radar positions along Γ_i and the pixels into Ω_i are equal for all the subdomains. Therefore, by taking into account equation [4.1], it is evident that the matrix $\underline{\underline{\mathbf{L}}}$ is the same for each couple of subdomains Γ_i and Ω_i. This means that such a matrix and its SVD must be evaluated for just the first subdomain, and the inversion in each subdomain Ω_i

involves only the multiplication of a matrix by a data vector. Accordingly, the SZ approach reduces the computational cost to:

$$\sim O\left(\frac{K^{2\frac{N}{\beta}}}{\alpha}\right) \quad [4.7]$$

where α and β are the scaling factors related to the reduced size of measurement line Γ_i and subdomain Ω_i, respectively. Therefore, an exponential reduction of the computational cost is obtained.

4.5. GPRT surveys in cultural heritage

To provide tangible proof of GPRT performance in cultural heritage surveys, the results referring to measurement campaigns carried out by the authors and their colleagues at some famous archaeological sites are provided in this section.

In cultural heritage applications, GPR surveys are either focused on subsoil exploration, related to preventive archeology (Masini et al. 2017; Masini et al. 2018), or aimed at characterizing historical artifacts and structures, in order to support technical choices for their conservation and restoration (Masini et al. 2010). Accordingly, the following examples have been selected in such a way as to demonstrate the GPRT capabilities in both imaging still-buried ruins and integrating archaeological knowledge for conservation purposes. Note that in the following examples, the spectral content and the PSF concepts have been exploited to foresee the achievable performance as a function of the parameters of the measurement configuration. On the other hand, the SZ has been used only in the case of large investigated areas.

4.5.1. *Discovery of a small temple – archeological park of Paestum and Velia*

This section deals with a case study wherein GPRT allowed the localization and basement reconstruction of a buried Doric temple at the archaeological site of Paestum, sited in the Campania region, south Italy.

Paestum, originally called Poseidonia, was one of the most important ancient Greek cities in Italy and its archeological site is famous for its monumental city walls and three well-preserved Greek temples. However, the area placed between the ceremonial zone and the city walls has been poorly investigated and preserves a great potential for archaeological research. In 2019, several accidental findings performed in a wide area usually devoted to agricultural activities and outside the management of the Sovrintendenza of the Archaeological Park of Paestum and Velia suggested the existence of a buried Doric temple of a small size. These unexpected archaeological findings motivated the use of non-invasive sensing technologies in order to confirm the suspected presence of the temple and restrict the area to be excavated as much as possible. Accordingly, a measurement campaign based on the joint exploitation of magnetometry (MAG) (Gaffney 2008) and GPRT was performed in the same year. As detailed in Capozzoli et al. (2020), MAG was used to survey the whole area, which is located in the northwest quarter of the ancient city and is about 2 hectares wide, while GPRT was exploited to conduct a detailed investigation of the area considered relevant according to the MAG results.

Figure 4.4 depicts the whole area investigated by the MAG, which is outlined in green, and the small portion that the GPRT was interested in, the area outlined in red. The latter is 600 m^2 wide and is a few meters away from the Porta Marina gate. The IDS manufactured HI-Mod GPR system, equipped with a dual frequency shielded antenna working at 200 and 600 MHz, was used. The data were processed by using the processing strategy described in section 4.2 and by exploiting the SZ approach to keep the computational burden feasible. Specifically, the subdomains Γ_i and Ω_i were 1 m wide along the data acquisition direction.

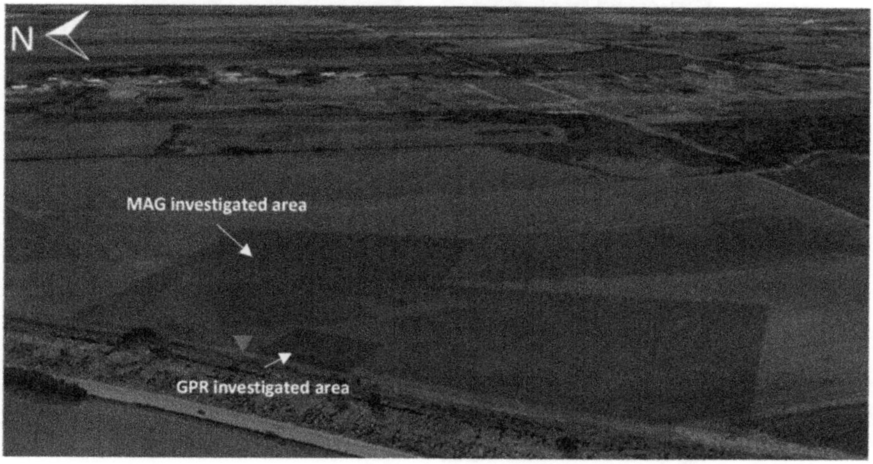

Figure 4.4. *Archaeological site of Paestum – surveyed area*

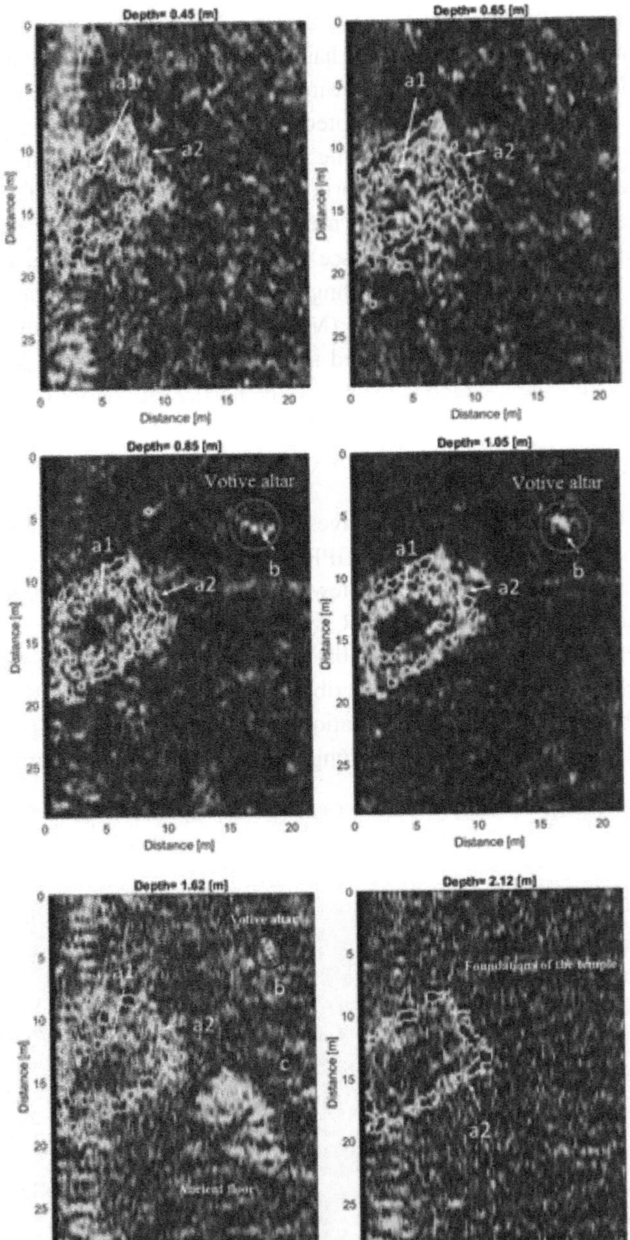

Figure 4.5. *GPRT results – MWT depth slices achieved by processing data collected at the central frequency of 200 MHz*

The results of the GPRT survey, achieved by assuming a soil relative permittivity equal to 9, are reported in Figure 4.5. This figure shows the most significant depth slices referring to the 200 MHz dataset, for which a spatial resolution around 40 cm was expected. Figure 4.5 shows the presence of two concentric rectangles; the outer rectangle (a2) has an extent of 6 m × 8 m, whereas the inner rectangle has an extent of 4 m × 6 m (a1). The tomographic images highlight the top of the structure, which is at a depth of about 0.6–1.0 m below the air–soil interface. Furthermore, two other interesting areas appear. The first one is marked as a red circle in the depth slices at 0.85 m and 1.05 m, and it could be associated with the presence of a votive altar, similar to the ones characterizing the other temples at the site. The second reflective area (red rectangle) is at a depth of 1.5–1.7 m and could be an ancient floor. Reflections are identified up to about 2 m; the temple foundations could be located at this depth.

The GPRT results allowed not only the temple localization and its size estimation, but also an impressive visualization of its basement, which was fully confirmed by the subsequent archeological excavation campaign (see Figure 4.6). Notably, the information provided by the GPRT investigation was also exploited by the Sovrintendenza of the Archeological Park of Paestum and Velia to realize a virtual reconstruction of the temple. The reconstructive hypothesis is a stylobate of 10.83 m × 6.80 m on which 4 × 7 columns were arranged, with a significantly increased intercolumniation on the short sides (2.02 m) compared to the long sides (1.68 m).

Figure 4.6. *Archaeological excavation finding*

4.5.2. *On the traces of Sir Arthur Evans restorations – Knossos Palace*

Knossos Palace is a Bronze Age archaeological site located in Crete, Greece, and is considered as Europe's oldest city, being the ceremonial, economic, social, political centre of the first civilization of the Mediterranean basin, that is, the Minoan civilization. The Old Palace was erected around 2000 BCE, but it was destroyed around 1,700 BCE and immediately rebuilt (New Palace) (Evans 1921/1928/1930/1935).

The Palace was partially excavated by M. Kalokairinos and fully excavated by Sir A. Evans, between 1900 and 1905. The need for preservation and restoration of the monument was obvious from the first years of the excavation. Since 1905, most of its parts were sheltered and in some cases restored. Sir A. Evans, indeed, reconstructed entire floors of the monument using reinforced concrete, created artistic copies of the original frescos, and reconstructed columns and pillars with concrete.

The Palace was under a complete conservation program and, in order to improve the available knowledge about Evans' restoration actions and the materials he used, a GPR measurement campaign was carried out in July 2018. The GPR surveys were performed in the frame of the Horizon 2020 HERACLES – HEritage Resilience Against CLimate Events on Site project (see: http://www.heracles-project.eu/), in collaboration with Ephorate of Antiquities of Heraklion, and regarded several pillars, columns and concrete roof-slabs of the King's room (Figure 4.7(a)) and Queen's room (Figure 4.7(b)). The data were collected by means of the IDS manufactured HI-Mod GPR system equipped with a 2 GHz shielded antenna and processed according to the processing chain in section 4.2. Due to the limited size of the investigated areas, the SZ was not applied. The relative permittivity of the concrete was fixed equal to 9; hence, by taking into account the working frequency, the expected resolution was 2 cm.

a) b)

Figure 4.7. *Knossos Palace: (a) King's room; (b) Queen's room*

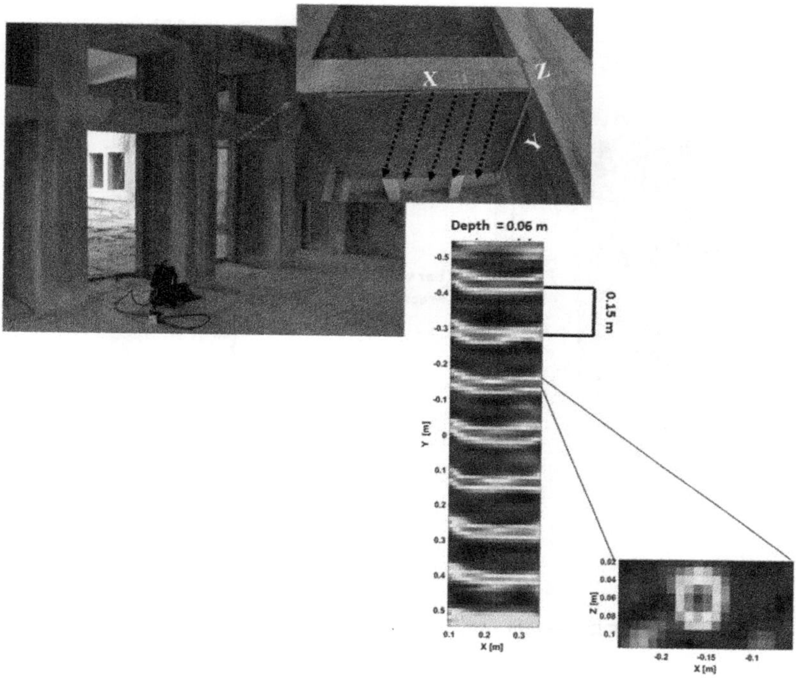

Figure 4.8. *GPRT result referring to a structure in the King's room*

GPRT surveys provided focused images, which made it possible to identify the iron girders and re-bars used by Evans during the first (i.e. before the First World War) and second (i.e. after the First World War) restoration periods.

As an example, Figure 4.8 shows one of the investigated areas in the King's room, as well as the relative GPRT images, which corroborate the presence of a reinforced concrete made up by rebar 15 cm, equally spaced, with a diameter of about 3 cm. The rebars clearly appear at a depth of 6 cm. In addition, Figure 4.9 shows one of the investigated areas in the Queen's room, as well as the relative GPRT image. These results corroborate the fact that the restorations of the King's and Queen's rooms occurred in different historical periods due to the use of two different kinds of iron bars. Specifically, the Queen's room was restored before World War I as the adopted rebars have a rectangular cross-section; conversely, the King's room was restored after the war, and in this case, the rebars have a circular cross-section.

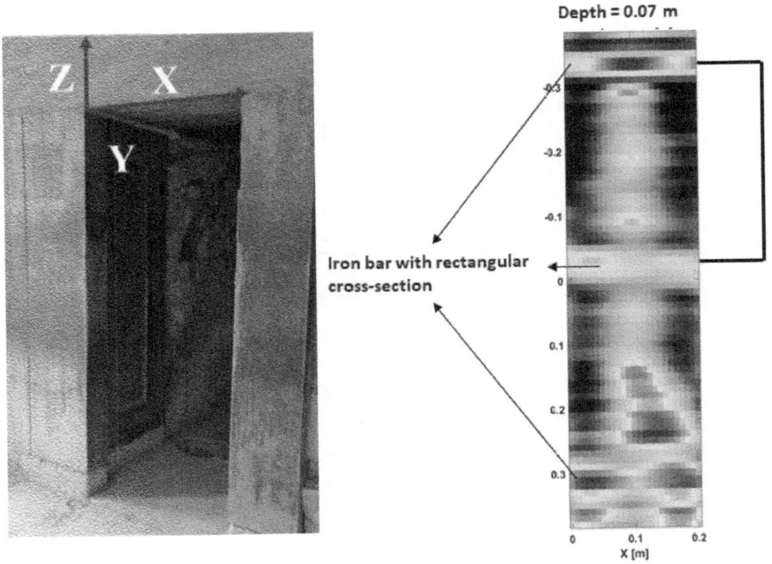

Figure 4.9. *GPRT result referring to a structure in the Queen's room*

4.5.3. *Surveys of columns – tombs of the Kings, Paphos, Cyprus*

The Tombs of the Kings is a necropolis in the Archaeological Park of Kato Paphos. It is one of the most important archaeological sites in Cyprus, dating back to the 4th century, and has been recognized as a UNESCO World Heritage site since 1980.

The necropolis is characterized by several tombs, which were carved out in solid rock and made up of an atrium with open air porticoes supported by monolithic columns and/or masonry pillars (see Figure 4.10).

GPRT surveys were carried out in the frame of the Horizon 2020 – ATHENA Remote Sensing Science Center for Cultural Heritage project (see: https://athena2020.eu/), in collaboration with the Eratosthenes Research Center, Remote Sensing and Geo-environment Research Lab, Cyprus University of Technology, and two institutes of the National Research Council of Italy: the Institute of Archaeological and Monumental Heritage and the Institute of Methodologies for Environmental Analysis. The surveys focused on several columns of the atrium of Tomb 3 and Tomb 4, and aimed to retrieve information on previous restoration actions and, in particular, discover the possible presence of invisible inner cracks and fractures filled with mortar and/or detect metallic reinforcement bars. The IDS HI-Mod GPR system equipped with a 2 GHz shielded antenna was used to perform both vertical scans (i.e. linear profiles) and horizontal ones (i.e. along the column circumference – circular profiles); the data were processed as described above. Note that the MWT approach allowed for the processing of data collected along the two types of scans, that is, linear profiles, as well as circular profiles.

Figure 4.10. *Tombs of the Kings Necropolis*

The GPRT images provided indications concerning the development/propagation of cracks and the existence of discontinuities within the rock material. As an example, Figures 4.11 and 4.12 summarize the results referring to one of the columns of Tomb 3 and Tomb 4, respectively.

The GPRT images in Figure 4.11 refer to two horizontal scans, named Profile a and b, at a height of 105 cm and 153 cm from the floor, and to four 90° spaced vertical scans, named Profiles 1–4. Based on these images, we can infer that:

1) the interior of the column is quite homogeneous, although several weak anomalies appear mainly in the upper part, see the images referring to horizontal scans and vertical scans number 2 and 3;

2) the back side of the column is slightly visible in the images of Profiles 1–4;

3) there is an inner interface 10 cm far from the column surface, which is clearly visible in the reconstruction referring to the Profiles a and b, and on the back side of those of Profiles 1–4; in all the images the air-column interface is not visible since it was removed by time.

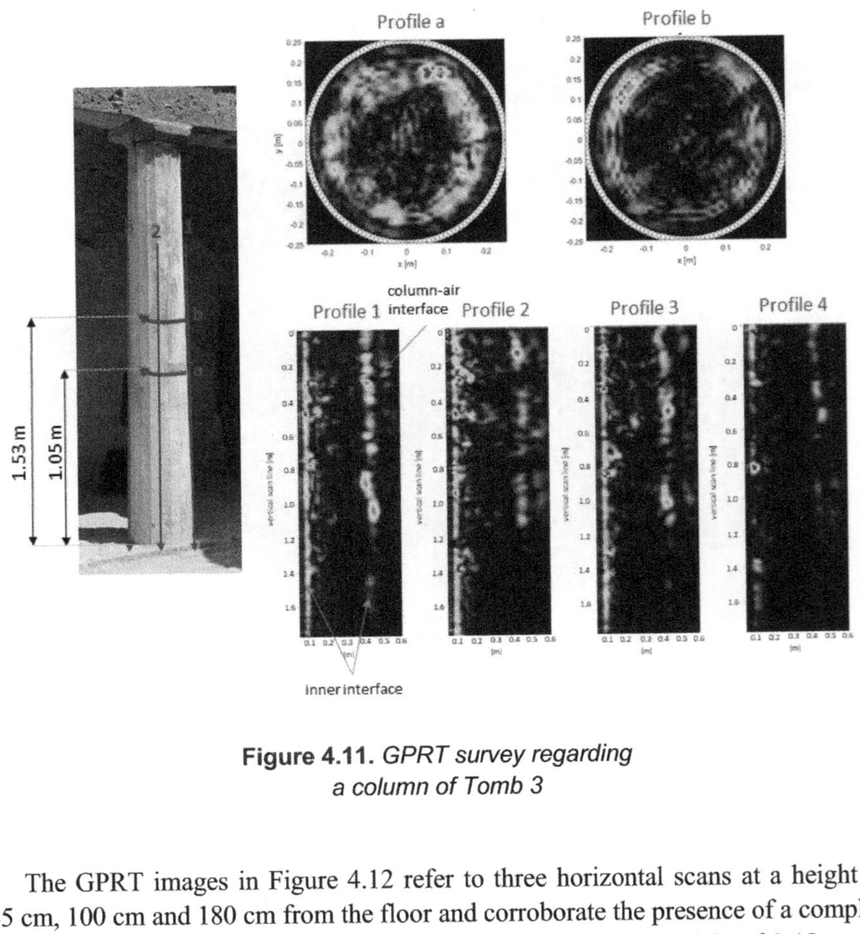

Figure 4.11. *GPRT survey regarding a column of Tomb 3*

The GPRT images in Figure 4.12 refer to three horizontal scans at a height of 45 cm, 100 cm and 180 cm from the floor and corroborate the presence of a complex system of fractures affecting the entire column. Specifically, at height of 0.45 m and 1.8 m, several strong anomalies cover about half of the column section and indicate that the fracturing level is very intense and pervades the entire column. The anomalies are less evident at 100 m, possibly due to surface discontinuities, which compromised data acquisition slightly.

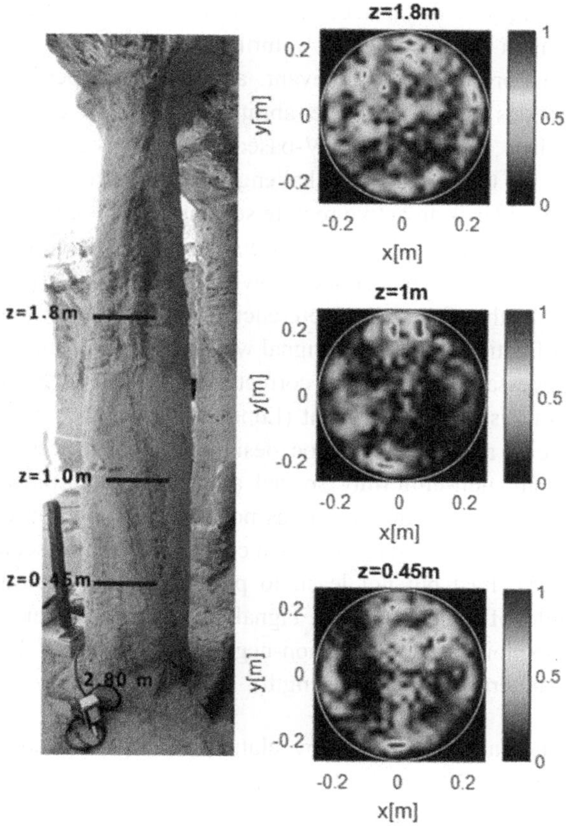

Figure 4.12. *GPRT survey regarding a column of Tomb 4*

4.6. Future perspectives: UAV-based GPRT

Today, UAV-based sensing technologies are widely considered in several fields, including archaeology and cultural heritage inspections. The main advantages offered by UAV as observation platforms concern the possibility of scanning large areas in a short time, the simplification of planning and executing measurement campaigns thus reducing their costs, and the increasing availability of autonomous systems that push more and more toward plug-and-fly solutions.

Mostly adopted UAV sensors are essentially optical, thermal and multi-spectral cameras, which are passive technologies designed to measure the signal emitted into the optical and (near- and far-) infrared portions of the electromagnetic spectrum. These technologies exploit techniques like aero-photogrammetry to get high-resolutions images of the surface features in the scanned area (Colomina and Molina 2014; Salamí et al. 2014). A further possibility is offered by radar technology, which provides the relevant advantage to perform surface and subsurface inspections because of the ability of microwaves to penetrate into non-metallic media. However, UAV-based radar systems are still under development due to the numerous challenges related to data acquisition and processing. Indeed, radar is an active remote sensing technology and this introduces not trivial issues related to payload and assets' constraints. Moreover, in order to obtain well-focused images, an accurate knowledge of the UAV position during its flight is required. In this frame, geo-referencing the UAV position with a spatial error lower than a fraction of the radar signal wavelength is a main challenge. At the same time, it is necessary to perform a correct association between the radar traces and the UAV positions during the flight (López et al. 2022; Noviello et al. 2022). Furthermore, there is a need for ad hoc designed signal processing strategies to manage data that are non-uniformly spaced along a non-rectilinear measurement domain. Note that such a requirement does not regard ground-based GPR surveys where an odometer typically assures a known constant spatial offset among the radar scans (Annan 2005). Last but not least, to perform subsurface surveys, imaging approaches capable of accounting for signal propagation in an inhomogeneous medium must be exploited, due to the non-negligible distance of the radar from the air–soil interface (in terms of the wavelength).

Additional issues are represented by clutter and signal attenuation problems, which typically concern GPR applications and become even more relevant when the sensor is mounted on an UAV (López et al. 2022; Noviello et al. 2022).

Accordingly, although some examples of UAV-based GPR systems are currently available (García-Fernández et al. 2018; García-Fernández et al. 2019; Vergnano et al. 2022; Wu et al. 2022), this technology has not yet been assessed and can be considered as the new and most challenging frontier of GPR development.

In this challenging frame, MWT is a flexible imaging tool, relatively easily adaptable to process data gathered by UAV-based GPR (Catapano et al. 2021a) . To corroborate this statement, let us observe that the integral relationship between the scattered field and the contrast function given by equation [4.1] can be specified according to the scattering model adopted to describe signal propagation. In other words, provided that Green's function and the incident field are properly determined, equation [4.1] can be used to face the imaging in the canonical homogeneous reference scenario, as well as in the case of an inhomogeneous medium (i.e. a half-space reference scenario). Therefore, whatever the observation platform, the imaging is faced by using the mathematical tools introduced in section 4.2.

As a further advantage, UAV positioning information provided by Global Navigation Satellite System (GNSS) techniques can be directly considered in the definition of the kernel of the scattering operator by means of the expression of the source and observation positions (see equation [4.1]). This means that the imaging can be performed regardless of the spatial distribution of the measurement points, without requiring a motion compensation strategy aimed at correcting the flight deviations with respect to the nominal one. On the other hand, the motion compensation strategy is necessary in order to fully exploit the computational advantages offered by the SZ (Noviello et al. 2021). Of course, reliable images are only possible if a sufficient amount of independent data samples are available (Catapano et al. 2021b).

It is also worth pointing out that different scattering models describing the signal propagation in presence of the air-medium interface under the Born approximation can be considered to face the imaging, and their performance has been investigated by the PSF (Ludeno et al. 2020). Such an analysis has corroborated that the equivalent permittivity model, originally proposed in Ludeno et al. (2018a), allows us to tackle the signal propagation in an inhomogeneous medium with the same computational complexity of a homogeneous background and an acceptable degradation of the reconstruction quality. Accordingly, it deserves attention to assure the trade-off between imaging accuracy and computational burden, as also demonstrated in Noviello et al. (2022), where it has been used to elaborate experimental data collected by means of the UAV-based Cobra GPR system available on the market.

To demonstrate the potential of UAV-GPRT, the authors have designed a UAV-based radar system prototype (Ludeno et al. 2018b), and MWT-based imaging strategies have been proposed and experimentally tested (Catapano et al. 2020; Noviello et al 2020, 2021). The prototype is made of a mini rotary wing UAV as the flying platform and the Pulson440 as the radar system. The latter is equipped with log-periodic antennas pointed at nadir and operates in the frequency range of 3.1–4.8 GHz. Moreover, for an accurate reconstruction of the UAV positions during the flight, the system exploits two GPS receivers (one onboard and one in the ground station) and the carrier-phase differential GPS (CDGPS) technique implemented in post-processing (Ward 1997; Catapano et al. 2020).

The prototype has been tested in an experimental campaign carried out at the archeological park of Paestum and Velia, specifically in the area of the ancient thermal baths. Due to the working frequency range, which does not allow sufficient penetration into the Paestum soil down to the expected depth of the buried structures, this proof-of-concept study was limited to imaging on the surface ruins. The data were gathered along multiple lines by planning the flight grid and flying in autonomous modality, and were processed by using an MWT approach that was suitable for processing data collected along multiple lines (Noviello et al. 2021). The inspected area, the acquisition grid and some examples of the obtained images are given in Figure 4.13, where the white line outlines the elevation of the detected anomalies. The GPRT results in Figure 4.13 and those published in Noviello et al. (2021) corroborate the ability to estimate the different elevations of the structures in the investigated area.

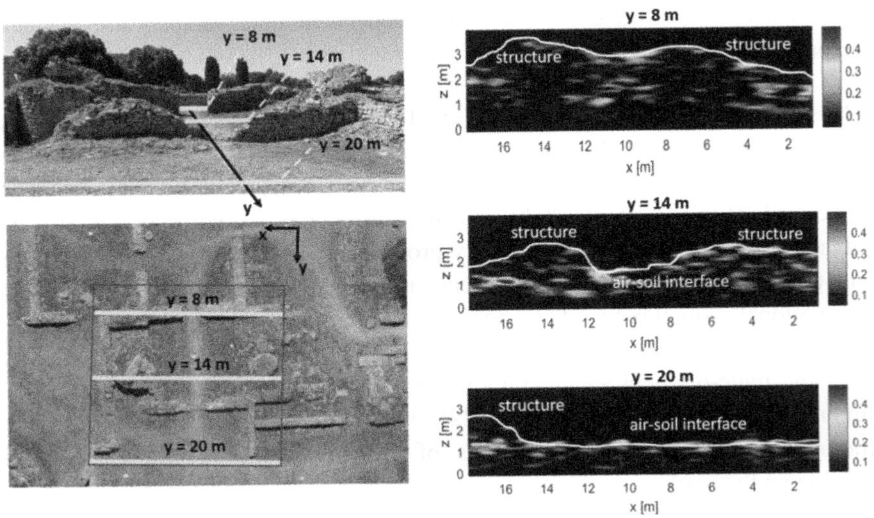

Figure 4.13. *UAV-based GPRT results – slices of investigated volume at constant y values: y = 8 m, y = 14 m, y = 20 m*

4.7. References

Álvarez López, Y., García-Fernández, M., Álvarez Narciandi, G., Las-Heras Andrés, F. (2022). Unmanned aerial vehicle-based ground-penetrating radar systems: A review. *IEEE Geoscience and Remote Sensing Magazine*, 10(2), 66–86.

Annan, A.P. (2005). Ground-penetrating radar. In *Near-Surface Geophysics*, Butler, D.K. (ed.). Society of Exploration Geophysicists, Tulsa, OK.

Balanis, C.A. (2012). *Advanced Engineering Electromagnetics*. John Wiley & Sons, Hoboken, NJ.

Bertero, M. (1989). Linear inverse and ill-posed problems. *Advances in Electronics and Electron Physics*, 75, 1–120.

Capozzoli, L., Catapano, I., De Martino, G., Gennarelli, G., Ludeno, G., Rizzo, E., Soldovieri, F., Uliano Scelza, F., Zuchtriegel, G. (2020). The discovery of a buried temple in Paestum: The advantages of the geophysical multi-sensor application. *Remote Sensing*, 12(17), 2711.

Catapano, I., Crocco, L., D'Urso, M., Isernia, T. (2009). 3D microwave imaging via preliminary support reconstruction: Testing on the Fresnel 2008 database. *Inverse Problems*, 25(2), 024002.

Catapano, I., Affinito, A., Gennarelli, G., di Maio, F., Loperte, A., Soldovieri, F. (2014). Full three-dimensional imaging via ground penetrating radar: Assessment in controlled conditions and on field for archaeological prospecting. *Applied Physics A*, 115, 1415–1422.

Catapano, I., Gennarelli, G., Ludeno, G., Soldovieri, F. (2019a). Applying ground-penetrating radar and microwave tomography data processing in cultural heritage: State of the art and future trends. *IEEE Signal Processing Magazine*, 36(4), 53–61.

Catapano, I., Gennarelli, G., Ludeno, G., Soldovieri, F., Persico, R. (2019b). Ground-penetrating radar: Operation principle and data processing. In *Wiley Encyclopedia of Electrical and Electronics Engineering*, Webster, J.G. (ed.). John Wiley & Sons, New York.

Catapano, I., Gennarelli, G., Ludeno, G., Noviello, C., Esposito, G., Renga, A., Fasano, G., Soldovieri, F. (2020). Small multicopter-UAV-based radar imaging: Performance assessment for a single flight track. *Remote Sensing*, 12(5), 774.

Catapano, I., Gennarelli, G., Ludeno, G., Noviello, C., Esposito, G., Soldovieri, F. (2021a). Contactless ground penetrating radar imaging: State of the art, challenges, and microwave tomography-based data processing. *IEEE Geoscience and Remote Sensing Magazine*, 10(1), 251–273.

Catapano, I., Noviello, C., Soldovieri, F. (2021b). Down-looking airborne radar imaging performance: The multi-line and multi-frequency. *Remote Sensing*, 13(23), 4897.

Chew, W.C. (1995). *Waves and Fields in Inhomogeneous Media*. IEEE Press, New York.

Colomina, I. and Molina, P. (2014). Unmanned aerial systems for photogrammetry and remote sensing: A review. *ISPRS Journal of Photogrammetry and Remote Sensing*, 92, 79–97.

Conyers, L.B. (2013). *Ground-Penetrating Radar for Archaeology*. AltaMira Press, Lanham, MD.

Daniels, D.J. (2005). *Ground Penetrating Radar*. John Wiley & Sons, Hoboken, NJ.

Economou, N., Benedetto, F., Bano, M., Tzanis, A., Nyquist, J., Sandmeier, K.J., Cassidy, N. (2017). Advanced ground penetrating radar signal processing techniques. *Signal Processing*, 132, 197–200.

Evans, A. (1921/1928/1930/1935). *The Palace of Minos at Knossos*, vols. I–IV. Macmillan, London.

Feng, X. and Sato, M. (2004). Pre-stack migration applied to GPR for landmine detection. *Inverse Problems*, 20(6), S99.

Gaffney, C. (2008). Detecting trends in the prediction of the buried past: A review of geophysical techniques in archaeology. *Archaeometry*, 50(2), 313–336.

Galli, A., Comite, D., Catapano, I., Gennarelli, G., Soldovieri, F., Pettinelli, E. (2013). 3D imaging of buried dielectric targets with a tomographic microwave approach applied to GPR synthetic data. *International Journal of Antennas and Propagation*, 10.

García-Fernández, M., Alvarez-López, Y., Arboleya-Arboleya, A., González-Valdés, B., Rodríguez-Vaqueiro, Y., Las-Heras Andrés, F., Pino García, A. (2018). Synthetic aperture radar imaging system for landmine detection using a ground penetrating radar on board a unmanned aerial vehicle. *IEEE Access*, 6, 45100–45112.

García-Fernández, M., Alvarez-López, Y., Las-Heras Andrés, F. (2019). Autonomous airborne 3D SAR imaging system for subsurface sensing: UWB-GPR on board a UAV for landmine and IED detection. *Remote Sensing*, 11(20), 2357.

Gennarelli, G., Catapano, I., Soldovieri, F., Persico, R. (2015). On the achievable imaging performance in full 3-D linear inverse scattering. *IEEE Transactions on Antennas and Propagation*, 63(3), 1150–1155.

Gennarelli, G., Catapano, I., Soldovieri, F. (2017). Reconstruction capabilities of down-looking airborne GPRs: The single frequency case. *IEEE Transactions on Computational Imaging*, 3(4), 917–927.

Gennarelli, G., Ludeno, G., Catapano, I., Soldovieri, F. (2020). Full 3-D imaging of vertical structures via ground-penetrating radar. *IEEE Transactions on Geoscience and Remote Sensing*, 58(12), 8857–8873.

Goodman, D. (1994). Ground-penetrating radar simulation in engineering and archaeology. *Geophysics*, 59(2), 224–232.

Küçükdemirci, M. and Sarris, A. (2022). GPR data processing and interpretation based on artificial intelligence approaches: Future perspectives for archaeological prospection. *Remote Sensing*, 14(14), 3377.

Lai, W.W.L., Derobert, X., Annan, P. (2018). A review of ground penetrating radar application in civil engineering: A 30-year journey from locating and testing to imaging and diagnosis. *NDT & E International*, 96, 58–78.

Lasaponara, R. and Masini, N. (2013). Remote sensing in archaeology: An overview. *Journal of Aeronautics and Space Technologies*, 6(1), 7–17.

Ludeno, G., Capozzoli, L., Rizzo, E., Soldovieri, F., Catapano, I. (2018a). A microwave tomography strategy for underwater imaging via ground penetrating radar. *Remote Sensing*, 10(9), 1410.

Ludeno, G., Catapano, I., Renga, A., Vetrella, A.R., Fasano, G., Soldovieri, F. (2018b). Assessment of a micro-UAV system for microwave tomography radar imaging. *Remote Sensing of Environment*, 212, 90–102.

Ludeno, G., Gennarelli, G., Lambot, S., Soldovieri, F., Catapano, I. (2020). A comparison of linear inverse scattering models for contactless GPR imaging. *IEEE Transactions on Geoscience and Remote Sensing*, 58(10), 7305–7316.

Masini, N. and Soldovieri, F. (2011). Integrated non-invasive sensing techniques and geophysical methods for the study and conservation of architectural, archaeological and artistic heritage. *Journal of Geophysics and Engineering*, 8(3), E01.

Masini, N., Persico, R., Rizzo, E., Calia, A., Giannotta, M.T., Quarta, G., Pagliuca, A. (2010). Integrated techniques for analysis and monitoring of historical monuments: The case of S. Giovanni al Sepolcro in Brindisi (Southern Italy). *Near Surface Geophysics*, 8(5), 423–432.

Masini, N., Capozzoli, L., Chen, P., Chen, F., Romano, G., Lu, P., Tang, P., Sileo, M., Ge, Q., Lasaponara, R. (2017). Towards an operational use of geophysics for archaeology in Henan (China): Archaeogeophysical investigations, approach and results in Kaifeng. *Remote Sensing*, 9(8), 809.

Masini, N., Capozzoli, L., Romano, G., Sieczkowska, D., Sileo, M., Bastante, J., Victoria, F.A., Ziolkowski, M., Lasaponara, R. (2018). Archaeogeophysical based approach for Inca archaeology: Overview and one operational application. *Survey in Geophysics*, 39, 1239–1262.

Negishi, T., Gennarelli, G., Soldovieri, F., Liu, Y., Erricolo, D. (2020). Radio frequency tomography for nondestructive testing of pillars. *IEEE Transactions on Geoscience and Remote Sensing*, 58(6), 3916–3926.

Noviello, C., Esposito, G., Fasano, G., Renga, A., Soldovieri, F., Catapano, I. (2020). Small-UAV radar imaging system performance with GPS and CDGPS based motion compensation. *Remote Sensing*, 12(20), 3463.

Noviello, C., Esposito, G., Catapano, I., Soldovieri, F. (2021). Multilines imaging approach for mini-UAV radar imaging system. *IEEE Geoscience and Remote Sensing Letters*, 19, 1–5.

Noviello, C., Gennarelli, G., Esposito, G., Ludeno, G., Fasano, G., Capozzoli, L., Soldovieri, F., Catapano, I. (2022). An overview on down-looking UAV-based GPR systems. *Remote Sensing*, 14(14), 3245.

Pastorino, M. (2010). *Microwave Imaging*. John Wiley & Sons, New York.

Persico, R. (2014). *Introduction to Ground Penetrating Radar: Inverse Scattering and Data Processing*. John Wiley & Sons, New York.

Persico, R., Bernini, R., Soldovieri, F. (2005). The role of the measurement configuration in inverse scattering from buried objects under the Born approximation. *IEEE Transactions on Antennas and Propagation*, 53(6), 1875–1887.

Persico, R., Ludeno, G., Soldovieri, F., De Coster, A., Lambot, S. (2017). Two-dimensional linear inversion of GPR data with a shifting zoom along the observation line. *Remote Sensing*, 9(10), 980.

Pierri, R. and Soldovieri, F. (1998). On the information content of the radiated fields in the near zone over bounded domains. *Inverse Problems*, 14, 321–337.

Pierri, R., Persico, R., Bernini, R. (1999). Information content of the Born field scattered by an embedded slab: Multifrequency, multiview, and multifrequency–multiview cases. *Journal of the Optical Society of America A*, 16(10), 2392–2399.

Salamí, E., Barrado, C., Pastor, E. (2014). UAV flight experiments applied to the remote sensing of vegetated areas. *Remote Sensing*, 6(11), 11051–11081.

Soldovieri, F., Hugenschmidt, J., Persico, R., Leone, G. (2007). A linear inverse scattering algorithm for realistic GPR applications. *Near Surface Geophysics*, 5(1), 29–41.

Stolt, R.H. (1978). Migration by Fourier transform. *Geophysics*, 43(1), 23–48.

Tejedor, B., Lucchi, E., Bienvenido-Huertas, J.D., Nardi, I. (2022). Non-destructive techniques (NDT) for the diagnosis of heritage buildings: Traditional procedures and futures perspectives. *Energy and Buildings*, 263, 112029.

Trinks, I., Johansson, B., Gustafsson, J., Emilsson, J., Friborg, J., Gustafsson, C., Nissen, J., Hinterleitner, A. (2010). Efficient, large-scale archaeological prospection using a true three-dimensional ground-penetrating radar array system. *Archaeological Prospection*, 17(3), 175–186.

Vaughan, C.J. (1986). Ground-penetrating radar surveys used in archaeological investigations. *Geophysics*, 51(3), 595–604.

Vergnano, A., Franco, D., Godio, A. (2022). Drone-borne ground-penetrating radar for snow cover mapping. *Remote Sensing*, 14(7), 1763.

Ward, N. (1997). Understanding GPS – Principles and applications. *The Journal of Navigation*, 50(1), 151–152.

Wu, S., Wang, L., Zeng, X., Wang, F., Liang, Z., Ye, H. (2022). UAV-mounted GPR for object detection based on cross-correlation background subtraction method. *Remote Sensing*, 14(20), 5132.

5

The Full-Wave Radar Equation for Wave Propagation in Multilayered Media and Its Applications

Sébastien LAMBOT[1], Kaijun WU[1], Arthur SLUŸTERS[2] and Jean VANDERDONCKT[2]

[1] *Earth and Life Institute, Université catholique de Louvain, Louvain-la-Neuve, Belgium*

[2] *Louvain Research Institute in Management and Organizations (LouRIM), Institute of Information and Communication Technologies, Electronics and Applied Mathematics (ICTEAM), Université catholique de Louvain, Louvain-la-Neuve, Belgium*

5.1. Introduction

A microwave radar such as an ultrawide band (UWB) ground-penetrating radar (GPR) can be used to non-destructively identify the electromagnetic properties of a medium. This requires a wave propagation model and an inversion scheme, which optimize the model parameters of interest by fitting the model to the radar data. Approaches for radar electromagnetic modeling range from simpler straight-ray propagation models (Huisman et al. 2003) to full-wave 3D models solving Maxwell's equations. While a few relatively simple geometric configurations have closed-form

For a color version of all the figures in this chapter, see www.iste.co.uk/serhir/radar.zip.

Ground Penetrating Radar,
coordinated by Mohammed SERHIR and Dominique LESSELIER.

analytical solutions, such as 3D planar or cylindrical layered media, full-wave solutions of Maxwell's equations require numerical methods, including the finite-difference time-domain (FDTD) method (Hyun et al. 2009; Meles et al. 2011; Millington et al. 2011; Warren and Giannopoulos 2011), the finite element method (FEM) (Venkatarayalu et al. 2008; Ilic et al. 2009) and the method of moments (MoM) (Craeye et al. 2009; Chen et al. 2010). However, the use of 3D numerical methods for full-wave modeling and inversion of the radar signal is limited by the required computing resources and the inherent differences between the conceptualized geometrical models and the real radar–antenna–medium system.

In this chapter, we present and explain the closed-form, far-field and near-field radar equations introduced by Lambot et al. (2004) and Lambot and André (2014) for full-wave modeling of radar data. Examples of inverse modeling applications are then given, with a focus on soil moisture mapping and hand gesture recognition.

5.2. Far-field radar equation

The far-field full-wave radar equation (Lambot et al. 2004; Lambot and André 2014) assumes a homogeneous field distribution for the back-scattered field over the antenna aperture (Figure 5.1(a)). This is applicable when the medium is far enough from the antenna, meaning the 3D field distribution from its aperture to the radar reference plane where the radar signal is measured and the corresponding current distribution on the antenna are constant and do not depend on the medium properties. Only the amplitude and phase of the field change as a function of the medium properties and distance. Along the field propagation paths within the antenna, continuous and discontinuous variations of impedance are encountered as a function of the antenna geometry, feed and coaxial line (Figure 5.1(b)). These variations of impedance lead to transmission and reflections within the antenna. The impedance profiles are not known exactly, but the following mathematical identity applies: any multilayered medium with variations of impedance (propagation paths in this case) can be replaced by an equivalent one-layer medium that is characterized with a so-called global impedance, or, in an equivalent way, global reflection and transmission coefficients (Figure 5.1(c), equations [5.18] and [5.20]). These global functions are complex valued and frequency dependent. Given this identity, the propagation of the fields within the antenna can be intrinsically and exactly described with only four quantities, namely, the global reflection coefficient of the antenna for fields incident from the radar reference plane onto the source point ($R_i(\omega)$), the global transmission coefficient for fields incident from the field point onto the radar reference plane ($T_s(\omega)$), the global transmission coefficient for fields incident from the radar reference plane onto the source point ($T_i(\omega)$) and the global reflection coefficient for fields incident from the layered medium onto the field point ($R_s(\omega)$), where ω is the angular frequency. Because of the superposition principle, these functions integrate all propagation paths within the antenna. The model concept as illustrated in Figure 5.1 can also be represented by the linear system in Figure 5.2, where the quantities $a(\omega)$ and $b(\omega)$ are the source and returned signals in the frequency domain, respectively. Wave propagation from the radar antenna to the multilayered medium can then be exactly described using the 3D multilayered media Green function $G_{xx}(\omega)$ (Chew 1990; Michalski and Zheng 1990; Slob and Fokkema 2002; Lambot et al. 2004).

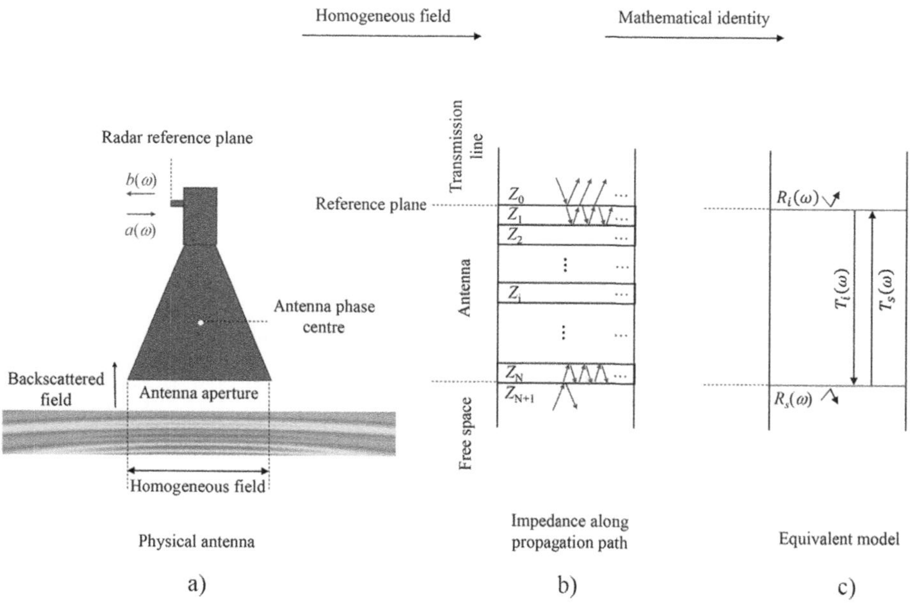

Figure 5.1. *Full-wave radar model concept of Lambot et al. (2004) and Lambot and André (2014), from the physical antenna (a), equivalent electric model based on the homogeneous field assumption and superposition principle (b), to the intrinsic antenna representation (c). In (b), the red arrows illustrate the infinite reflections occurring within the antenna*

The equivalent linear block diagram shown in Figure 5.2 can be solved as follows:

$$b(\omega) = a(\omega)R_i(\omega) + z(\omega)T_s(\omega) \quad [5.1]$$

with

$$z(\omega) = [a(\omega)T_i(\omega) + z(\omega)R_s(\omega)]\, G_{xx}(\omega) \quad [5.2]$$

which can be solved as

$$z(\omega) = \frac{a(\omega)T_i(\omega)G_{xx}(\omega)}{1 - R_s(\omega)G_{xx}(\omega)} \qquad [5.3]$$

Hence, we obtain

$$b(\omega) = a(\omega)R_i(\omega) + \frac{a(\omega)T_i(\omega)G_{xx}(\omega)}{1 - R_s(\omega)G_{xx}(\omega)}T_s(\omega) \qquad [5.4]$$

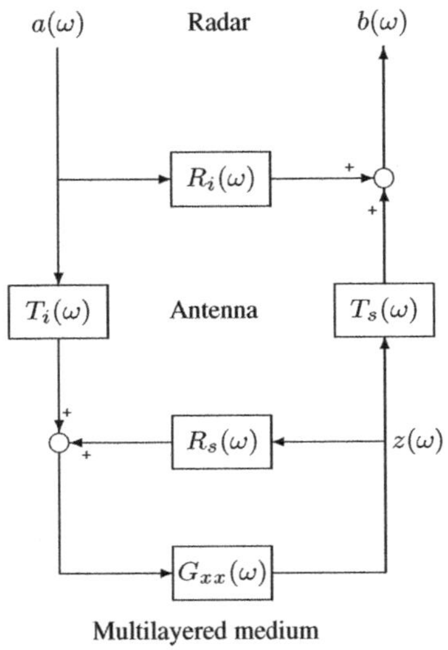

Figure 5.2. *Equivalent linear block diagram of the radar–antenna–medium system to which corresponds the far-field radar equation (Lambot et al. 2004)*

Considering the ratio $S(\omega) = b(\omega)/a(\omega)$, for example, the quantity S_{11} measured by a vector network analyzer, and the product $T(\omega) = T_i(\omega)T_s(\omega)$, equation [5.4] can be written as (Lambot et al. 2004):

$$S(\omega) = R_i(\omega) + \frac{T(\omega)G_{xx}(\omega)}{1 - R_s(\omega)G_{xx}(\omega)} \qquad [5.5]$$

For a pulse radar, the quantity $s(t)$ is measured, with t being time, and $b(\omega)$ is obtained using the Fourier transform:

$$b(\omega) = \mathcal{F}\{s(t)\} = \int_{-\infty}^{\infty} s(t)e^{-2\pi j t\omega}dt \qquad [5.6]$$

Then, defining $H_i(\omega) = a(\omega)R_i(\omega)$, $H(\omega) = a(\omega)T(\omega)$ and $H_s(\omega) = R_s(\omega)$, equation [5.4] becomes:

$$\mathcal{F}\{s(t)\} = H_i(\omega) + \frac{H(\omega)G_{xx}(\omega)}{1 - H_s(\omega)G_{xx}(\omega)} \tag{5.7}$$

Whether the radar provides the ratio $b(\omega)/a(\omega)$ or the signal $b(\omega)$, the radar–antenna can be fully described by only three independent functions.

It is worth noting that $R_i(\omega)$ or $H_i(\omega)$ corresponds to a free-space measurement, for which $G_{xx}(\omega) = 0$, and that $R_s(\omega)$ is responsible for the multiple reflections between the antenna and the target medium.

Once the radar–antenna characteristic functions are known, from [5.5], the Green's function can be directly calculated from the radar measurements as:

$$G_{xx}(\omega) = \frac{S(\omega) - R_i(\omega)}{T(\omega) + S(\omega)R_s(\omega) - R_i(\omega)R_s(\omega)} \tag{5.8}$$

From the measured radar data, this equation allows us to obtain a normalized quantity, the Green function, that is independent of the radar and antenna. The undesired radar–antenna effects such as internal multiple reflections or antenna–medium interactions are eliminated. For time-domain radars, the so-called time-zero becomes precisely defined and does correspond to the antenna phase center. A practical application of the far-field radar equation is therefore to enhance radar imaging (e.g. Lopera et al. (2007)).

5.3. Determination of the radar–antenna characteristic functions

The characteristic functions $R_i(\omega)$, $T(\omega)$ and $R_s(\omega)$ (or $H_i(\omega)$, $H(\omega)$ and $H_s(\omega)$) can be calculated by solving a system of equations [5.5] for a minimum of three different model configurations (denoted k). Our practice is to use well-defined model configurations, with the antenna situated at varying heights above a metal sheet, representing an infinite perfect electric conductor (PEC). Knowing the properties of the medium, $G_{xx,k}(\omega)$ can be computed and the corresponding $S_k(\omega)$ can be measured.

The initial system of equations is changed to a linear form by rewriting equation [5.5] in the following manner (frequency dependence ω is assumed):

$$S_k = R_i + S_k G_{xx,k} R_s + G_{xx,k}\left(T - R_i R_s\right) \tag{5.9}$$

Hence, the resulting linear system of equations can be written in the matrix form as:

$$\mathrm{b} = \mathbf{A}\mathbf{x} \tag{5.10}$$

where

$$\mathrm{b} = \begin{bmatrix} S_1 \\ \vdots \\ S_k \\ \vdots \\ S_n \end{bmatrix} \tag{5.11}$$

$$\mathbf{A} = \begin{bmatrix} 1 & S_1 G_{xx,1} & G_{xx,1} \\ \vdots & \vdots & \vdots \\ 1 & S_k G_{xx,k} & G_{xx,k} \\ \vdots & \vdots & \vdots \\ 1 & S_n G_{xx,n} & G_{xx,n} \end{bmatrix} \tag{5.12}$$

$$\mathbf{x} = \begin{bmatrix} R_i \\ R_s \\ T - R_i R_s \end{bmatrix} \tag{5.13}$$

The vector of unknowns is then determined using the least squares approach:

$$\mathbf{x} = \left(\mathbf{A}^{\mathrm{H}}\mathbf{A}\right)^{-1}\mathbf{A}^{\mathrm{H}}\mathrm{b} \tag{5.14}$$

where symbol H denotes the Hermitian or conjugate transpose ($\mathbf{A}^{\mathrm{H}} \equiv \overline{\mathbf{A}^T}$).

In general, it is recommended to use more than three different configurations, for example, five configurations, to ensure that the linear system is well-posed for all frequencies, depending on the measurement distances and frequency range. As per Tran et al. (2013a), the far-field radar equation is applicable when the distance to the medium is greater than 1.2 times the antenna aperture. To achieve an optimal signal-to-noise ratio, the calibration distances should begin from this threshold and span at least one wavelength. In order for the assumption of the infinite plane to be valid, the size of the metal sheet should be greater than five times the maximum wavelength at these distances. For low-frequency radars, the required size of the calibration plane may be relatively large (e.g. 15 m at 100 MHz). In this case, calibrating the radar–antenna over a large, deep and smooth water body may be more practical than using a PEC (Mahmoudzadeh Ardekani and Lambot 2014; Wu et al. 2022).

5.4. Planar multilayered media Green's function

The dyadic Green's function for wave propagation in planar multilayered media is an invaluable mathematical tool in the field of electromagnetic field theory, with countless applications ranging from antenna design to microwave remote sensing. Considering a classical GPR configuration, we define the Green's function $G_{xx}(\omega)$ as the back-scattered x-directed electric field $E_x(\omega)$ at the field point for a unit-strength x-directed electric source J_x situated at the source point. Depending on the antenna configuration, other components from the dyadic Green's function can be considered as well. Assuming the radar is situated in the air above the medium, the source and field points are set at a distance of h_1 away from the first medium interface, as illustrated in Figure 5.3. The distance between the source point and the field point in the xy-plane is denoted ρ. When the same antenna simultaneously operates as transmitter and receiver, $\rho = 0$ and the source and field point corresponds to the antenna phase center. The Green function can be derived using a recursive scheme to compute the global reflection coefficients of the multilayered medium in the spectral domain (Chew 1990; Michalski and Zheng 1990; Slob and Fokkema 2002; Lambot et al. 2004). The transformation back to the spatial domain is performed by evaluating numerically a semi-infinite integral, for which a fast procedure is applied (Lambot et al. 2007).

More precisely, the spatial-domain Green's function in [5.5] is computed as

$$G_{xx} = \frac{1}{8\pi} \int_0^{+\infty} \tilde{G}_{xx}(k_\rho) k_\rho dk_\rho \qquad [5.15]$$

with $\tilde{G}_{xx}(k_\rho)$ being the spectral Green's function, that is, the Green function in the Fourier domain with respect to the x and y coordinates:

$$\tilde{G}_{xx}(k_\rho) = \left[J_0(k_\rho \rho) \left(\frac{\Gamma_1 R_1^{TM}}{\sigma_1 + \jmath\omega\varepsilon_1} - \frac{\jmath\omega\mu_1 R_1^{TE}}{\Gamma_1} \right) - \right.$$
$$\left. J_2(k_\rho \rho) \cos(2\theta) \left(\frac{\Gamma_1 R_1^{TM}}{\sigma_1 + \jmath\omega\varepsilon_1} + \frac{\jmath\omega\mu_1 R_1^{TE}}{\Gamma_1} \right) \right] \exp(-2\Gamma_1 h_1) \qquad [5.16]$$

where J_0 and J_2 are, respectively, the first kind zero- and second-order Bessel's functions, θ is the angle in the xy-plane between the field and source points, k_ρ is the spectral-domain counterpart of ρ, subscript 1 refers to the upper half-space (free-space), $\jmath = \sqrt{-1}$, R_1^{TM} and R_1^{TE} are, respectively, the transverse magnetic (TM) and transverse electric (TE) global reflection coefficients accounting for all reflections in the multilayered medium, Γ is the vertical wavenumber defined as $\Gamma = \sqrt{k_\rho^2 - k^2}$, while $k^2 = \omega^2\mu(\varepsilon - (\jmath\sigma/\omega))$ with the magnetic permeability μ, dielectric permittivity ε and electrical conductivity σ.

The global reflection coefficients at interface n are recursively calculated as:

$$R_n^{TM} = \frac{r_n^{TM} + R_{n+1}^{TM}\exp(-2\Gamma_{n+1}h_{n+1})}{1 + r_n^{TM}R_{n+1}^{TM}\exp(-2\Gamma_{n+1}h_{n+1})} \quad [5.17]$$

$$r_n^{TM} = \frac{(\sigma_{n+1} + \jmath\omega\varepsilon_{n+1})\Gamma_n - (\sigma_n + \jmath\omega\varepsilon_n)\Gamma_{n+1}}{(\sigma_{n+1} + \jmath\omega\varepsilon_{n+1})\Gamma_n + (\sigma_n + \jmath\omega\varepsilon_n)\Gamma_{n+1}} \quad [5.18]$$

$$R_n^{TE} = \frac{r_n^{TE} + R_{n+1}^{TE}\exp(-2\Gamma_{n+1}h_{n+1})}{1 + r_n^{TE}R_{n+1}^{TE}\exp(-2\Gamma_{n+1}h_{n+1})} \quad [5.19]$$

$$r_n^{TE} = \frac{\mu_{n+1}\Gamma_n - \mu_n\Gamma_{n+1}}{\mu_{n+1}\Gamma_n + \mu_n\Gamma_{n+1}} \quad [5.20]$$

where r_n^{TM} and r_n^{TE} denote, respectively, the local TM- and TE-mode reflection coefficients at interface n ($n = 1 \cdots N-1$). The recursion is initiated with $R_{N-1}^{TM} = r_{N-1}^{TM}$ and $R_{N-1}^{TE} = r_{N-1}^{TE}$ as there are no incident waves coming from the lower half-space.

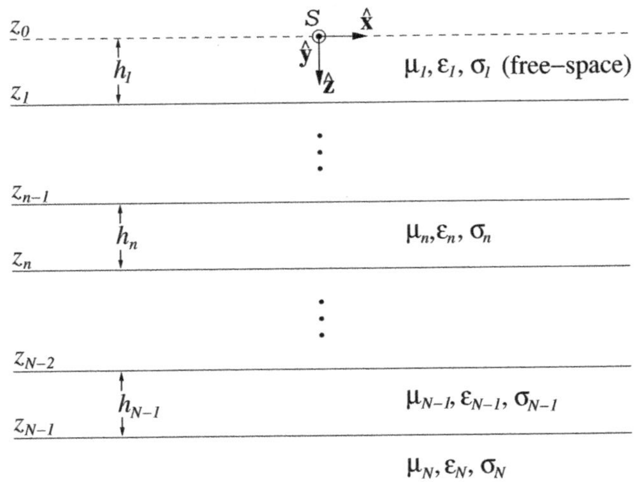

Figure 5.3. *Three-dimensional N-layered medium with a point source-receiver at the origin S of the coordinate system ($\rho = 0$ in this sketch). The medium of the nth layer is characterized by homogeneous magnetic permeability (μ_n), dielectric permittivity (ε_n), electric conductivity (σ_n) and thickness (h_n)*

The spectral Green's function [5.16] contains various singularities, namely poles and branch points. These poles occur due to denominators vanishing, which can be physically interpreted as modes guided by the layered medium (surface wave poles). Branch points arise from the Γ dependence, as it is a square root of a complex number, and thus has two possible values, which creates a branch cut when Im(Γ)=0 intersects the plane of integration. Each branch cut terminates in a branch point at $k_\rho = \pm k$. In the case of a lossless medium, these singularities are located on the real k_ρ axis, whereas for a conductive medium, they are situated in the negative imaginary k_ρ plane, as the integration runs from zero to infinity. A branch cut exists for every layer n (Chew 1990).

To ensure efficient and quick convergence of the integral, the integration path should be designed to avoid any singularities in the integrand and minimize oscillations of the function. It is recommended to deform the path in the positive part of the complex k_ρ plane, as this avoids the need to identify all poles and branch points that occur for negative values of $\text{Im}(k_\rho)$. Since the free-space exponential in [5.16] has a very significant effect on the spectral Green's function for a far-field model configuration, the path can be chosen to minimize its oscillating behavior, resulting in a constant imaginary part of the square root in the exponential, equal to the value at $k_\rho = 0$, where the integration has started.

Lambot et al. (2007) provide an optimal integration path in the complex k_ρ plane for far-field zero-offset ($\rho = 0$) radar. Defining k_ρ as the complex number $(x + \jmath y)$ in Γ_1, and changing variables so that $a = x\frac{c}{\omega}$ and $b = y\frac{c}{\omega}$, with $c = 1/\sqrt{(\mu_0\varepsilon_0)}$ being the free-space velocity, we find:

$$\text{Im}(\Gamma_1) = \text{Im}\left(\sqrt{(a+\jmath b)^2 - 1}\right) = \frac{1}{2}\sqrt{2 + 2\sqrt{a^4 + 2a^2b^2 - 2a^2 + b^4 + 2b^2 + 1} - 2a^2 + 2b^2} \quad [5.21]$$

For $k_\rho = 0$, we find $\text{Im}(\Gamma_1) = 1$. Substituting this value in [5.21], we obtain the following relationship for the constant phase integration path:

$$y(x) = \frac{x}{\sqrt{\left(\frac{xc}{\omega}\right)^2 + 1}} \quad [5.22]$$

Then, using Cauchy's integral theorem, we can calculate Green's function in the spatial domain as follows:

$$G_{xx} = \int_0^{x_{max}} \tilde{G}_{xx}(x + \jmath y(x)) \cdot [x + \jmath y(x)] \frac{\partial (x + \jmath y(x))}{\partial x} dx \qquad [5.23]$$

with

$$\frac{\partial (x + \jmath y(x))}{\partial x} = 1 + \jmath \left(\frac{1}{\sqrt{\left(\frac{xc}{\omega}\right)^2 + 1}} - \frac{x^2}{\sqrt{\left(\left(\frac{xc}{\omega}\right)^2 + 1\right)^3}} \frac{c^2}{\omega^2} \right) \qquad [5.24]$$

and the upper limit of the integral, x_{max}, can be determined by considering the exponential damping factor d (e.g. $d = 10^{-6}$):

$$x_{max} = \sqrt{\left(\frac{-\ln d}{2h_1}\right)^2 + \left(\frac{\omega}{c}\right)^2} \qquad [5.25]$$

This value represents the point at which the integral becomes negligible.

5.5. Near-field radar equation

When the medium is situated in the near-field of the antenna, the homogeneous field assumption over the antenna aperture no longer holds, and the scattered field distribution depends on the antenna–medium distance and the medium's properties. An efficient approach for producing the correct antenna radiation field and capturing the scattered field distribution is the use of an equivalent set of infinitesimal electric dipoles ($J_{x,n}$, $n = 1 \cdots N$) for the source and a set of points ($E_{x,m}$, $m = 1 \cdots N$) where the field is sampled for the receiver (Gentili and Spagnolini 2000; Serhir et al. 2010). Here, N does not refer to the multilayered medium. This approach relies on the superposition principle. Figure 5.4 shows an example of such a model configuration, with a bowtie antenna pair, as an example, and equivalent infinitesimal electric sources and receiving field points. When the same antenna is used as a transmitter and receiver, the source and field points are the same. The number of points to consider depends on the complexity of the scattered field distribution, which is intrinsically decomposed into a number of elementary distributions, that is, plane waves.

The near-field radar equation can be derived in a similar fashion to the far-field equation. This equation accounts for wave propagation between the point sources or field points and the radar transmission line reference plane by means of complex, frequency-dependent global reflection and transmission coefficients. The field points act as point sources within infinite reflection loops due to the scattered field antenna reflection coefficients. Although the medium-dependent part of mutual antenna coupling is not explicitly considered in the model (only free-space coupling is described through T_0), it is implicitly accounted for because of the linearity of Maxwell's equations. The near-field radar equation is expressed as follows (Lambot and André 2014):

$$S(\omega) = \frac{b(\omega)}{a(\omega)} = T_0(\omega) + \mathbf{T_s}\left(\mathbf{I_N} - \mathbf{G^0R_s}\right)^{-1}\mathbf{GT_i} \qquad [5.26]$$

with

$$\mathbf{T_i} = \left[T_{i,1}(\omega)\; T_{i,2}(\omega)\; \cdots\; T_{i,N}(\omega)\right]^T \qquad [5.27]$$

$$\mathbf{T_s} = \left[T_{s,1}(\omega)\; T_{s,2}(\omega)\; \cdots\; T_{s,N}(\omega)\right] \qquad [5.28]$$

$$\mathbf{R_s} = \text{diag}\left(\left[R_{s,1}(\omega)\; R_{s,2}(\omega)\; \cdots\; R_{s,N}(\omega)\right]\right) \qquad [5.29]$$

$$\mathbf{G} = \begin{bmatrix} G_{11}(\omega) & G_{12}(\omega) & \cdots & G_{1N}(\omega) \\ G_{21}(\omega) & G_{22}(\omega) & \cdots & G_{2N}(\omega) \\ \vdots & \vdots & & \vdots \\ G_{N1}(\omega) & G_{N2}(\omega) & \cdots & G_{NN}(\omega) \end{bmatrix} \qquad [5.30]$$

and

$$\mathbf{G^0} = \begin{bmatrix} G^0_{11}(\omega) & G^0_{12}(\omega) & \cdots & G^0_{1N}(\omega) \\ G^0_{21}(\omega) & G^0_{22}(\omega) & \cdots & G^0_{2N}(\omega) \\ \vdots & \vdots & & \vdots \\ G^0_{N1}(\omega) & G^0_{N2}(\omega) & \cdots & G^0_{NN}(\omega) \end{bmatrix} \qquad [5.31]$$

where the different quantities are defined as in [5.5] and following Figure 5.4, with $T_0(\omega)$ being the global transmission or reflection coefficient of the antenna in free space (referred to by subscript 0) for non-zero or zero-offset ($T_0(\omega) = R_i(\omega)$) source-receiver, respectively, $G^0_{..}(\omega)$ being the layered medium Green's functions for fields incident from the field points onto the field points themselves, $\mathbf{I_N}$ being the N-order identity matrix, and superscript T denoting transpose. When the number of source and receiver points is equal to 1, [5.26] reduces to the far-field radar equation [5.5].

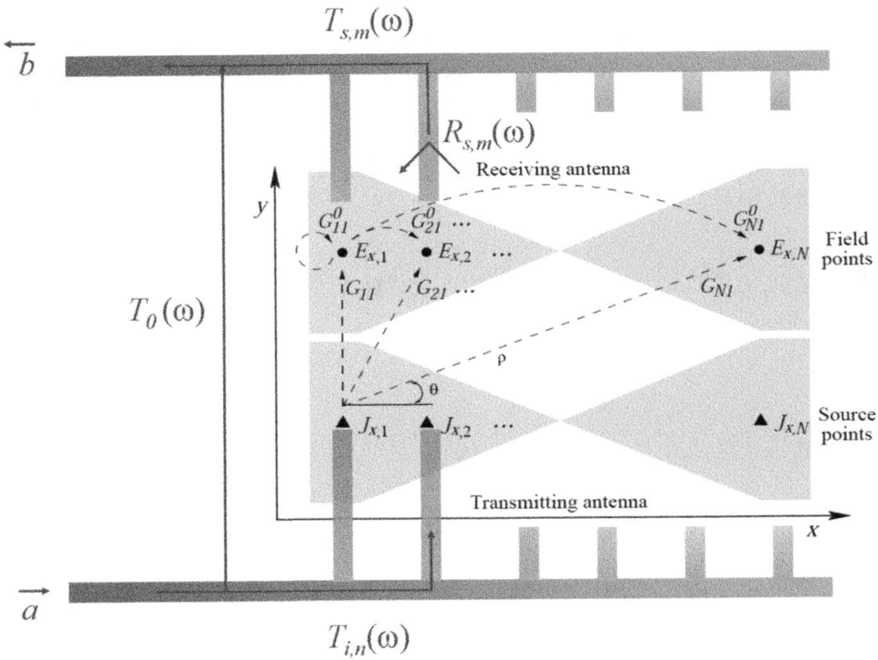

Figure 5.4. *Distribution of equivalent infinitesimal electric dipole point sources ($J_{x,\cdot}$) and field points ($E_{x,\cdot}$) in the xy-plane of a 3-D Cartesian system illustrating the near-field radar equation. Gray outlines represent bowtie antennas as an example to show the interaction paths between antenna sources and the layered medium. Dashed lines represent layered medium Green's functions from the sources to the receivers ($G_{..}$) and from the receivers, acting as secondary sources, to the receivers themselves ($G^0_{..}$). Blue outlines indicate the model wave propagation paths along which variations of impedance are encountered and to which correspond global transmission and reflection coefficients*

The global transmission and reflection coefficients of the antenna can be numerically determined by solving an inverse problem with a larger number of unknowns. To obtain accurate modeling results, it is necessary to carry out a relatively large number of measurements at different distances from the calibration medium. Generally, utilizing 6- or 8-point sources and field points is sufficient, even when the antenna is just a centimeter away from the medium.

Figure 5.5 shows an example of near-field calibration data, with measured and modeled signals for a pulse radar (StructureScan Mini HR, Geophysical Survey Systems, Inc. (GSSI)). Measurements were performed from the near-field to the far-field at 100 distances over a PEC. The near-field radar model agrees very well with the measurements over the entire range of distances. From a distance of 0.30 m and greater, the direct coupling between the transmitter and receiver is clearly distinguishable from the reflection of the PEC. At around 0.10 m, both the direct coupling and PEC reflection interfere with each other, resulting in a low recorded signal amplitude due to destructive interference. As the distance further decreases, the interference intensifies, causing the signal center frequency to shift to lower frequencies, as observed from the larger wavelength. Additionally, the multiple reflections occurring at later times are also visible in the images.

An example of measured and modeled near-field radar data with the same radar over dry sand is illustrated in Figure 5.6. Measurements were taken at two different distances, 1.6 cm and 14.0 cm, from the sand with a relative permittivity of approximately 2.6. At the greater distance, the model accurately reflects the measurements, with a slight time shift observed for the direct coupling between the transmitter and receiver. This time drift is a common result of pulse GPR systems. The reflection from the sand surface is visible at about 3.5 ns. For the shorter distance, the model still performs quite well, but larger discrepancies are observed. This can be attributed to the model's higher sensitivity to measurement and calibration errors in close proximity and may suggest that the calibration could be improved. This discrepancy is also to be partly attributed to the amplitude drift of the radar system. Generally, better results are yielded from calibrated vector network analyzer-based radars. Additional near-field model comparisons, including with 3D FDTD, are presented by Tran et al. (2013b).

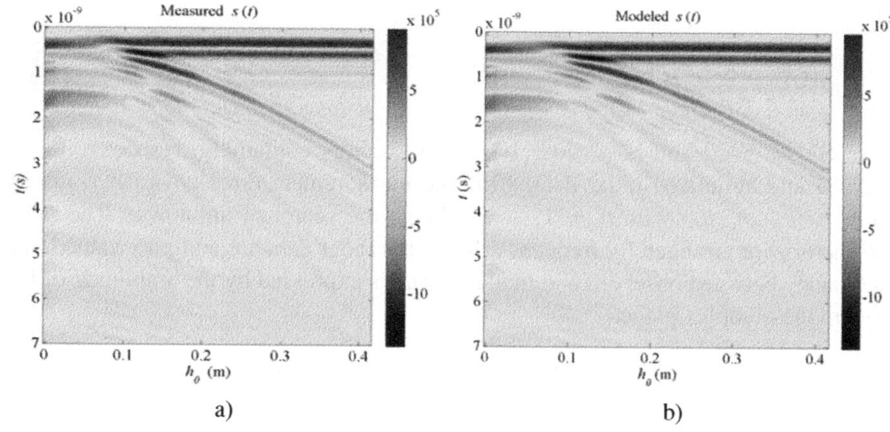

Figure 5.5. *Measured (a) and modeled (b) calibration data for a pulse radar (StructureScan Mini HR from Geophysical Survey Systems, Inc. (GSSI)). Measurements were performed from the near-field to the far-field at 100 distances from a PEC*

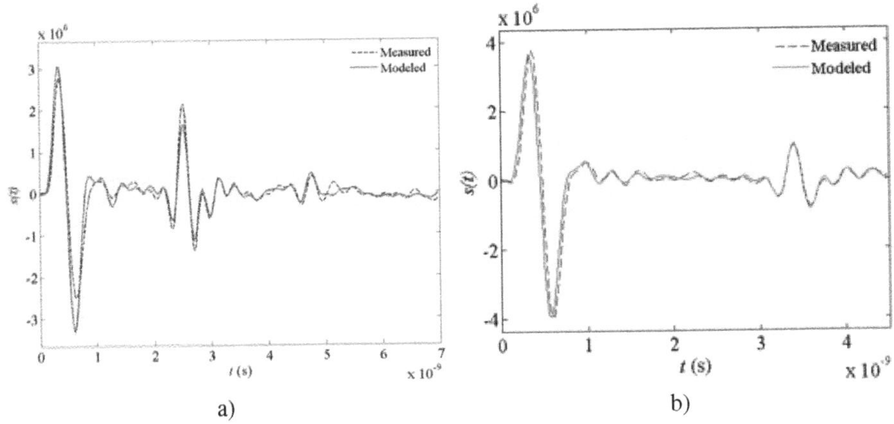

Figure 5.6. *Measurements and near-field modeling of pulse radar data using the StructureScan Mini HR from Geophysical Survey Systems, Inc. (GSSI) in near-field conditions over dry sand with a relative permittivity of approximately 2.6 at two different distances: 1.6 cm (a) and 14.0 cm (b)*

5.6. Full-wave inversion

The properties of the layered medium can be determined through full-wave inversion, which minimizes an objective function formulated in the least squares sense, namely:

$$\phi(\mathbf{b}) = |\mathbf{S}^* - \mathbf{S}|^T \mathbf{C}^{-1} |\mathbf{S}^* - \mathbf{S}| \qquad [5.32]$$

where $\mathbf{S}^* = S(\omega)$ and $\mathbf{S} = S(\omega, \mathbf{b})$ are the vectors containing, respectively, the observed and simulated radar data, $\mathbf{C}$ is the measurement error covariance matrix and $\mathbf{b} = [\varepsilon_n, \sigma_n, h_n]$ with $n = 1, \cdots, N$ being the vector of unknowns. The radar data vectors are arranged by frequency. Since the radar data are complex valued, the discrepancy between observed and modeled data is expressed by the amplitude of the errors in the complex plane.

In practice, when the far-field model is employed, the antenna effects are filtered out from $S(\omega)$ and the objective function is expressed as a function of Green's functions. This method provides a more precise evaluation of the fit quality and a better understanding of the results. On the other hand, if the near-field model is used, the inversion is normally performed on $S(\omega) - H_i(\omega)$, as Green's functions cannot be derived directly from this approach (De Coster and Lambot 2018).

It is also beneficial to perform inversion in the time-domain in order to focus on a particular time window, such as the medium surface reflection (Lambot et al. 2006).

This reduces the inverse model configuration, for example, to a half-space medium, thereby limiting the number of unknowns in the inverse problem. To convert the frequency domain data into the time domain, the inverse Fourier transform is applied:

$$g_{xx}(t) = \mathcal{F}^{-1}\left\{G_{xx}(\omega)\right\} = \int_{-\infty}^{\infty} G_{xx}(\omega)e^{2\pi j\omega t}d\omega \qquad [5.33]$$

Then, a maximal time t_{max} can be defined to focus on the first reflectors of interest such as the surface reflection (see Figure 5.7). The objective function becomes:

$$\phi(\mathbf{b}) = \left(\mathbf{g}^*_{\mathbf{xx}} - \mathbf{g}_{\mathbf{xx}}\right)^T \mathbf{C}^{-1}\left(\mathbf{g}^*_{\mathbf{xx}} - \mathbf{g}_{\mathbf{xx}}\right) \qquad [5.34]$$

where

$$\mathbf{g}^*_{\mathbf{xx}} = g^{\uparrow *}_{xx}(t)\Big|_0^{t_{max}} \quad \text{and} \quad \mathbf{g}_{\mathbf{xx}} = g_{xx}(t, \mathbf{b})\Big|_0^{t_{max}} \qquad [5.35]$$

are the vectors containing, respectively, the observed and simulated time-domain windowed Green functions.

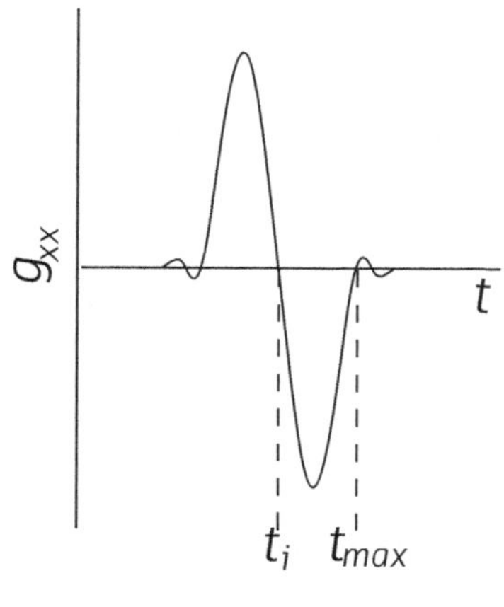

Figure 5.7. *Sketch of the radar signal in the time domain, $g_{xx}(t)$, with antenna effects removed. The time window for the inversion, denoted as t_{max}, is chosen to focus, as an example, on the surface reflection. t_i denotes the approximate time corresponding to the surface interface for the Green's function*

Depending on the number of unknowns to estimate and the amount of information contained in the radar data, the topography of the objective function to optimize can be quite complex, and in some cases the inverse problem can even be ill-posed. When dealing with one or two unknowns, local optimization using a method such as the Levenberg–Marquardt algorithm or a look-up table approach, which calculates the objective function across the entire parameter space, is both reliable and efficient. However, for more complex problems, global optimization is necessary. In this case, the Global Multilevel Coordinate Search algorithm (Huyer and Neumaier 1999) combined sequentially with the classic Nelder–Mead simplex algorithm (Lagarias et al. 1998) has been found to produce good results in previous studies.

5.7. Soil moisture mapping application

The characterization of the soil water content has been an important task in hydrology, meteorology, environmental science and agriculture. For example, in hydrology, soil moisture plays a vital role at various spatial and temporal scales, and its temporal patterns can provide insight into vegetation growth and root water uptake (Hupet and Vanclooster 2005, 2002). Additionally, agricultural irrigation accounts for a significant portion of water usage (e.g. 70% (Podimata and Yannopoulos 2015)). Knowledge of soil moisture at a field scale with a high spatial resolution would lead to more efficient use of water resources.

Remote sensing has become a valuable tool for mapping soil characteristics, including soil moisture. Optical remote sensing can provide wide coverage but is limited in measurement depth and can be affected by clouds and vegetation (Peng et al. 2017). Microwave remote sensing has also been shown to be promising for soil characterization, as it can measure deeper and is sensitive to soil water content (Kornelsen and Coulibaly 2013; Bertoldi et al. 2014). It is based on differences in dielectric properties between dry soil and water, and there are various models that describe the relationship between microwave surface emissivity, radar back-scattering coefficient and soil moisture (Wagner et al. 1999; Baghdadi et al. 2011; Al-Yaari et al. 2017). However, estimates can be affected by factors such as surface roughness and vegetation. Additionally, the resolution of microwave remote sensing is typically limited to 10 m × 10 m, which may not be precise enough for agricultural applications. For this reason, time-domain reflectometry (TDR) is commonly used in agriculture to measure soil dielectric permittivity, though it is less efficient.

GPR is considered a promising technique for measuring soil water content as it is sensitive to changes in electromagnetic medium properties. Like the TDR method, the GPR measures the relative dielectric permittivity of the soil and uses petrophysical relationships such as Topp's and Ledieu's models to convert this permittivity to the volumetric water content (Topp et al. 1980; Ledieu et al. 1986). It is found that it is easier to obtain soil permittivity from the surface reflected waveform than from the electromagnetic wave propagation velocity. The full-wave inversion method is considered as more accurate and efficient in comparison with the surface reflection coefficient method (Lambot et al. 2006; Pramudita et al. 2022).

Therefore, an air-coupled GPR system can be used flexibly on mobile platforms such as quads, drones, etc., to measure soil water content using the full-wave inversion method (Minet et al. 2012; Wu et al. 2019).

5.7.1. *Test sites*

In our study, we used GPR systems on a drone and a quad to measure the soil water content at two test sites located within the central Belgian loess belt. The soils at these sites consist mainly of sandy and silt-loam. Drone-GPR measurements were conducted in the Château de Bousval vineyard on March 3, 2021, when the grape trellises were low. A 0.5-m dipole antenna was used and the operating frequency range was set to 235–395 MHz. Quad-GPR measurements were performed at the Ferme de Lauzelle in Louvain-la-Neuve on April 30, 2021, when the land was covered with short grass. We utilized a log-periodic antenna and the operating frequency range was set to 300–850 MHz.

5.7.2. *Radar system*

Pictures of the radar system are shown in Figure 5.8. It is composed of a handheld vector network analyzer (VNA), a micro-computer, an antenna, a GPS receiver, a power bank and a smartphone. The VNAs shown in Figure 5.8 are, respectively, Planner R60 and R54 from Copper Mountain technologies, in Indianapolis, USA. The R60 and R54 are both 50 Ω 1-port devices capable of measuring S_{11} from 1 MHz to 6 GHz and 85 MHz to 4.8 GHz, respectively. The R60 has a weight of 0.35 kg, with dimensions of 130 × 65 × 28 mm. The R54 has a weight of only 0.25 kg and dimensions of 120 × 43 × 23 mm. In the system, the VNA connects directly to the antenna, acting as a frequency domain radar. The micro-computer is an Intel Compute Stick that provides a platform for controlling the radar system. An application corresponding to the VNA is installed in the micro-computer to control the VNA. The overall system is controlled by the gprSense application (Sensar Consulting, Belgium, see www.gprsense.com). The microcomputer is connected to a smartphone via a Wi-Fi hotspot. The application can be accessed through the smartphone browser using the corresponding IP address. The gprSense application, which controls the VNA systems, allows users to preset and control various measurement settings, such as operating frequency range, frequency steps and other relevant parameters. This provides users with the flexibility to customize the measurement to their specific antennas. The software also permits automated calibration of the radar and real-time full-wave inversion, for example, for real-time soil moisture mapping.

Figure 5.8. *Lightweight radar system with a dipole antenna mounted on a drone (left) and a log-periodic antenna mounted on a quad (right) for soil moisture mapping*

5.7.3. *Radar calibration*

Both the dipole and log-periodic antennas were calibrated on a copper sheet measuring 3 m $\times$ 3 m, with the objective of calculating the characteristic functions $R_i(\omega)$, $T(\omega)$, and $R_s(\omega)$ as described in equation [5.5]. In particular, when the dipole was calibrated, the drone was also included in the process to account for the drone–radar interactions, which would be present when the antenna was in use. The quad was positioned at a distance far enough away, such that these interactions could be disregarded. Figure 5.9 displays the characteristic antenna functions for the dipole on the drone. An antenna is considered to operate well if $|R_i| < 0.5$, that is, more than 50% of the input field are radiated. In this case, the antenna was found to be more effective at 288 MHz, which is its resonant frequency. The characteristic function T, which displays the antenna gain in the direction of the medium, was maximal at 280 MHz. The maximal gain is slightly shifted with respect to the resonant frequency due to the non-homogeneous radiation pattern of the antenna, including the influence of the drone. The antenna gain remains at more than half its maximum throughout the frequency band of interest. Finally, the reflection function for the incident fields from the medium, $R_s(\omega)$, is below 10^{-3}, which is relatively good for the conditions in the far field. It is essential to recalibrate the system if any changes have been made to the radar, drone or their relative positions, such as modifications to the drone's body or wings. Such modifications could affect the antenna's radiation pattern and should not be overlooked.

5.7.4. *Radar images*

Figures 5.10 and 5.11 illustrate the application of the far-field radar equation [5.8] to filter out the effects of the antenna from raw data. The example shown involves 2,612 measurements conducted at the Ferme de Lauzelle, with the GPR system mounted on a quad. The inverse Fourier transform [5.33] was applied to convert the signals to the time domain. Figure 5.10(a) shows the amplitude of $S_{11}(f)$ for all measurements, while Figure 5.10(b) displays the corresponding time domain counterpart $s(t)$. These data contain significant reflections from the antenna itself, which greatly affect the image. For example, a continuous and stable reflection can be observed at around 7 ns throughout the entire measurement, which is represented by R_i in equation [5.5]. After applying the radar equation using [5.8], Figure 5.11 displays the resulting Green function. The surface reflection can now be identified between 4–5 ns, and the time zero corresponds to the antenna phase centre. As the quad moves, the height of the radar above the ground is altered, which is reflected in the varying time for the surface reflection. Slight reflections occurring at a later time may originate from deeper soil layers, reflections from the quad or calibration artifacts from the laboratory conditions. However, these reflections are not considered here for the inversion due to time gating.

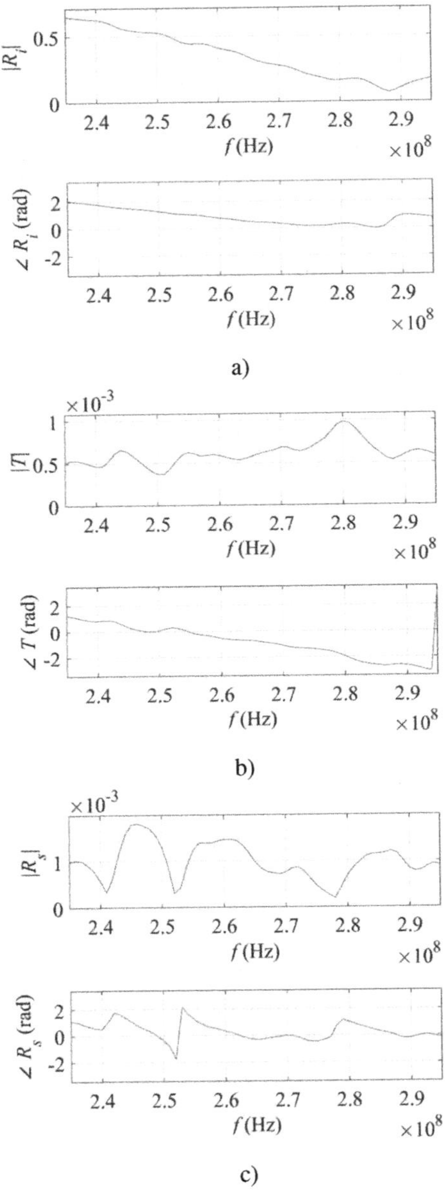

Figure 5.9. *Characteristic functions of the dipole antenna mounted on the drone for soil moisture mapping in the vineyard. Both amplitude and phase are shown*

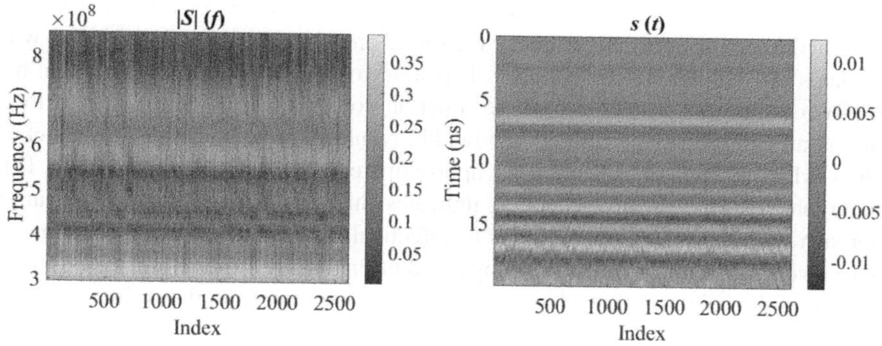

Figure 5.10. *Raw radar images in the frequency (a) and time domains (b)*

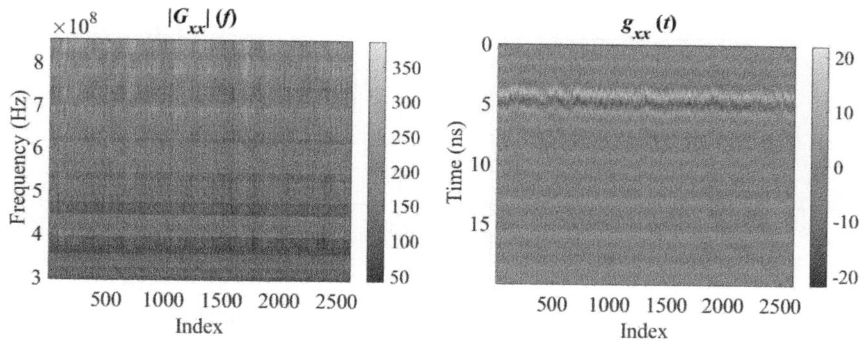

Figure 5.11. *Radar images in both the frequency (a) and time (b) domains obtained after applying the far-field radar equation, thus allowing for the filtering out of antenna effects*

5.7.5. *Inversion focused on the surface reflection*

We employed a Look-Up-Table (LUT) approach to conduct full-wave inversion, which focused on surface reflection in the time domain. The parameter vector of unknowns b in equation [5.34] includes the relative permittivity ε_r and height h. We set ε_r to range from 2 to 22 with an increment of 0.5, and h to range from 0.5 to 5 m with an increment of 0.005 m. This resulted in 36,941 computation iterations (901 x 41) to generate the LUT containing the simulated time-domain signals for all possible parameter combinations. We only generated the LUT once and compared the measured Green functions with all the LUT simulations to calculate the objective functions and identify its minimum.

Figure 5.12 shows an example of full-wave inversion. When the global minimum is reached, the best fitting between the measured and simulated data is achieved, with estimates for the parameters ε_r and h. It is noteworthy that the inverse problem has only two unknowns, yet the objective function contains local minima. These local minima are caused by the limited bandwidth, resulting in similar objective function values when the height is shifted by approximately one average wavelength. This topography of the objective function indicates that a LUT or global optimization approach must be employed in order to effectively solve this inverse problem. An alternative approach could be to employ an additional sensor to provide a good initial estimate of the height, followed by rapid local optimization from that value.

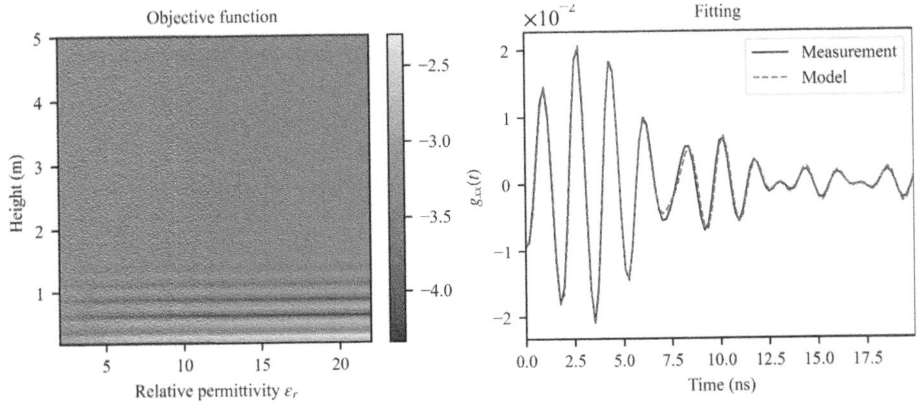

Figure 5.12. *Example of objective function to minimize (left) and corresponding fit in the time domain (right)*

We used Topp's equation (Topp et al. 1980) to calculate the volumetric water content of the soil θ from the relative permittivity ε_r:

$$\theta = -5.3 \times 10^{-2} + 2.92 \times 10^{-2}\varepsilon_r - 5.5 \times 10^{-4}\varepsilon_r^2 + 4.3 \times 10^{-6}\varepsilon_r^3 \quad [5.36]$$

Subsequently, kriging interpolation was conducted using the GPS position of each radar measurement to generate the soil moisture maps.

Figure 5.13 presents the soil moisture map obtained with quad-GPR in the Ferme de Lauzelle. Volumetric moisture levels range from approximately 11% to 33%, which are values that are within the expected range. In particular, the wetter area of the field corresponds to a small valley, whereas the drier parts are located on the slopes. At the south-west and north-east parts of the field, where the terrain is flatter, the soil moisture values are intermediate. The quad-GPR-derived soil map for the field is highly consistent with the topography. This is further supported by the Google Earth orthophoto from July 2019, when the weather was particularly dry, which displays similar patterns. In fact, the humid areas allowed the vegetation to remain lush and green, while the drier areas saw the vegetation thin and fade away.

Figure 5.13. *Test site located in the Ferme de Lauzelle in Belgium. The orthophoto (left) is from Google Earth Pro. The elevation model (right) was obtained from drone imaging using the DJI Phantom 4 Pro and processed with the Agisoft Metashape software using photogrammetry. The soil moisture map was obtained using quad-GPR and full-wave inversion (right)*

Figure 5.14 shows the soil moisture map obtained with the drone-GPR in the vineyard. Soil moisture ranges from about 12% to 41%. In general, the eastern part of the field is drier than the western part. The correlation with the orthophoto remains limited, although some correspondences can be observed, especially in the north-west part of the plot. It should be noted that the metallic structures of the vineyard may have influenced the radar measurements.

5.7.6. *Conclusions and perspectives*

In this section, the full-wave radar equation is used alongside lightweight GPR systems, both on quad and on drone, to map soil moisture with high spatial resolution. From a technical perspective, improvements can be made to further increase the accuracy of the proposed method. The accuracy of the estimates is heavily dependent on the calibration of the antenna with the far-field model, particularly for low-frequency antennas. Moreover, the relative orientation of the antenna must also be considered during the calibration process (Wu and Lambot 2022b). This becomes especially important with the drone, as the antenna pointing in relation to the soil surface is ever-changing.

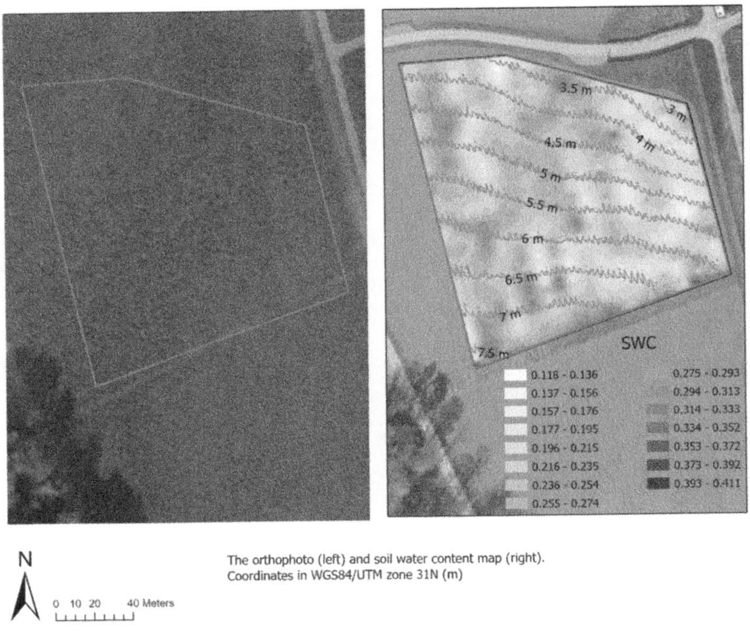

Figure 5.14. *Test site located in Bousval, Belgium. The orthophoto and elevation models are obtained from DJI Phantom 4 Pro and processed with the Agisoft Metashape software*

In agriculture, the GPR system has the potential to be used for automated and real-time determination of soil moisture for precise and automatic irrigation. For example, a trial was conducted on trench-hill potato fields where the GPR was used on an irrigation robot (Wu et al. 2022). The GPR has also been shown to be a valuable decision-support tool for tree planting in variable edaphic contexts (Mahieu et al. 2020). Using a drone as a platform, the system can be used to map soil moisture in a non-destructive manner during the growing season. Furthermore, it can be used in more complex and wild fields such as swamps and deserts. GPR technology can be used to calibrate global-scale soil moisture data obtained from satellite sensing and can also be used to characterize soil electrical conductivity using lower frequencies (Wu and Lambot 2022a). In conclusion, GPR technology and advanced data processing enrich remote sensing methods and drone applications for digital soil mapping by providing more information on soil properties and allowing for the production of soil maps with higher resolution and efficiency, for example, to support precision agriculture and environmental monitoring.

5.8. Radar-based interaction

In this section, we will discuss how the concepts of radar antenna characterization and full-wave inversion can be applied to gesture recognition. In the field of human–computer interaction (HCI), gestural interaction remains a popular modality for interacting with interactive applications, such as multimedia browsing (Sluÿters et al. 2022c), because of its natural and intuitive character to users (Wigdor and Wixon 2011; Huang et al. 2019). Until now, gestures were mostly captured on a wide variety of devices, such as touchable surfaces (e.g. touchscreens and tablets), vision-based sensors (e.g. cameras and Leap Motion Controller (Sluÿters et al. 2022b)) and wearable devices (e.g. armbands). While touchable and wearable devices induce physical contact and some intrusiveness and privacy concerns, vision-based sensors are contactless but are dependent on ambient conditions, such as light, occlusion, and weather.

In contrast, radar sensors are particularly interesting for gestural interaction (Yeo and Quigley 2017) in environments subject to changes in weather and lighting conditions. They can seamlessly integrate into everyday objects, such as smartphones, desks, and doors, pose fewer privacy concerns than other sensors, and can be combined with other modalities, such as vocal interaction (Parthiban et al. 2022).

However, radar data are complex, high-dimensional and difficult to interpret, making radar less attractive compared to other types of sensors. Our approach based on radar–antenna characterization and full-wave inversion aims at normalizing the radar signal and extracting two physically significant features: the hand-radar distance and the apparent relative permittivity of the medium defined by the hand. These features are straightforward to interpret and very small in size.

This section is structured as follows: section 5.8.1 explains how the concepts of radar antenna characterization and full-wave inversion were applied to gesture recognition in the form of a signal processing pipeline, section 5.8.2 describes the radar sensor used in our experiments, and section 5.8.3 discusses the performance of the pipeline on a dataset of 16 mid-air hand gestures.

5.8.1. *Processing pipeline*

Based on the concepts of antenna characterization and full-wave radar inversion, a seven-stage pipeline was designed to process the radar signal for gesture recognition.

1) *Raw data capture (and fast Fourier transform)* (Figure 5.15(a)): the pipeline retrieves data from one or more receiver antennas. If the signal is in the time domain, it is first converted to the frequency domain using equation [5.6], as required by the next processing stages. The frequency range is set to achieve a trade-off between bandwidth and signal-to-noise ratio.

2) *Radar source and antenna effects removal* (Figure 5.15(b)): this stage aims to remove the effects of the radar source and antenna (e.g. internal reflections and transmissions, antenna-target interactions) by applying the radar equation (equation [5.8]) to the raw signal. In the case of antenna arrays, this stage requires us to generate a far-field model of each pair of emitter and receiver antennas by computing their characteristic functions. The far-field model is less resource-intensive to compute than the near-field one and is valid as long as the users stays sufficiently far away from the radar. The resulting signal from this stage is a Green's function that is independent of the radar, and thereby normalized in the operating frequency range of the radar. Undesired echoes are removed, thus making users' reflections stand out.

3) *Background scene removal* (Figure 5.15(c)): based on the superposition principle, the signal from the surrounding environment (e.g. walls, plants, furniture) is subtracted from the normalized signal. The resulting signal only contains the reflections of moving objects and persons, including the user, thus improving the accuracy and efficiency of full-wave inversion.

4) *Inverse fast Fourier transform* (Figure 5.15(d)): the signal is converted to the time domain using equation [5.33] to simplify computations in the next processing stage.

5) *Time gating* (Figure 5.15(e)): the time domain signal is truncated to keep only reflections within a distance from the radar (see Figure 5.7 and equation [5.35]). The maximum distance is carefully selected to ensure a sufficient signal-to-noise ratio while maintaining a sufficient distance between the user and the radar. The benefits of this stage are twofold: (1) moving objects behind the user are ignored, thus improving the accuracy of the next stage, and (2) the number of data to process in the full-wave inversion is reduced, thereby increasing the efficiency of the next stage.

6) *Full-wave inversion* (Figure 5.15(f)): the radar signal is reduced to two physically meaningful dimensions by the process of full-wave inversion, namely the hand-radar distance and the relative permittivity of an infinite plane equivalent to the hand. Therefore, the retrieved permittivity is an apparent permittivity and not the true permittivity of the hand. Distance gives an idea of the position of the hand, whereas apparent permittivity provides information on the pose of the hand. The inversion process consists of adjusting the distance and apparent relative permittivity parameters of a model to find the best fit between the modeled and measured signals

(equation [5.34]). This process is sped up by the use of a pre-calculated LUT. The drastic data reduction, while keeping essential information, resulting from the full-wave inversion enables the use of simple template matching algorithms for gesture recognition.

7) *Filtering* (Figure 5.15(g)): the result of full-wave inversion is subject to the quality of the radar data. A low signal-to-noise ratio or moving objects behind the user may result in an incorrect estimation of the distance and apparent permittivity. In an attempt to clean up the resulting signal, we discard the distance and permittivity values if they are under a given threshold and replace them with default values. In addition, large variations in the estimated distance and permittivity are smoothed out by applying a moving average filter. The need for filtering may decrease with radar systems that can achieve a higher signal-to-noise ratio.

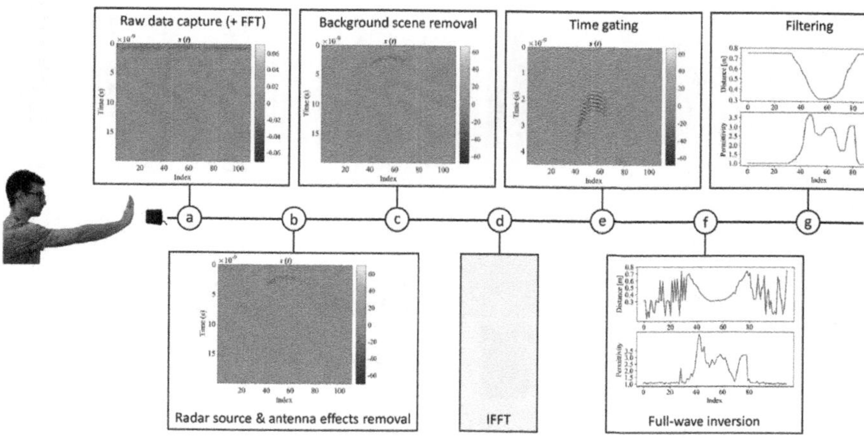

Figure 5.15. *Radar data processing pipeline for gesture recognition*

Data from each stage can serve as input to gesture recognition algorithms, such as deep learning or template matching. For instance, data from stages b, c and e are suited to image-processing algorithms, such as CNNs.

5.8.2. *Radar system*

The Walabot Developer (Figure 5.16) is an affordable, readily available frequency-modulated continuous wave (FMCW) radar array featuring 18 antennas. Three of those antennas are not used, four act as emitters and receivers, and the rest only act as receivers. Depending on its version, the Walabot Developer operates between 3.3 and 10 GHz (US/FCC) or between 6.3 and 8 GHz (EU/CE). It is a good candidate to evaluate our processing pipeline, as it provides challenges in the quality of its data (lower signal-to-noise ratio, more internal reflections and transmissions than more expensive radar systems) and in the quantity of data (a large number of receiving antennas). It provides easy access to the raw time-domain signal through its API and can be configured with various profiles, which impact the emitter/receiver pairs used, the sampling rate, and the frequency range. Our processing pipeline based on radar–antenna characterization and full-wave inversion should allow us to get the best out of this device.

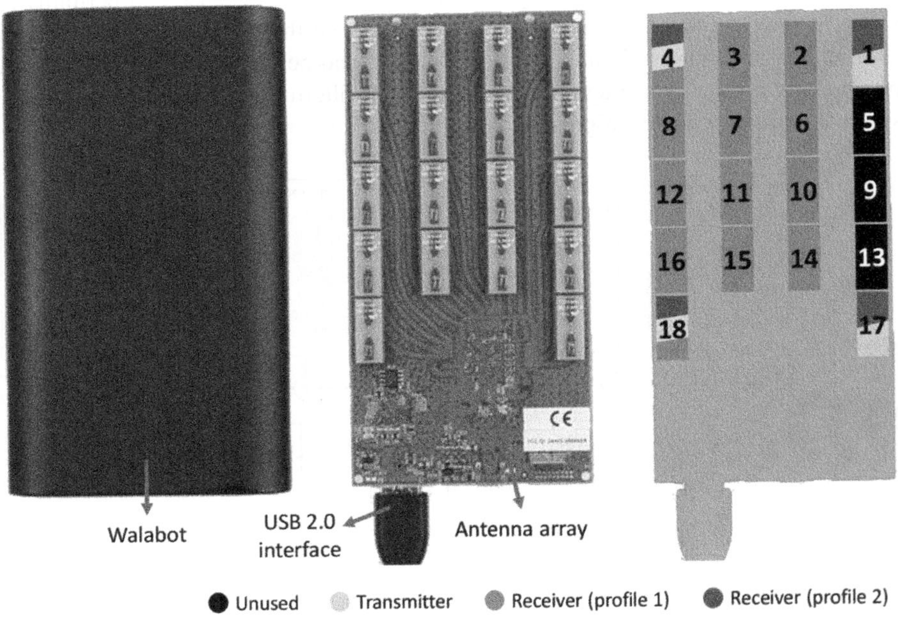

Figure 5.16. *Case (left), PCB (middle) and antenna layout (right) of the Walabot Developer*

5.8.3. *Results*

The system was evaluated on a set of 16 hand gestures (Sluÿters et al. 2022a) with Jackknife, a popular modality-agnostic gesture recognition algorithm (Taranta II et al. 2017). Each gesture was performed 10 times in front of a Walabot by three different participants, resulting in 480 gesture recordings. Figure 5.18 (respectively, Figure 5.19) shows the confusion matrices obtained with data from two different stages of the pipeline in a user-dependent (respectively, user-independent) testing scenario.

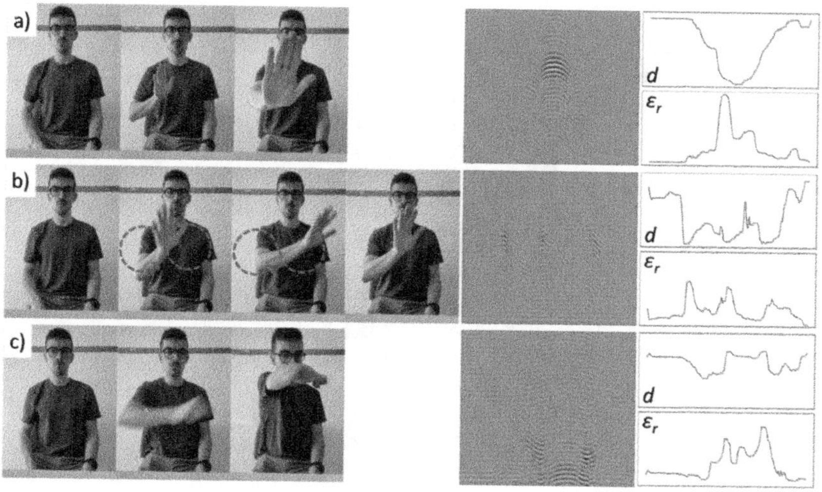

Figure 5.17. *Examples of gestures (left) and the corresponding signal after background scene removal (middle) and filtering (right)*

In the user-dependent scenario, the Jackknife recognizer (Taranta II et al. 2017) was trained and tested with gestures from the same participant. This simulates a scenario where users interact with a system that was trained on their own gestures.

The system performed extremely well on data from the "background scene removal" stage, reaching 98% accuracy (Figure 5.18(a)). The accuracy was lower with data from the "filtering" stage, at 67%, which can be explained by the loss of information and the greater sensitivity to noise incurred by the data inversion, making distinguishing between similar fine-grained gestures (e.g. showing one vs. two fingers) more difficult. However, the smaller size of the data greatly sped up gesture recognition.

In the user-independent scenario, the gesture recordings used for training were performed by different participants than the recordings used for testing. This simulates a more challenging scenario in which users interact with a system that does not know them.

The system performance dropped drastically in the user-independent scenario, reaching 12% on data from the "background scene removal stage". It performed slightly better on data from the "filtering stage", with 19% precision. The low performance in this scenario highlights a need for data normalization with respect to the characteristics of the users (e.g. hand size, height, weight or body shape). Accuracy was better in the "filtering" stage as some information related to the user was filtered out from the data, such as the reflection of their body.

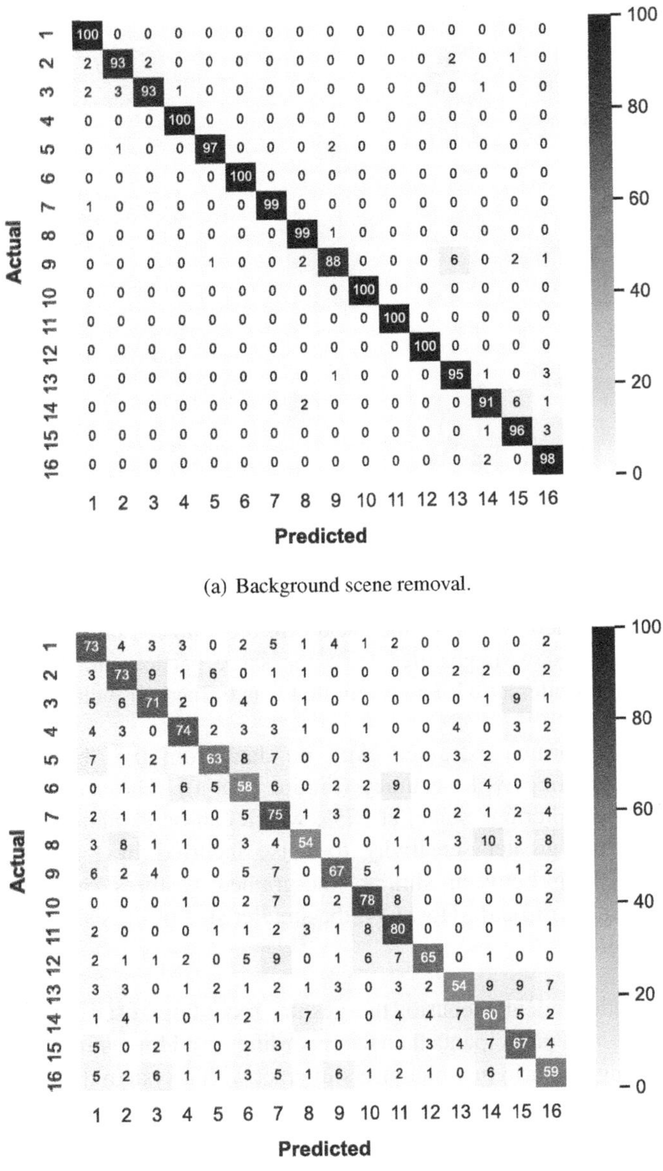

(a) Background scene removal.

(b) Filtering.

Figure 5.18. *Normalized confusion matrices for the user-dependent testing scenario with gesture data from two stages of the pipeline*

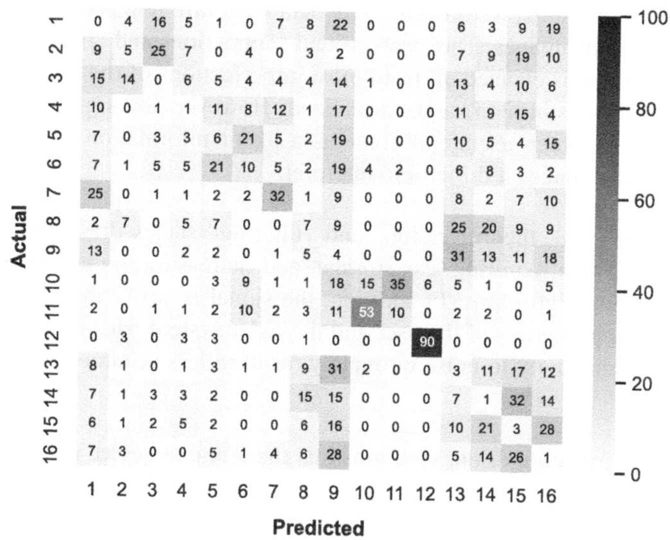

(a) Background scene removal.

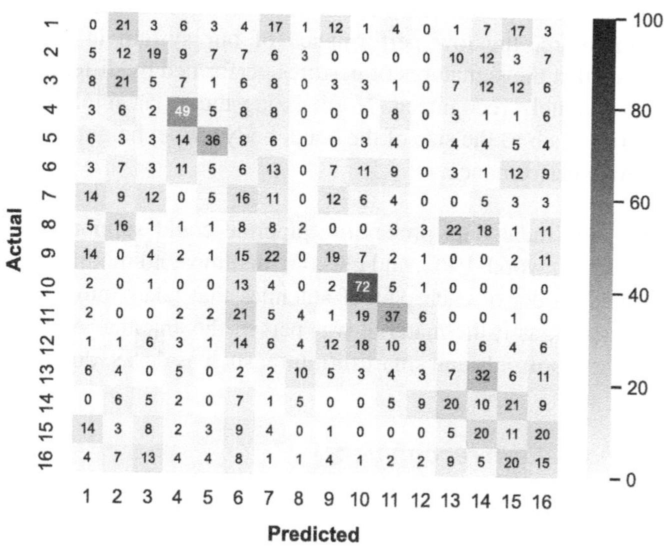

(b) Filtering.

Figure 5.19. *Normalized confusion matrices for the user-independent testing scenario with gesture data from two stages of the pipeline*

Based on these results, we observed that data from different stages of the pipeline provide different advantages. The "background subtraction" and "time gating" stages provide a signal that is cleaned up from most imperfections of the radar system (e.g. interference and transmission) and normalized. However, the signal is sensitive to small variations between users and contains a lot of information, which slows the execution time and makes it difficult to interpret.

On the other hand, the "inversion" and "filtering" stages drastically reduce the quantity of data, thus requiring less storage, and increasing the execution speed of gesture recognition algorithms. In addition, the signal is easier to interpret and less dependent on the properties of the user and the radar system. However, the reduction in data and the sensitivity to noise ultimately result in less accuracy on large gesture sets.

One current limitation of our system is that the Walabot device used in our testing provides data at a very high frame rate from multiple receive channels, resulting in a large amount of data to process which makes achieving real-time execution challenging. The execution speed of the pipeline could be improved by introducing optimizations (e.g. pruning, multithreading, and data dimension reduction) and comprises (e.g. reduced frame rate, lower resolution, and use of fewer receive channels). Furthermore, the low performance of our system in user-independent scenarios suggests that the signatures of gestures performed by different users are too different to be accurately recognized. Normalizing the radar signal with respect to user characteristics, such as the size of the hand, body shape, height, and weight, may drastically improve performance.

In the future, the radar data processing pipeline could support gesture sensing through materials (e.g. wood, PVC, and glass) with almost no modification other than building a far-field model of each pair of antennas that takes into account both the radar and the material, as if the material was part of the antenna. Applications could include interactive shopfronts and smart furniture, such as desks, doors, and cabinets.

5.8.4. *Breathing pattern recognition*

Breathing pattern recognition consists of identifying and recognizing specific patterns in the regular breathing of a person. It has applications in various contexts, including healthcare, as a tool for monitoring breathing exercises and as a non-invasive way for people with disabilities (e.g. tetraplegia) to interact with computers and other objects (e.g. a wheelchair). Breathing pattern recognition could also serve as a discreet method of interaction in situations where other methods may not be appropriate (too loud, too visible), such as for controlling a smart home from the bed without waking up other people in the room or calling for help without alerting an aggressor. Finally, it could allow users to unconsciously modify their environment and improve their quality of life. For example, lights could change color depending on mood (tired, stressed), music playlists could change with the intensity of exercise, and smart assistants could provide suggestions to relieve anxiety.

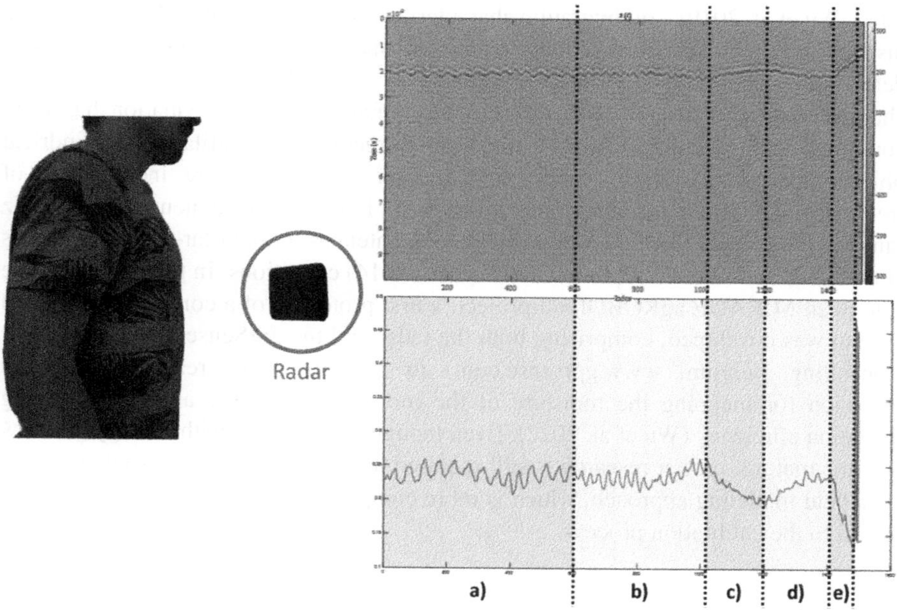

Figure 5.20. *Breathing patterns processing. At the top, the time-domain signal after background scene removal. At the bottom, the result of full-wave inversion*

The radar data processing pipeline defined in section 5.8.1 could be applied to the recognition of breathing patterns by extracting the small distance variations between the radar and the user caused by the expansion and contraction of their lungs. Figure 5.20 shows the time-domain radar signal and the result of full-wave inversion for five different scenarios: (a) normal breathing, (b) faster breathing, (c) deep breathing, (d) breathing out, and (e) moving closer to the radar. Each pattern can be clearly identified in the signal, which suggests the feasibility of breathing pattern recognition with radar.

5.9. Conclusion and perspectives

Provided that the modeling assumptions are accurately respected, the closed-form full-wave radar equations introduced by Lambot et al. (2004) and Lambot and André (2014) have demonstrated a high level of both modeling accuracy and computational efficiency, making the approach highly practical for non-destructive testing in the field of ultrawide-band radar applications. This method has been successfully employed for a variety of purposes, including soil moisture mapping (Minet et al. 2012; Wu et al. 2019), soil hydraulic characterization (Jadoon et al. 2008, 2011), road inspection (De Coster et al. 2018) and the enhancement of landmine imaging and detection (Lopera et al. 2007; Soldovieri et al. 2011), as well as concrete characterization (Patriarca et al. 2011). Additionally, the radar equation has been combined with cylindrical Green's functions to reconstruct multilayered cylindrical objects (Gao et al. 2021). Since the equation assumptions are independent of frequency, it is also applicable to electromagnetic induction at frequencies in the kHz range, with adapted Green's functions for loop antennas, in both far-field (Moghadas et al. 2010) and near-field (André and Lambot 2014) conditions. In the context of the European MIRAGE agROBOfood project, a first prototype of a complete automated system was developed, comprising both the radar and the gprSense software (Sensar Consulting, Belgium, www.gprsense.com), in order to provide real-time full-wave inversion for mapping the moisture of the root zone, with the aim of improving irrigation efficiency (Wu et al. 2022). Even though the far-field methodology has now become mature, further research is still required in order to take full advantage of the near-field modeling approach, which is more complicated to employ, especially with regard to the calibration process.

5.10. References

Al-Yaari, A., Wigneron, J.-P., Kerr, Y., Rodríguez-Fernández, N., O'Neill, P., Jackson, T., De Lannoy, G.J.M., Al Bitar, A., Mialon, A., Richaume, P. et al. (2017). Evaluating soil moisture retrievals from ESA's SMOS and NASA's SMAP brightness temperature datasets. *Remote Sensing of Environment*, 193, 257–273.

André, F. and Lambot, S. (2014). Intrinsic modeling of near-field electromagnetic induction antennas for layered medium characterization. *IEEE Transactions on Geoscience and Remote Sensing*, 52(11), 7457–7469.

Baghdadi, N., Abou Chaaya, J., Zribi, M. (2011). Semiempirical calibration of the integral equation model for SAR data in C-band and cross polarization using radar images and field measurements. *IEEE Geoscience and Remote Sensing Letters*, 8(1), 14–18.

Bertoldi, G., Della Chiesa, S., Notarnicola, C., Pasolli, L., Niedrist, G., Tappeiner, U. (2014). Estimation of soil moisture patterns in mountain grasslands by means of SAR RADARSAT2 images and hydrological modeling. *Journal of Hydrology*, 516, 245–257.

Chen, Y., Yang, S., He, S., Nie, Z. (2010). Fast analysis of microstrip antennas over a frequency band using an accurate MoM matrix interpolation technique. *Progress in Electromagnetics Research – PIER*, 109, 301–324.

Chew, W.C. (1990). *Waves and Fields in Inhomogeneous Media*. Van Nostrand Reinhold, New York.

Craeye, C., Gilles, T., Dardenne, X. (2009). Efficient full-wave characterization of arrays of antennas embedded in finite dielectric volumes. *Radio Science*, 44, 1–12.

De Coster, A. and Lambot, S. (2018). Full-wave removal of internal antenna effects and antenna-medium interactions for improved ground-penetrating radar imaging. *IEEE Transactions on Geoscience and Remote Sensing*, 57(1), 93–103. doi: 10.1109/TGRS.2018.2852486.

De Coster, A., Van der Wielen, A., Grégoire, C., Lambot, S. (2018). Evaluation of pavement layer thicknesses using GPR: A comparison between full-wave inversion and the straight-ray method. *Construction and Building Materials*, 168, 91–104.

Gao, L., Dachena, C., Wu, K., Fedeli, A., Pastorino, M., Randazzo, A., Wu, X., Lambot, S. (2021). Full-wave modeling and inversion of UWB radar data for wave propagation in cylindrical objects. *Remote Sensing*, 13, 2370. doi: 10.3390/rs13122370.

Gentili, G.G. and Spagnolini, U. (2000). Electromagnetic inversion in monostatic ground penetrating radar: TEM horn calibration and application. *IEEE Transactions on Geoscience and Remote Sensing*, 38(4), 1936–1946.

Huang, J., Jaiswal, P., Rai, R. (2019). Gesture-based system for next generation natural and intuitive interfaces. *Artificial Intelligence for Engineering Design, Analysis and Manufacturing*, 33(1), 54–68.

Huisman, J.A., Hubbard, S.S., Redman, J.D., Annan, A.P. (2003). Measuring soil water content with ground penetrating radar: A review. *Vadose Zone Journal*, 2, 476–491.

Hupet, F. and Vanclooster, M. (2002). Intraseasonal dynamics of soil moisture variability within a small agricultural maize cropped field. *Journal of Hydrology*, 261(1), 86–101.

Hupet, F. and Vanclooster, M. (2005). Micro-variability of hydrological processes at the maize row scale: Implications for soil water content measurements and evapotranspiration estimates. *Journal of Hydrology*, 303(1), 247–270.

Huyer, W. and Neumaier, A. (1999). Global optimization by multilevel coordinate search. *Journal of Global Optimization*, 14(4), 331–355.

Hyun, S.Y., Kim, S.Y., Kim, Y.S. (2009). An equivalent feed model for the FDTD analysis of antennas driven through a ground plane by coaxial lines. *IEEE Transactions on Antennas and Propagation*, 57(1), 161–167.

Ilic, M.M., Djordjevic, M., Ilic, A.Z., Notaros, B.M. (2009). Higher order hybrid FEM-MoM technique for analysis of antennas and scatterers. *IEEE Transactions on Antennas and Propagation*, 57(5), 1452–1460.

Jadoon, K., Slob, E.C., Vanclooster, M., Vereecken, H., Lambot, S. (2008). Uniqueness and stability analysis of hydrogeophysical inversion for time-lapse ground penetrating radar estimates of shallow soil hydraulic properties. *Water Resources Research*, 44, W09421.

Jadoon, K.Z., Lambot, S., Slob, E.C., Vereecken, H. (2011). Analysis of horn antenna transfer functions and phase-center position for modeling off-ground GPR. *IEEE Transactions on Geoscience and Remote Sensing*, 49(5), 1649–1662.

Kornelsen, K.C. and Coulibaly, P. (2013). Advances in soil moisture retrieval from synthetic aperture radar and hydrological applications. *Journal of Hydrology*, 476, 460–489.

Lagarias, J.C., Reeds, J.A., Wright, M.H., Wright, P.E. (1998). Convergence properties of the Nelder–Mead simplex method in low dimensions. *SIAM Journal on Optimization*, 9(1), 112–147.

Lambot, S. and André, F. (2014). Full-wave modeling of near-field radar data for planar layered media reconstruction. *IEEE Transactions on Geoscience and Remote Sensing*, 52(5), 2295–2303.

Lambot, S., Slob, E.C., van den Bosch, I., Stockbroeckx, B., Vanclooster, M. (2004). Modeling of ground-penetrating radar for accurate characterization of subsurface electric properties. *IEEE Transactions on Geoscience and Remote Sensing*, 42(11), 2555–2568.

Lambot, S., Weihermüller, L., Huisman, J.A., Vereecken, H., Vanclooster, M., Slob, E.C. (2006). Analysis of air-launched ground-penetrating radar techniques to measure the soil surface water content. *Water Resources Research*, 42, W11403.

Lambot, S., Slob, E.C., Vereecken, H. (2007). Fast evaluation of zero-offset Green's function for layered media with application to ground-penetrating radar. *Geophysical Research Letters*, 34, L21405.

Ledieu, J., De Ridder, P., De Clerck, P., Dautrebande, S. (1986). A method of measuring soil moisture by time-domain reflectometry. *Journal of Hydrology*, 88(3–4), 319–328.

Lopera, O., Slob, E.C., Milisavljevic, N., Lambot, S. (2007). Filtering soil surface and antenna effects from GPR data to enhance landmine detection. *IEEE Transactions on Geoscience and Remote Sensing*, 45(3), 707–717.

Mahieu, A., Ponette, Q., Mounir, F., Lambot, S. (2020). Using GPR to analyze regeneration success of cork oaks in the Maâmora forest (Morocco). *NDT & E International*, 115, 102297.

Mahmoudzadeh Ardekani, M.R. and Lambot, S. (2014). Full-wave calibration of time- and frequency-domain ground-penetrating radar in far-field conditions. *IEEE Transactions on Geoscience and Remote Sensing*, 52(1), 664–678.

Meles, G., Greenhalgh, S., van der Kruk, J., Green, A., Maurer, H. (2011). Taming the non-linearity problem in GPR full-waveform inversion for high contrast media. *Journal of Applied Geophysics*, 73(2), 174–186.

Michalski, K.A. and Zheng, D. (1990). Electromagnetic scattering and radiation by subsurfaces of arbitrary shape in layered media: Parts I and II. *IEEE Transactions on Antennas and Propagation*, 38, 335–352.

Millington, T.M., Cassidy, N.J., Nuzzo, L., Crocco, L., Soldovieri, F., Pringle, J.K. (2011). Interpreting complex, three-dimensional, near-surface GPR surveys: An integrated modelling and inversion approach. *Near Surface Geophysics*, 9(3), 297–304.

Minet, J., Bogaert, P., Vanclooster, M., Lambot, S. (2012). Validation of ground penetrating radar full-waveform inversion for field scale soil moisture mapping. *Journal of Hydrology*, 424, 112–123.

Moghadas, D., André, F., Vereecken, H., Lambot, S. (2010). Efficient loop antenna modeling for zero-offset, off-ground electromagnetic induction in multilayered media. *Geophysics*, 75(4), WA125–WA134.

Parthiban, V., Maes, P., Sellier, Q., Sluÿters, A., Vanderdonckt, J. (2022). Gestural-vocal coordinated interaction on large displays. In *EICS '22 Companion of the 2022 ACM SIGCHI Symposium on Engineering Interactive Computing Systems*, 26–32.

Patriarca, C., Lambot, S., Mahmoudzadeh, M.R., Minet, J., Slob, E.C. (2011). Reconstruction of sub-wavelength fractures and physical properties of masonry media using full-waveform inversion of proximal penetrating radar. *Journal of Applied Geophysics*, 74(1), 26–37.

Peng, J., Loew, A., Merlin, O., Verhoest, N.E.C. (2017). A review of spatial downscaling of satellite remotely sensed soil moisture. *Reviews of Geophysics*, 55(2), 341–366.

Podimata, M. and Yannopoulos, P. (2015). Evolution of game theory application in irrigation systems. *Agriculture and Agricultural Science Procedia*, 4, 271–281.

Pramudita, A.A., Wahyu, Y., Rizal, S., Prasetio, M.D., Jati, A.N., Wulansari, R., Ryanu, H.H. (2022). Soil water content estimation with the presence of vegetation using ultra wideband radar-drone. *IEEE Access*, 10, 85213–85227.

Serhir, M., Besnier, P., Drissi, M. (2010). Antenna modeling based on a multiple spherical wave expansion method: Application to an antenna array. *IEEE Transactions on Antennas and Propagation*, 58(1), 51–58.

Slob, E.C. and Fokkema, J. (2002). Coupling effects of two electric dipoles on an interface. *Radio Science*, 37(5), 1073.

Sluÿters, A., Lambot, S., Vanderdonckt, J. (2022a). Hand gesture recognition for an off-the-shelf radar by electromagnetic modeling and inversion. In *27th International Conference on Intelligent User Interfaces, IUI '22*. Association for Computing Machinery, New York.

Sluÿters, A., Ousmer, M., Roselli, P., Vanderdonckt, J. (2022b). QuantumLeap, a framework for engineering gestural user interfaces based on the leap motion controller. In *Proceedings of the ACM on Human-Computer Interaction*, 6(EICS), 47. Association for Computing Machinery, New York. doi: 10.1145/3532211.

Sluÿters, A., Sellier, Q., Vanderdonckt, J., Parthiban, V., Maes, P. (2022c). Consistent, continuous, and customizable mid-air gesture interaction for browsing multimedia objects on large displays. *International Journal of Human-Computer Interaction*, 38, 1–32. doi: 10.1080/10447318.2022.2078464.

Soldovieri, F., Lopera, O., Lambot, S. (2011). Combination of advanced inversion techniques for an accurate target localization via GPR for demining applications. *IEEE Transactions on Geoscience and Remote Sensing*, 49(1), 451–461.

Taranta II, E.M., Samiei, A., Maghoumi, M., Khaloo, P., Pittman, C.R., LaViola Jr., J.J. (2017). Jackknife: A reliable recognizer with few samples and many modalities. In *Proceedings of the 2017 CHI Conference on Human Factors in Computing Systems, CHI '17*. Association for Computing Machinery, New York.

Topp, G., Davis, J.L., Annan, A.P. (1980). Electromagnetic determination of soil water content: Measurements in coaxial transmission lines. *Water Resources Research*, 16, 574–582.

Tran, A.P., André, F., Craeye, C., Lambot, S. (2013a). Near-field or far-field full-wave ground penetrating radar modeling as a function of the antenna height above a planar layered medium. *Progress in Electromagnetics Research – PIER*, 141, 415–430.

Tran, A.P., Warren, C., André, F., Giannopoulos, A., Lambot, S. (2013b). Numerical evaluation of a full-wave antenna model for near-field applications. *Near Surface Geophysics*, 11(2), 155–165.

Venkatarayalu, N.V., Gan, Y.B., Lee, R., Li, L.W. (2008). Application of hybrid FETD-FDTD method in the modeling and analysis of antennas. *IEEE Transactions on Antennas and Propagation*, 56(9), 3068–3072.

Wagner, W., Lemoine, G., Rott, H. (1999). A method for estimating soil moisture from ERS scatterometer and soil data. *Remote Sensing of Environment*, 70(2), 191–207.

Warren, C. and Giannopoulos, A. (2011). Creating finite-difference time-domain models of commecrical ground-penetrating radar antennas using Taguchi's optimization method. *Geophysics*, 76(2), G37–G47.

Wigdor, D. and Wixon, D. (2011). *Brave NUI World: Designing Natural User Interfaces for Touch and Gesture*, 1st edition. Morgan Kaufmann Publishers, Burlington, MA.

Wu, K. and Lambot, S. (2022a). Analysis of low-frequency drone-borne GPR for root-zone soil electrical conductivity characterization. *IEEE Transactions on Geoscience and Remote Sensing*, 60, 1–13.

Wu, K. and Lambot, S. (2022b). Effect of radar incident angle on full-wave inversion for the retrieval of medium surface permittivity for drone-borne applications. *IEEE Transactions on Geoscience and Remote Sensing*, 60, 1–10.

Wu, K., Rodriguez, G.A., Zajc, M., Jacquemin, E., Clément, M., De Coster, A., Lambot, S. (2019). A new drone-borne GPR for soil moisture mapping. *Remote Sensing of Environment*, 235, 111456.

Wu, K., Desesquelles, H., Cockenpot, R., Guyard, L., Cuisiniez, V., Lambot, S. (2022). Ground-penetrating radar full-wave inversion for soil moisture mapping in trench-hill potato fields for precise irrigation. *Remote Sensing*, 14(23), 1–16.

Yeo, H.-S. and Quigley, A. (2017). Radar sensing in human-computer interaction. *Interactions*, 25(1), 70–73.

6

Assessment of Flexible Pavements by GPR: 20 Years of R&D in France

Xavier DÉROBERT[1], Amine IHAMOUTEN[1], Vincent BALTAZART[1], David GUILBERT[2], Shreedhar Savant TODKAR[1], Grégory ANDREOLI[2], Bachir TCHANA-TANKEU[2], Jean-Michel SIMONIN[1] and Cyrille FAUCHARD[2]

[1] Université Gustave Eiffel, Nantes Campus, France

[2] Cerema, ENDSUM Research Team, France

6.1. Context

Roadways are one of the most widely spread networks of transportation across the world. In France alone, roadways stretch over a million kilometers. Most of these roads are made up of multilayered bituminous concrete (BC) and were laid over three decades ago. The French pavement design method consists of optimizing the choice of materials and their thickness. Stresses and strains are calculated at different depths and compared to limits obtained by laboratory fatigue tests. Over time, traffic and environmental factors may lead to material or interface damage. On the road surfaces, this damage induces permanent deformation in the transverse (rutting) or longitudinal profile (evenness), or cracking, usually in the wheelpath. Cracks promote the entry of water into the structure, which facilitates degradation of interfaces (debonding) and materials (disaggregation or stripping). This damage affects the residual life of the pavement substantially.

Ground Penetrating Radar,
coordinated by Mohammed SERHIR and Dominique LESSELIER.

Pavement management systems aim to maintain good road conditions. They require regular road monitoring by a measurement method. They merge the monitoring results to plan and budget road maintenance for several years. Structural conditions assessment requires information from surface conditions monitoring (cracking, longitudinal and transverse profiles), and also from monitoring with non-destructive techniques (NDT) such as deflection measurement and ground penetrating radar (GPR). At the project level, other local techniques can be added to refine diagnosis and obtain a robust reinforcement solution. Thus, coring, which is not appreciated by road managers, can be required to confirm a diagnosis from NDT measurement. Additional information will be given on certain techniques in the "complementary method" section.

Fifty years ago, road monitoring used devices performing static measurements or at least at reduced speed. These methods are not effective for surveying large road networks. In addition, they generate problems of traffic congestion and safety of users and agents carrying out the measurement. To avoid these issues, equipment manufacturers are seeking to develop traffic speed monitoring devices. GPR was the first NDT method to offer this opportunity at the beginning of the 1990s.

Since the 1980s, the growth of GPR in geophysics and civil engineering applications has resulted in numerous scientific works, including several books for the professional public. We can mention the examples of Daniels (1996, 2004), Dérobert (2003), Jol (2009), Benedetto and Pajewski (2015) and Carrik Utsi (2021). The recent trends are moving toward special issues in scientific journals, and among the most recent ones, we can cite Lambot et al. (2016), Solla and Lagüela (2019) and Solla et al. (2021).

In the frame of pavement inspections, either for periodical ones or those initiated by visible damage requiring some diagnostic, a GPR survey has been automatically integrated into the process of evaluation of the structures for about the last 20 years (Saarenketo and Scullion 2000; Laï et al. 2018; Elseicy et al. 2022). Indeed, GPR technological developments have resulted in various improvements: increased acquisition speeds, availability of very high frequency antennas for better resolution, availability of antenna arrays for a density of measurements guaranteeing accurate mapping, etc. In the same way, computer developments have brought to the fore data processing softwares, optimized for the application of pavements, through automatic picking of the interfaces, or the design of maps, with time-saving processing and improved interpretation of the data.

One of the first applications of GPR in road engineering was to determine road layer thicknesses (Saarenketo and Scullion 2000; Al-Qadi and Lahouar 2005; Loizos and Plati 2007). An extended application of GPR has concerned the cracks as GPR technology improved (Diamanti and Redman 2012; Solla et al. 2014). More recently, research has shifted toward debonding, compaction and tack coat evaluation; these topics of interest are detailed later on in this chapter.

Independent to this research, different standards or guidelines have been established for quality GPR inspections on pavements. The most known standards come from ASTM, which promote general use of GPR (ASTM D6432-11), specific information for the thickness of asphalt layers (ASTM D4748-10) and particular points for asphalt-covered concrete bridges (ASTM 6087-08). These standards are accompanied by several guidelines. We can mention the AASHTO (2013) Designation 37-04 from North America, recommendations produced in Europa, such as the Mara-Nord Project Recommendations (2011), and more recently, guidelines from the European GPR Association GS1601 (EuroGPR Guideline 2016) and the Belgian Road Research Centre (BRRC 2016). Lastly, the European community has gathered in a COST Action TU1208 (2013–2017), producing some guidelines, one being dedicated to GPR surveys on flexible pavement.

In this context, the scientific and technical network (RST) of the French Ministry of the Ecological Transition has been heavily involved in the development of radar methodology for the past 40 years, carrying out actions along two axes: a field approach, validating some survey methodology – including impulse and stepped-frequency radar (SFR) technologies – and a research and development approach, studying the poorly solved or unresolved problems, like the measurement of the thickness of very thin wearing courses and the evaluation of the compactness and bonding of asphalt layers.

6.2. GPR systems and acquisition

The major information presented in the following two sections come mainly from the COST Action TU1208 (Dérobert et al. 2019), the ASTM standard (D4748-10) and the AASHTO (2013) R37-04 guideline.

Different GPR systems can be used for road investigations, from single- to multi-antenna systems. Generally, GPR systems for road applications are not shielded, as the antennas are air-coupled, hung on the front or the back of a vehicle. As they work in the time domain, the antennas are required to be ultra-wide band (UWB) in the frequency domain, inducing central frequencies from about 400–2,000 MHz, for the subgrade-sub-base or the wearing course investigations, respectively.

Most of the radar systems work in bistatic mode, with antennas for transmission and reception being separate, allowing the removal of any internal multi-reflexions from the transmitter. Some systems can promote 3D investigations while using antenna arrays. This last kind of technology enables the collection of several parallel scans at once, accurately positioned near each other, producing 3D data.

Radar technologies propose either transmission of short-time duration pulses or successive linearly increasing frequencies in discrete steps. For the latter, an inverse fast Fourier transform (IFFT) is used to provide time domain radar data (A-scans), comparable to impulse systems. Each has its advantages, with the impulse systems being said to be faster in acquisition but more sensitive to radio frequency interference (RFI), while the stepped-frequency ones are electronically more stable and less sensitive to RFI.

Moreover, the antennas are developed to operate either at a limited distance above the pavement (part of air-coupled antennas) or in close proximity to the soil (part of the ground-coupled antennas). Most air-coupled antennas are horns or the Vivaldi kind, working in the range of 500–2,000 MHz, with the main advantage being a constant coupling, translated into a stable radio-frequency emission insensitive to change in pavement properties. On the other hand, ground-coupled antennas provide information about deeper pavement layers, but the surface coupling authorize only qualitative information from the near surface (Figure 6.1).

Figure 6.1. *Example of radar system acquisition using a 900 MHz ground-coupled antenna (on the left) and a 2 GHz air-coupled antenna on the right, towed by a vehicle (courtesy of Cerema). For a color version of this figure, see www.iste.co.uk/serhir/radar.zip*

The complete acquisition system generally includes an inspection vehicle, an odometer trigging data acquisition for an accurate location of the radar data along the acquisition profile, and potentially a geo-referencing system (differential GNSS), which enables the relation of the survey coordinates to a geographical information system (GIS).

Concerning the planning of the survey and the data acquisition, several levels of inspections have to be distinguished. At a network level, the inspections are used for evaluation or classification, providing a general overview of the pavement's structure, while dividing the global road linear into homogeneous sections for further investigation and analysis. At this level, only the layers thickness assessment is generally performed, with a limited number of scans per meter. The testing is performed at high speed, with almost no interference with the traffic. Generally, only one profile is measured along one-wheel path of a lane using air-coupled antennas.

General and detailed project levels require a more detailed analysis of the pavement. The density of measurement (scans per meter) is higher and more longitudinal profiles along the road are performed. For detailed information, such as debonding and material properties, the density of scans increase toward some grid in order to obtain a C-scan. An array antenna system is recommended for this level, to the detriment of a partial traffic closure.

For better preparation and a more accurate analysis, all information coming from the client is required, as well as the objectives of the client as technical information on construction and maintenance (e.g. number of elementary layers that make up the base course or the sub-base, which could be visible on B-scans and wrongly interpreted), or as indirect information from the neighbourhood (e.g. utilities). This step enables the creation of an evaluation framework that can integrate the accessibility and safety of the surveys, the GPR investigation – checking the EM environment and climate, cores' extraction for calibration, and complementary investigations, if required (detailed in a later section).

Lastly, concerning the setting parameters, we can just mention the following points:

– 2–4 scans/m for thickness measurements, or 20–40 scans/m for detailed investigations;

– a frequency bandwidth coherent with respect to the antenna and a time gain constant as far as possible;

– a distance of odometer calibration superior to 200 m;

– a time range 1.5 times double the expected travel time of echoes from the deepest layer.

To complement this, all of the notes, photos and drawings due to visual observations should be recorded in a log book.

6.3. GPR processing and interpretation

Several GPR-basic processing methods exist for application to flexible pavements. The most common methods require a signal picking in the time domain, which can be manual or automatic. The measured two-way travel times can then be converted into thicknesses through some calibration, while drilling cores or using single- or multi-offset reflection methods. Beyond these techniques, inverse modeling methods are becoming popular as no calibration cores are required for the assessment of the relative dielectric permittivity of the layers (see the following sections).

Remaining in a basic approach, air-coupled antennas configuration, in a bistatic mode, presents some advantages. The first one is to remain under the hypothesis of a mono-dimensional configuration with vertical wave propagation, perpendicular to the layers. This induces a simple relation between the picking measurement Δt and the estimated thickness h (equation [6.1]):

$$2h = v.\Delta t \qquad [6.1]$$

where v is the radar velocity in the material defined by:

$$v = \frac{c}{\sqrt{\varepsilon}} \qquad [6.2]$$

where c is the wave propagation velocity in the free space and ε is the relative dielectric permittivity of the medium.

Usually, the velocity is estimated using the thickness of each layer measured by coring, the latter being positioned along a B-scan at specific location(s) representative to the surveyed homogeneous section or above abnormal zones.

The second advantage of air-coupled antennas configuration is to rely on the proportionality between the intensity of the reflections and the value of the contrast between the dielectric properties of the pavement layers (Saarenketo and Scullion 2000). The methodology requires a static measurement on a large metallic plate placed on the surface of the pavement. The reference-reflected amplitude A_m is measured for different heights of the antennas while moving them vertically. Then, the dielectric permittivities of the course and base layers can be computed from the amplitudes of the considered reflections. We present only the equation for the first asphalt layer (A_{al}), considering that the other equations from Saarenteko and Scullion (2000) do not present sufficient reliability for road managers (equation [6.3]):

$$\varepsilon_{al} = \left[\frac{\left(1+\frac{A_{al}}{A_m}\right)}{1-\left(\frac{A_{al}}{A_m}\right)}\right]^2 \qquad [6.3]$$

Even if such a relation cannot be used for ground-coupled radar acquisitions, amplitudes coming from the direct and reflected echoes are sensitive to the EM nature and states of the media. To illustrate these standard approaches and processing on road structures, some results of a controlled experimental campaign are presented herein. Gustave Eiffel University has performed a GPR monitoring to survey debonding areas within pavement structure during accelerated pavement tests (APT) conducted on a full-scale fatigue carrousel. Thirteen artificial defect sections composed of three types of defects (tack-free, geotextile, and sand-based) were embedded during the construction phase between the top and the base layers. The data were collected in two stages, covering the entire lifecycle of the pavement structure, using four GPR systems: an air-coupled ultra-wideband GPR (SF-GPR) using the frequency band (0.8–10.8 GHz), two wideband 2D ground coupled GPRs (a SIR-4000 with a 1.5 GHz antenna and a 2.6 GHz-StructureScan from a GSSI manufacturer) and a wideband 3D GPR (from 3D-radar that became the KONTUR manufacturer) composed of 21 ground-coupled antennas working in the frequency band (40 MHZ–3 GHz).

The first stage of the experiments took place in 2012–2013 and lasted up to 300,000 loadings. During this stage, the pavement structure presented no clear degradation. The second stage of experiments was conducted in 2019 and continued until the pavement surface demonstrated a strong degradation, which was observed at 800,000 loadings. At the end of the GPR experiments, several trenches were cut at various sections to get the ground truth of the pavement structure. Finally, the GPR data were processed using the conventional amplitude ratio test (ART) to study the evolution of the echoes coming from the debonded areas. All of the results and analysis from this study are described in detail in Dérobert et al. (2021), and raw data are available in two dataverses (Dérobert et al. 2020a, 2020b).

Gustave Eiffel University's pavement fatigue carrousel is a full-scale road traffic simulator, with accelerated pavement testing, composed of four arms carrying the moving heavy loads (65 kN on twin wheel for this experiment) at a maximum speed of 90 km/h (Simonin et al. 2012). The test track associated with this experiment is a

25 m structure of road composed of two bituminous layers (6 cm thick wearing course and 8 cm thick base layer) over a granular sub-base.

Thirteen rectangular patches of materials (sand-based, geotextile and tack-coat free inter-face), simulating debonded areas, were inserted at the interface between the two asphalt layers (Simonin et al. 2014). The three types of defect were chosen to represent at least two levels of interface damage that can be observed by coring during road expertises: simple debonding due to tack-coat free and some sand layer simulating aggregate stripping, both being completed by a geotextile. Figure 6.2 presents the positioning of the 13 defects on the test track.

a) b)

Figure 6.2. *Test section (a) with the defects made during construction and (b) in the form of a site plan (Dérobert et al. 2021). For a color version of this figure, see www.iste.co.uk/serhir/radar.zip*

Figure 6.3 presents a longitudinal profile along the center of the global test section at 2.6 GHz, with every defect being detected (the I-1 location is more ambiguous). Defects I-4, I-9, I-10 and I-12 (but I-2) show stronger echoes than the others due to stronger dielectric contrasts. We can note several embedded sensors to monitor the stress under the heavy traffic located in the center of defect I-13 and between I-12 and I-13.

During the campaign, periodically longitudinal and transversal B-scans were acquired above the effects: at around 50,000, 10,000, 200,000, 250,000 and 300,000 loadings in 2012, and at about 310,000, 396,000, 420,000, 500,000, 600,000, 720,000 and 800,000 loadings in 2019–2020 (end of the APT).

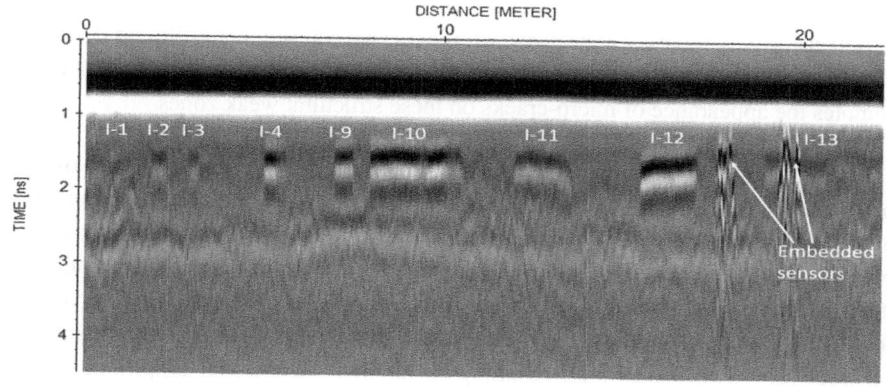

Figure 6.3. *Example of longitudinal B-scan at 2.6 GHz along the test section*

A comparison of raw B-scan data is shown in Figure 6.4, performed by a 2.6 GHz impulse radar (Figure 6.4(a)) and a 4.5 GHz home-made air-coupled SFR (Figure 6.4(b)) on the defect I-11 (sand) at 720,000 loadings. The surface echo at the top showing the strongest amplitude is called direct wave (DW) for both data types. The second strongest echo is the reflected wave (RW) from the sand-based debonded interface, with the echo being weaker in the healthy area, at roughly the same time-depth.

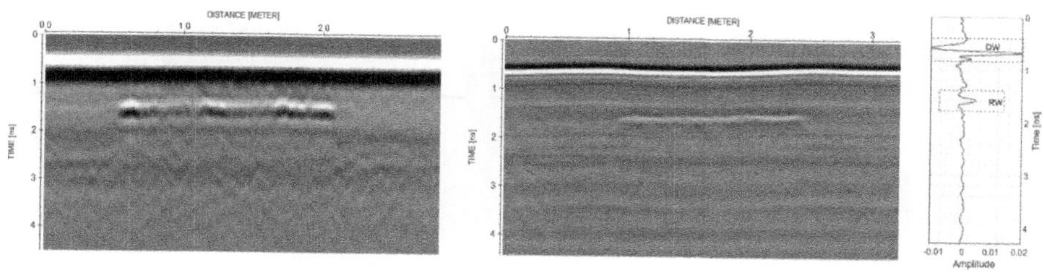

a) b) c)

Figure 6.4. *B-scans performed at 720,000 loadings on the defect I-11 (sand) by (a) a ground-coupled 2.6 GHz impulse system and (b) an air-coupled 4.5 GHz stepped-frequency system. (c) A-scan extracted from (b) and showing the direct wave (DW) and the reflected wave (RW) (Dérobert et al. 2021)*

As the defects I-11 (sand-based), I-12 (geotextile) and I-13 (tack coat free) are the largest ones, they were studied in more detail with four GPR systems during the global campaign. Figure 6.5 shows some examples of GPR maps performed by the 3D-radar system and extracted from the 1.5 GHz C-scan acquisitions. It is interesting to note that the variations of the reflected echoes amplitudes are logically visible above the defects, giving their accurate sizes. Moreover, we can see that GPR can detect the damage evolution of the surface asphalt course itself, due to heavy traffic, which promotes the appearance of micro-cracks on these structural weak zones.

The general processing methodology consists of the computation of the conventional ART, for each A-scan, as the ratio between the RW and the DW, namely ART = RW/DW (Todkar et al. 2019). Then, the data processing basically consists of automatically picking the maximum (or minimum) amplitude of the two latter echoes with some commercial radar software. In terms of pavement monitoring, any debonding provides additional echoes that mostly interact constructively with each other, resulting in an increased signal amplitude of the reflected signal (RW) compared to the one of the healthy zones. The pavement monitoring then turns to analyzing the amplitude ratio variation function of the traffic loading. Concerning the study of the DWs, pickings are normalized by a DW value taken from a sound zone outside the traffic path.

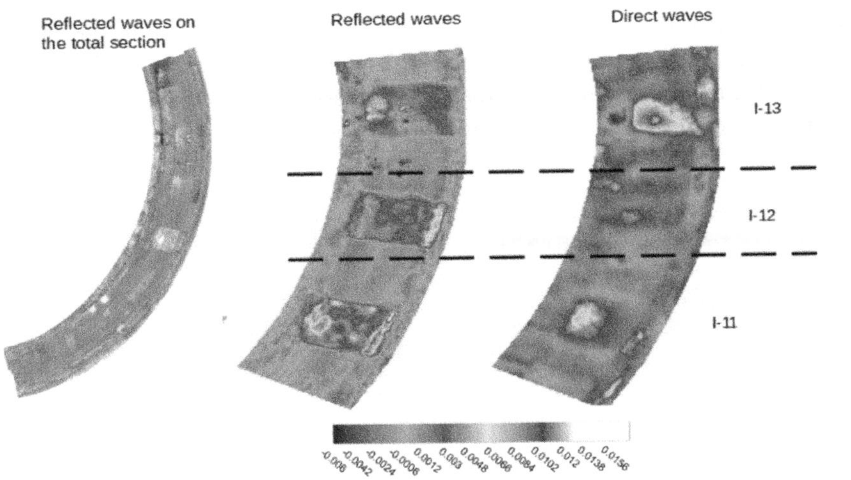

Figure 6.5. *Example of amplitude maps done at 720k loadings extracted from 1.5 GHz C-scans. For a color version of this figure, see www.iste.co.uk/serhir/radar.zip*

From over 500 B-scans performed during the global campaign, 2.6 GHz has shown to be the best frequency for debonding detection, when studying the ART of the database, compared to the other two. Figure 6.6 presents, as an example, some results from the defect I-11 (sand-based). At the end of the campaign, the major observations are as follows:

– In the healthy zones out of and under traffic, normalized RW remains stable. Apart from a natural ageing of the pavement layers, heavy traffic has had no effect on the road structure from a GPR point-of-view. Moreover, due to low EM contrasts, echo picking was not always easy and could sometimes lead to errors in their detection/location.

– Defects are clearly visible with a high level of RW amplitude, but with strong local variations. These variations could be due to local heterogeneities enhanced by internal moisture content and by the fact that GPR antennas did not pass exactly on the same path. Nevertheless, we cannot make a direct link between amplitude evolution versus traffic. Indeed, variations mainly come from moisture content inside the defect due to the weather conditions from the previous days – favored by micro-crack increases – and perhaps in a second step, from variations of temperature.

– Transverse data present a general trend of RW decrease, while moving toward the outside of the carrousel due to the road topography promoting inward water migration and possible lateral water gradient. It should be mentioned that this trend does not seem to be correlated with the lateral traffic density, which is mainly concentrated in the center of the path with a lateral V shape.

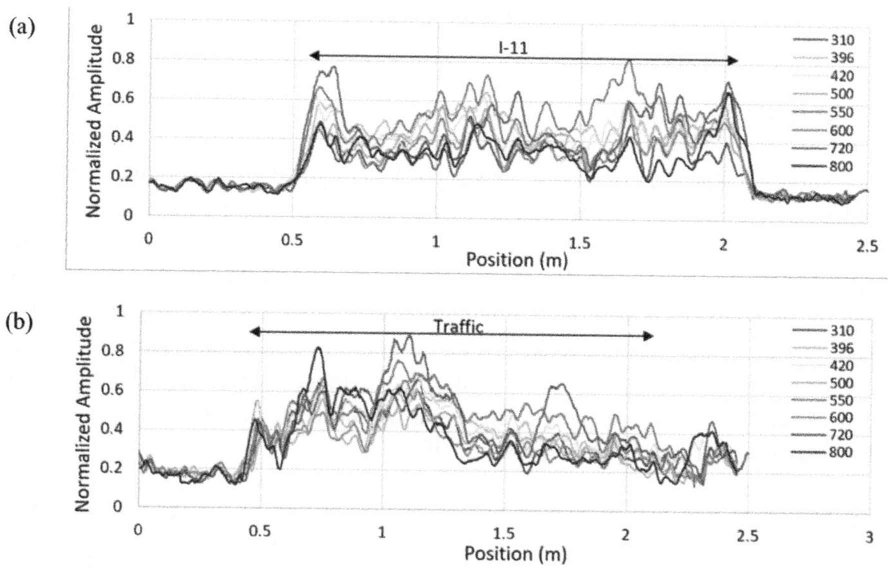

Figure 6.6. *2.6 GHz norm. RW amplitudes on sand (I-11) in the (a) longitudinal and (b) transversal directions versus kloadings (Dérobert et al. 2021). For a color version of this figure, see www.iste.co.uk/serhir/radar.zip*

6.4. Complementary ND techniques

Whatever their objective, the interpretation of radar measurements needs to be confirmed using other information. This can be the road history, from the design to the last monitoring. Results from quality control of the layered material (nature and thickness) make it possible to verify that the interface echoes are correctly identified. Several monitoring techniques can also be used to obtain a more robust diagnosis like coring, surface conditions monitoring, or deflection measurement (or other mechanical-based NDT methods).

6.4.1. *Coring*

Coring constitutes an on-site reference. It allows the calibration of GPR measurement and is necessary for each homogeneous section to adjust wave velocities. The GPR B-scans may help to choose the best location for coring and minimize the cores number. The quality of material from the sample, but also in the coring, also has to be observed. They can give information about interface quality (well bonded or not, presence of disaggregate material). In addition, samples can be tested in the laboratory to estimate the compaction (using gamma measurement), dielectric permittivity or Young modulus of the different layer.

6.4.2. *Surface conditions*

The survey of surface conditions consists of locating and characterizing "visible" degradation on the road surface. This includes permanent deformations in the longitudinal and transverse direction, but also the presence of cracks and the departure of materials. For example, wide radius rutting is generally associated with fatigue of the subsurface material. Cumulative traffic then leads to fatigue cracking of materials and interfaces between layers. Alteration of radar echoes at the same location reinforce the assumption of interface damage. Localized damage also influences surface condition parameters, such as short wavelength deformation of the longitudinal profile and/or the presence of cracking.

Several techniques are used to monitor surface conditions. Visual survey is for historic reasons the most common method. New methods appeared using either accurate but expensive sensors, or cheap sensors with lower performance. Devices of the first type use laser sensors or high-resolution cameras with a high acquisition frequency. They make it possible to reconstruct the 3D surface of the roadway in order to deduce the usual parameters of the surface condition. This expensive apparatus is generally only used on the structural network. The second survey technique benefits from the development of smartphones. They are used as sensors (accelerometer, camera) and acquisition systems, eventually coupled with external sensors located at a specific point in the vehicle. Data are pre-processed by the smartphone and then automatically transmitted, together with the GPS coordinates, to a data server. The road network evaluation benefits from the measurements made by different vehicles. Their data are combined by data fusion techniques to obtain a more robust evaluation (Martin 2015).

6.4.3. *Deflection*

Deflection measurement is the worldwide ND method used to assess the structural condition of roads. It consists of measuring the deflection of the roadway under the action of a heavy load. This mechanical test directly estimates the state of fatigue of the structure, which has been designed with respect to mechanical fatigue criteria. Traditional devices are static or slow moving such as the falling weight deflectometer (FWD) (stationary), Benkelman beam (static) or curviameter and deflectograph (slow-moving 18 and 8 km/h, respectively). They induce operational limitations such as lane closure, traffic congestion and potential workplace risks. To overcome these limits, new devices operating at traffic speed have recently appeared, such as the TSD in the early 2010s (Baltzer 2010) and more recently, the RAPTOR.

Deflection is very sensitive to pavement damage. Several indexes can be calculated from deflection to estimate the bearing capacity of different parts of the structure. More recently, a method has been developed to calculate optimized indicators, which focus on different parts of the structure, including interface behavior. These indexes assume a homogeneity of the structure. GPR makes it possible to confirm this assumption or define the limits of the different homogeneous structure. Another example of the complementarity of the two methods can be based on observing the maximum deflection only. Some variations show probable structural damage located where the deflection is highest. Radar B-scans can help identify the type of damage. This may be a reduction in the thickness of the materials, revealing a problem in the construction of the structure, or a damaged interface between two layers, which may be visible on the variations of the radar echoes of these interfaces.

Deflection and radar are clearly two complementary methods for structural assessment of pavements. The new traffic speed deflectometers (very expensive) will reinforce this complementarity. They already include an optional radar system to perform the two measurements simultaneously. This will produce very large amounts of data that will require new processing techniques.

6.4.4. *Other mechanical NDT methods*

Pavement structures are sized to resist traffic and the mechanical fatigue that these loads generate. Mechanical methods are therefore appropriate to monitor the fatigue evolution of the structure. Apart from the deflection measurement, which uses a static or quasi-static load, other mechanical methods apply based on dynamic stresses. These methods are associated with different physical phenomena such as mechanical wave propagation (Rayleigh, Lamb, compression, shear), plate vibrations or echo impact. The application of these methods for pavement monitoring remains in the field of research because their practice on road pavement is difficult and their efficiency is very low. Use for operational purposes is not (yet) possible.

Wave propagation methods have two limitations, which are difficult to overcome. First, bituminous materials have visco-elastic properties and are temperature sensitive. Viscosity can greatly attenuate wave propagation. To promote wave propagation, tests can be carried out at low temperatures, thus attenuating the viscous effect. Unfortunately, in the field it is often difficult to choose the measurement conditions. The temperature sensitivity of bituminous materials also has to be taken into account when interpreting the results. Thus, to estimate thicknesses by the echo impact method, it is necessary to consider the temperature of the materials, which will strongly modulate the propagation wave velocity. (Simonin et al. 2013). Practice on bituminous materials constitutes a second limitation. The generation of waves and the measurement of the surface response are particularly tricky to achieve. Waves are generated either by small balls or by transducers in contact with spikes. The operator must be very attentive to the positioning to generate the impulse on an aggregate and not on the bituminous binder. Indeed, in the latter case, the coupling is very bad, making the test unsuccessful (not enough energy) or unusable (bad frequency band).

Among the different methods, the Colibri prototype was specifically developed for road monitoring of research test sites. An original data processing allows us to calculate a "B-scan" in the frequency domain, which allows us to detect, identify and qualify the presence of damage, such as interface defects. (Simonin et al. 2012, 2014). A full-scale experiment on the fatigue carousel at the Nantes campus of Gustave Eiffel University, used several non-destructive methods (GPR, FWD, ultrasonic pulse echo and mechanical wave propagation) to explore debonded interfaces (Simonin et al. 2013). It shows the complementary between the different methods.

6.4.5. *Gammadensimetry*

The estimation of the compaction is another parameter required to check the mechanical characteristics of asphalt over the durability and maintenance study of flexible roads. Concerning new pavement courses, in the frame of quality control of acceptance of work, gammadensimetric techniques can be performed.

These techniques indirectly estimate either continuously or punctually the bulk density of the pavement courses, while measuring the scattering phenomenon of gamma rays inside materials. Their parameters (distance between the source and the receiver, emission angle and photon detection angle) are selected so as to obtain good measurement sensitivity, for an investigation depth which can vary from about 5–14 cm.

The measurement is non-destructive, independent of the aggregate chemical characteristics and works punctually or until 1.8 km/h, depending on the system. Laboratory calibration can be performed on some slabs extracted from the site to correlate measurements and bulk density. Otherwise, when measurements indicate values outside the thresholds, cores are drilled and analysed in a laboratory, using a gammadensimeter or a hydrostatic weighting, in order to validate these values and locate the zones out of contract.

6.5. Research and innovations

Numerous applied research has been performed over the last 20 years while studying different software or hardware approaches for specific road application. In the following sections, these approaches are detailed technique by technique.

The following table enables us to summarize the possible applications (challenges to solve) related to this research. Bold crosses indicate the applications that will be presented as examples in the following sections.

	Thickness	Debonding	Tack-coat	Compaction	Moisture
SFR	**X**	X	x	**X**	
High resolution processing	x	**X**	x		
FWFI processing	**X**	X	**X**		**X**
ML processing	x	**X**	**X**		

Table 6.1. *Applied research associated with road applications*

6.5.1. *High-frequency stepped-frequency radar approach*

The interest of studying and developing a SFR technique comes from the limitation of impulse commercial systems for very thin course layers inspections. There is also the technological possibility of working in very high-frequency ranges above the impulse equivalent ones.

The principle herein is to be able to utilize a commercial network analyzer as both a generator and radar receiver. A synthetic pulse is reconstructed from a stepped-frequency signal, transmitted to the antenna, using an inverse Fourier transform. The major value of such a system stems from the potential to control a UWB around the central frequency for a given application and obtain a higher dynamic range than an equivalent impulsive radar.

The first experiments were performed with an HP8753C network analyzer and UWB antennas from the Vivaldi family working in the (500 MHz – 10 GHz) band, enabling the reconstruction of a pulse centered above 3 GHz (Dérobert et al. 2001). These experiments were done on controlled test sites, including one particularly devoted to GPR measurements (Figure 6.7). It is composed of three different structures, with local metallic plates making it possible to confirm some specific interfaces. The materials used are BC, bitumen-bound granular material (BBGM), cement-bound granular material (CBGM) and unbound granular material (UGM). Figure 6.8 presents the corresponding GPR profiles, done at 1.5 GHz with a GSSI system and with the 3 GHz SFR. We note the increase in resolution and the decrease in depth penetration in the latter system. Thus, the results coming from the different test sites have demonstrated the ability to detect both very thin asphalt layers (2.5 cm averaged thickness) and a second asphalt course in the base structure.

The SFR technology continued to be used in further research done by the RST, related to very thin asphalt layers (Fauchard et al. 2008), high-resolution processing (Todkar et al. 2018; Le Bastard et al. 2019) and full-wave form inversion developments (Ihamouten et al. 2018; Andreoli et al. 2022). Some of them are discussed in the following sections.

Another advantage of the SFR technology relies on the electronic stability which enables us to assess HMA (hot-mix asphalt) pavement compactness, one of the key parameters to control newly paved roads. The compactness, or degree of compaction, is classically obtained from a measurement of the density of the material (the NF EN 12697-6 standard (AFNOR 2012) and the ASTM D1188-07e1 standard (ASTM International 2007)) or from nuclear gauges (according to the NF P94-062 standard (AFNOR 1997) and the ASTM D2950/D2950M standard (ASTM

International 2011)), with an error of less than 1% on the reference values given by the weighting standard method on newly paved roads.

Work carried out at Cerema (Fauchard et al. 2009; Fauchard et al. 2015) showed that the use of an SFR made it possible to estimate the compactness of new pavement mixes with an accuracy equivalent to nuclear methods (Fauchard 2015). As previously mentioned, equation [6.3] is used to estimate the permittivity of the newly paved HMA in situ. Only the permittivity of the first layer, and thus its compactness, is estimated.

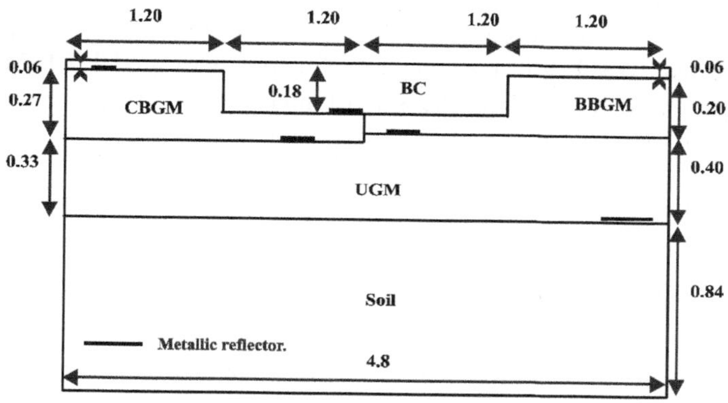

Figure 6.7. *Layout of a GPR test site for pavement layer thickness measurements (Dérobert et al. 2001)*

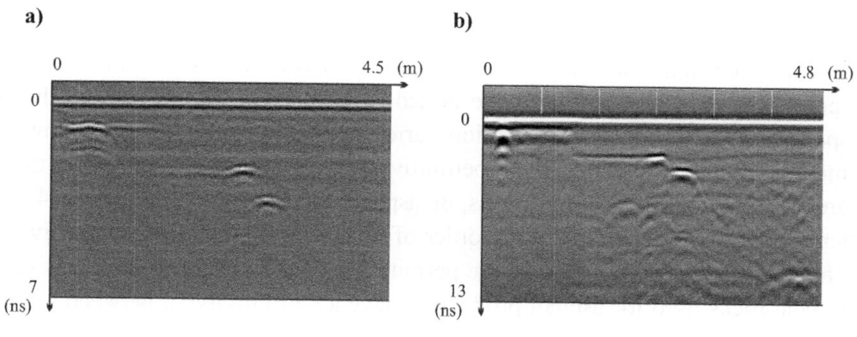

Figure 6.8. *Layout profiles of the GPR test site (Figure 6.7): (a) SFR profile at 3 GHz; (b) impulse profile done at 1.5 GHz (Dérobert et al. 2001)*

The compactness C of HMA is given by:

$$C = 1 - C_{air} = \frac{\rho_{HMA}}{\rho_{HMAM}} \quad [6.4]$$

where ρ_{HMA} is the specific gravity of the HMA and ρ_{HMAM} is the maximum specific gravity, that is, without air. We developed a physical model, based on the CRIM formula (Sihvola et al. 1998), giving the permittivity of a mix $\varepsilon_{HMA}^{\alpha}$ as a function of the volume proportions and dielectric characteristics of each component of the mix:

$$\epsilon_{HMA}^{\alpha} = \sum_i C_i \epsilon_i^{\alpha} \qquad [6.5]$$

where i is the component considered and can be f for filler, b for bituminous binder, a for aggregate, rec for recycled aggregates and air for air. C_i and ε_i^{α} are, respectively, the volume concentration and the permittivity of the ith component and α is a coefficient ($\alpha = 1/2$ for the complex refractive index model (CRIM), $\alpha = 1/3$ for the Birchak model, $\alpha = \pm 1$ for the Sihvola model and $\alpha > 0$ for the Lichtenecker and Rother approximation).

By combining both equations [6.4] and [6.5], our model is as follows:

$$C = \frac{\varepsilon_{HMA}^{\alpha} - 1}{\frac{1}{Z}\varepsilon_a^{\alpha} + \frac{C_{rrec}}{Z}\varepsilon_{rec}^{\alpha} + \frac{C_{rf}}{Z}\varepsilon_f^{\alpha} + \frac{C_{rb}}{Z}\varepsilon_b^{\alpha} - 1} \qquad [6.6]$$

where Z is equal to $Z = 1 + C_{rb} + C_{rf} + C_{rrec}$ and C_{ri} is the relative volume concentration of the ith component (see Araujo et al. (2017) for more details). Other models (Araujo et al. 2017) have been used and compared. In particular, the Böttcher and Rayleigh models also appeared to be very accurate for estimating compactness.

The method therefore requires precise knowledge about the permittivity of the aggregates, the bituminous binder and the air, the latter being given as ε_{air}. Conventional bituminous binders are a mixture of hydrocarbons from the distillation of petroleum to which additives are added to give them the desired mechanical properties. Their chemical composition varies according to the oil used. However, numerous studies show that their permittivity ε_b is approximately equal to 2.5. Moreover, their proportion by mass in asphalt mixes is low (generally 4–6%), whereas that of aggregates is of the order of 92%. It is therefore the permittivity of the aggregate that matters most in the permittivity of the asphalt. It has been shown that most rocks used for asphalt production have a permittivity ε_a between 4 and 8. Finally, the filler component is often made of limestone. Its weight concentration is about 1% and its permittivity is considered equal to 5.5.

Cerema has compiled a permittivity database of the main quarries supplying aggregates to road builders. The estimation of the compactness, knowing the mass contents of the constituents (given by the manufacturer's formulation), is therefore possible by considering the values listed in the permittivity database (Araujo et al. 2016). In the context of material reuse, the asphalt used for new pavements increasingly includes recycled aggregates whose origin is generally unknown, and consequently, the permittivity of the aggregates is unknown. Two solutions are then available as follows:

– coring (destructive survey) is necessary, allowing the measured permittivity to be recalibrated on the basis of a density measurement (hydrostatic weighing, gamma bench measurement);

– a complementary measurement in a cylindrical cavity of the aggregates used is carried out. This assumes that the manufacturer provides a sample of the aggregates and the dielectric characterisation in a cylindrical cavity is done in the laboratory (it could be done in situ). This approach does not require coring of the pavement.

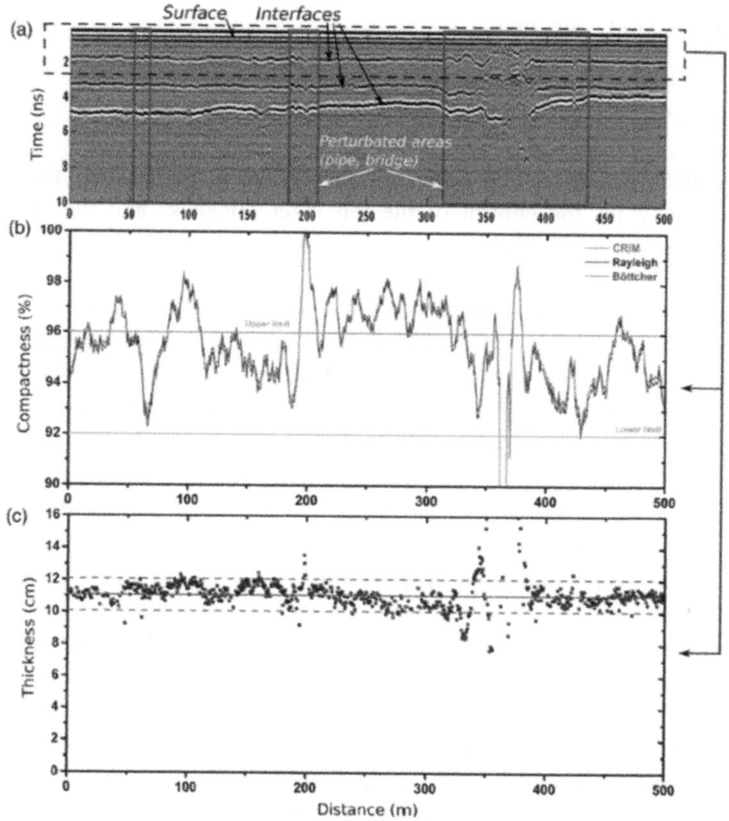

Figure 6.9. *(a) SFR B-scan on a newly paved road; (b) compactness assessment on the whole profile (2 A-scan/m, 7.5 km/h) according to Böttcher, Rayleigh and CRIM models; (c) calculated thickness as additional results of the first layer permittivity estimation (Araujo et al. 2017). For a color version of this figure, see www.iste.co.uk/serhir/radar.zip*

The latter approach has been successfully tested on site on a newly paved road layer containing 38% of recycled aggregates, and Araujo et al. (2017) showed that the assessment of compactness is below the required 1% accuracy, taking into account the nuclear testing as reference. Figure 6.9 shows as an example a SFR B-scan performed on a new pavement. Compactness assessment on the whole profile (2 A-scan/m, 7.5 km/h) is obtained according to Böttcher, Rayleigh and CRIM models. Figure 6.9(c)) gives complementary information on the wearing course layer thickness resulting from its permittivity estimation.

It also shows how this approach enables the evaluation of the compactness on the whole linear of the profile, contrary to the nuclear or low-frequency electromagnetic methods that are only punctual.

6.5.2. *Super/high time resolution processing*

As recalled in section 6.3, GPR has mainly been used within the scope of pavement survey for the control of the top layer thickness and the detection of interface debonding within the pavement structures.

The conventional radar configuration includes air-coupled antenna devices at nadir associated with commercial GPR pulse devices, in which the bandwidth has evolved from 1.2 to 2.6 GHz in the past 20 years. Data interpretation is usually performed offline by specialists (Lai et al. 2018). It is mostly based on time picking of backscattered echoes (in time and amplitude) on either 2D B-scan images or successive 1D A-scan signals. Some processing may provide users with the automatic time picking of detectable echoes along the B-scan. Time and amplitude picking allows the specialist to estimate the permittivity and layer thickness and then fully characterize the layered pavement structure.

The latter conventional processing is relevant only when the pavement structure shows thick enough layers, namely with a thickness above 5 cm (labeled the conventional layer thickness range in Figure 6.10), such that each pavement interface is clearly distinguished in time radar signal by unambiguous individual backscattered echoes (providing dielectric contrast between layers). The condition $B \times \Delta T \geq 1$ defines the time resolution limit for conventional processing, which relates the effective bandwidth B to the differential time shifts ΔT between adjacent echoes. The condition $B \times \Delta T = 1$ in Figure 6.10 defines the time resolution limit for conventional processing, which relates the effective bandwidth B to the differential time shifts ΔT between adjacent echoes. The conventional time resolution limit $B \times \Delta T = 1$ is displayed in Figure 6.10 as a reference.

To cope with thinner pavement surfacing techniques of about 1–2 cm in thickness, namely, the thin asphalt surfacing (TAS) and ultra-thin asphalt surfacing (UTAS) thickness range in gray in Figure 6.10, the solution which was first explored in Dérobert et al. (2001) relies on the use of both larger bandwidth B and UWB microwave antennas with higher central frequency. UWB antennas were coupled with the network analyser to form the step-frequency-radar (SFR) device. Originally, the SFR cannot support data acquisition at traffic speed. Recent technological evolution now achieves costly effective trade-off between dynamic range and acquisition time. According to Figure 6.10, the effective bandwidth must be larger than 3.5 GHz to control TAS and UTAS thickness layers with conventional data processing.

The second solution which has been explored for this application relies on signal processing techniques with a high time resolution capability (Le Bastard et al. 2007). This solution compensates for the limited bandwidth of commercial pulse GPR devices and may be easily adapted to both pulse and step-frequency radar technologies. It may at least improve the time resolution by a factor 10 at medium SNR. The new achievable time resolution limit $B \times \Delta T = 1$ has been shown[1] in Figure 6.10.

Within the context of a GPR-based pavement survey, Yamada et al. (1999) paved the way to combine the use of a network analyser with high-resolution time delay algorithms, while Wu et al. (2002) showed the real-time capability of super resolution TDE algorithms for the control of pavement thickness. Le Bastard et al. (2007) demonstrated the successful application of different super resolution TDE algorithms for measuring 1–2 cm thick layers with the two existing GPR technologies.

Roughly, three families of signal processing techniques can afford super time resolution capability: subspace-based methods (Le Bastard et al. 2007), regularized deconvolution techniques (He et al. 2011; Chahine et al. 2011) and machine learning methods (Le Bastard et al. 2014) at lower extent. Subspace-based time-delay estimation (TDE) techniques, for example, MUSIC (MUltiple SIgnal Classification), ESPRIT (Estimation of Signal Parameters via Rotational Invariance Techniques) and MPM (Matrix Pencil Method), have shown better time resolution capability compared to the others (Le Bastard et al. 2007; Chahine et al. 2011). The

1 The super time resolution limit is computed with the same permittivity as for the conventional resolution limit, namely, 5. The surrounding shadow orange area represents the resolution limit variations when the permittivity of the pavement debonding is filled with two extreme permittivity values, that is, $\varepsilon^r = 1$ (air) or $\varepsilon^r = 20$ (wet material).

performance assessment of these TDE techniques have been demonstrated from simulated data in the literature to reach the well-known Cramer-Rao bound at high SNR.

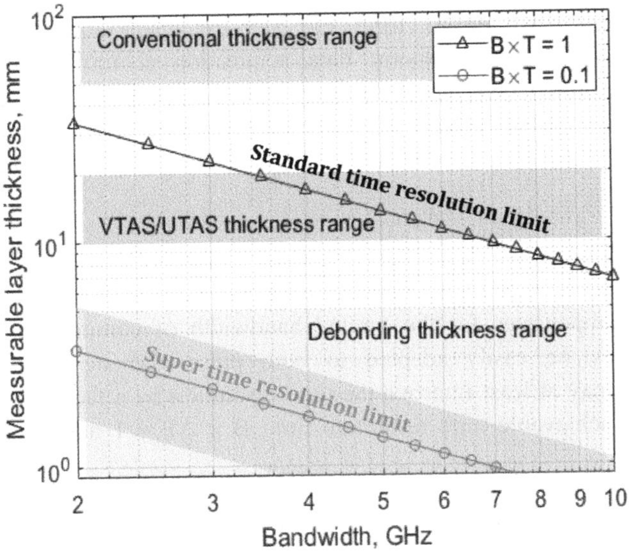

Figure 6.10. *GPR ability to control three thickness ranges of pavement layers with regard to the effective bandwidth and the time resolution limit of both conventional and advanced signal processing techniques. For a color version of this figure, see www.iste.co.uk/serhir/radar.zip*

All these methods rely on a simplified analytical GPR data model in the frequency domain, which is the sum of attenuated and shifted radar pulses, and assuming single electromagnetic scattering within the layered pavement structure. In practice, the strong correlation between backscattered echoes and possible multiple scattering within layered pavement structures may limit the performance of HR TDE algorithms. Frequency sub-band averaging techniques have been used in the literature to recover the full-time resolution capability. Among the existing techniques which have been compared in Sun (2016), MSSP (modified spatial smoothing processing) has shown a good trade-off between computer burden and TDE performance, while the SSS (signal subspace smoothing) technique in Sun et al. (2017) enables an improved robustness against noise.

In the past 10 years, high-resolution TDE algorithms have been associated with UWB SFR-based GPR to probe debonded pavement structures. As illustrated in Figure 6.11, the debonding area is observed between the two top layers of the pavement structure and may be a few mm thick. Such embedded defects weaken the overall pavement structure and may give rise to reflected cracks up to the top surface. In a GPR survey, debonding areas induce strongly overlapping backscattered echoes in time, where the differential time delay ΔT is much smaller than the conventional time resolution limit 1/B, and the overall magnitude has increased because of constructive interferences.

The latter magnitude enhancement allows the debonding areas to be detected using the widespread ART technique, which shows increased dynamic variation at larger bandwidth (Simonin et al. 2012; Todkar et al. 2019). According to Figure 6.10, the refined GPR signal analysis over the debonding areas requires the combination of both larger bandwidth and super resolution TDE algorithms. Once the time delays are estimated, the data model also allows the full characterization of the debonding layer, namely, the permittivity and the thickness, for monitoring purposes.

The data collected during the long-term experiment on the pavement fatigue carousel in Simonin et al. (2016) have been used to test both the conventional and high-resolution processing techniques over three types of artificial debondings, namely, uncoating, sand and geotextile areas. As reported in Dérobert et al. (2021), the conventional processing techniques allowed us to easily detect the defects by the two GPR configurations and technologies, namely, ground-coupled versus air-coupled and pulse commercial GPR and SFR-based GPR. The ongoing analysis in Tankeu (2023) with high-resolution processing techniques is expected to provide further insights into the monitoring of the three types of artificial debondings (Figure 6.11).

In parallel to the development of GPR application, further theoretical investigations have been carried out to reduce the computer burden of high-resolution TDE algorithms and improve the performance of TDE algorithms within a strong correlation multipath background. Then, the "linear" family of HR techniques, namely, OPM, BEWE and SWEDE, have been adapted to GPR data processing and compared to the polynomial version of subspace-based TDE algorithms (Le Bastard et al. 2008). The polynomial PUMA algorithm, which has recently been adapted to TDE for GPR application in Tankeu et al. (2021) achieves another step in reducing the computer burden and robustness against the correlation between echoes.

In the meantime, TDE algorithms were extended to a more realistic data model, including frequency-dependent propagation effects at UWB. In this situation, the GPR pulse shape spreads out with depth, thus reducing the time resolution for deeper echoes. TDE algorithms then become TDQE algorithms for time delay and Q factor estimations, where the Q factor can be interpreted differently according to the physical origin of the frequency dependence. For example, Sun et al. (2018) addressed the problem of the frequency-dependent decay amplitude of backscattered echoes, which is induced by the interlayer pavement roughness, and proposed some adaptations for the MUSIC algorithm to this aim. Chahine et al. (2010) proposed different MPM-based TDQE algorithms using the causal Jonscher parametrization of the medium permittivity, which then obeys to coupled amplitude and phase frequency dependences. Within the latter scope, MPM has shown its capability to be adapted to absorbing/dispersive medium, while being robust to the cross-correlation between echoes beyond medium SNR.

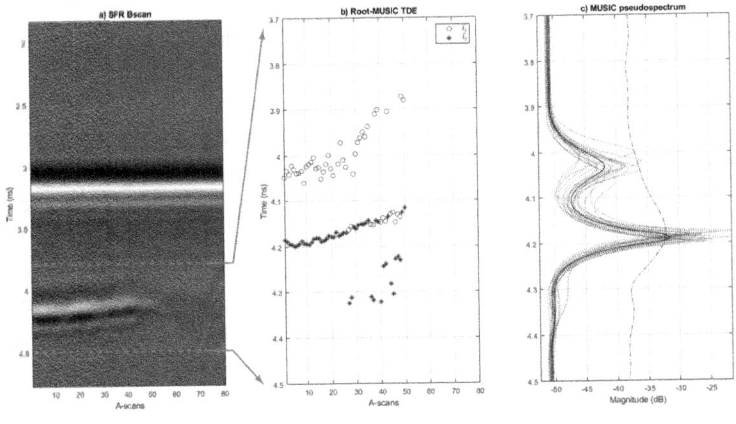

Figure 6.11. *Illustration of the GPR survey of debonded pavement structures during the long-term experiment in Dérobert et al. (2021). (a) SFR-based GPR Bscan image showing the echo over the artificial geotextile-based debonding (at roughly 4.2 ns); (b) time delay estimation obtained by the MUSIC algorithm over the (3.7, 4.5 ns) time interval; (c) MUSIC pseudospectra of the first 20 A-scans (in black) showing the two peaks from the debonding interfaces; the dashed red line displays the A-scan signal in magnitude. For a color version of this figure, see www.iste.co.uk/serhir/radar.zip*

6.5.3. *Full waveform inversion processing*

The application of the complete durability process for the physical and geometrical 3D mapping of civil engineering and building structures requires the estimation of non-destructive observables. Recent work has focused on the development of radar full waveform inversion (FWI), based on robust forward analytical and numerical models in front of high dispersive media, while including the global radar signal information. This method has already demonstrated its efficiency and remarkable performance for the dielectric and geometric characterization of materials in geophysics (Lambot et al. 2004; van der Kruk et al. 2014; Lambot and André 2014). Since the FWI method requires the radiation characteristics of the transmitting and receiving sources, the developments have mainly focused on the configuration of air-coupled radar antennas. However, some studies have opted for a processing methodology with near-field conditions so as to improve the resolution of the inverted signals, while allowing a better penetration depth of the waves in the medium. The choice to work with the strong assumption of near-field conditions was motivated by the desire to avoid edge effects due to multiple reflections not considered in the inverse models.

6.5.3.1. *Forward model and inverse problem: EM wave propagation models applied to dispersive multilayered media*

Lambot et al. (2004) and Guan et al. (2017) proposed an analytical model of antennas combined with the zero-offset Green's function to describe the propagation of EM waves in planar-stratified media. This model is valid under the conditions of a constant distribution of the backscattered field with respect to the aperture of the antenna, which is particularly true under far field conditions (homogeneous field). Subsequently, this model was generalized to near-field conditions for which the field distribution is variable (Lambot and André 2014). In this case, the multi-offset Green functions are used. Furthermore, Guan et al. (2017) used a power-frequency-law model to describe the frequency dependence of the dielectric permittivity of the dispersive material. However, the results of the work by Ihamouten et al. (2012) showed the interest of using the variants of Jonscher's model, which proved its efficiency in describing the dielectric response of highly dispersive materials with a limited number of parameters (2–4).

Through the various forward analytical or numerical models (Lambot et al. 2004; Guan et al. 2017; Liu et al. 2018) developed in recent years, it is possible to describe the propagation of radar waves in multilayered and dispersive media with physical and hydrical gradients. However, a robust inverse approach is needed to determine the problem parameters in terms of ND observables from the experimental data. The objective of the inversion is to reach the best set of the model parameters, which minimizes the difference between the analytical/numerical signals and the reference signals (experimental), while keeping a computational and especially physical consistency. There are three main types of inversion: amplitude inversion, phase inversion, complete waveform inversion. The EM waveform inversion has been studied and shown remarkable performances during the last few years (Lambot et al. 2004; van der Kruk et al. 2014; Lambot and André 2014). It makes full use of all the waveform information and can therefore provide a better result in front of coupling effects (even sometimes confusion) between the dielectric and the geometric characteristics of the propagation media. In addition, inversion in the frequency domain has most often been adopted because the complex experimental data are obtained in frequency and are raw processed directly without the contribution of filtering biases and/or transformations. However, at the end of the inversion process, the inverse Fourier transform is only used to help analyze the time signals for specialists in the GPR technical community.

The studied inverse approach requires a robust and stable algorithm for the optimal search of the (Green's function) forward model parameters minimizing the objective function. It is a global nonlinear optimization problem that could be solved by many numerical resolution algorithms. Conventional algorithms (gradient method, search for the extremum by climbing, ...) are fast, but have the disadvantage of converging prematurely toward local minima following the inappropriate choice of initial values. Indeed, it has been noted in various research that these algorithms are easily trapped in local areas that do not correspond to the solution. Some improved variants of these conventional search algorithms were also tested (Kirkpatrick et al. 1983; Glover 1986), but the convergence toward the global minima was still strongly dependent on the boundary conditions and the initialization of the iterative computation. However, the genetic algorithm (GA), a search algorithm that corresponds to the process of natural selection, has proven to be very effective in solving inverse problems. It generates solutions to nonlinear optimization problems using inheritance, mutation, selection, and crossover. The calculation process of this algorithm is naturally parallel, which considerably reduces the overall time cost of calculation in comparison with conventional algorithms. Moreover, its mutation process avoids a premature convergence toward the local minima, despite a non-optimized parametrization of the initial values. Recent research has therefore opted for the integration of this GA in the full waveform inversion method. For more details on the parametrization of this algorithm, we suggest that the reader refers to the PhD thesis by Guan (2017).

Finally, Figure 6.12 illustrates the general process of radar full waveform inversion, using the example of the zero-offset Green's function to describe the propagation of EM waves in planar-stratified media at the radar frequency band.

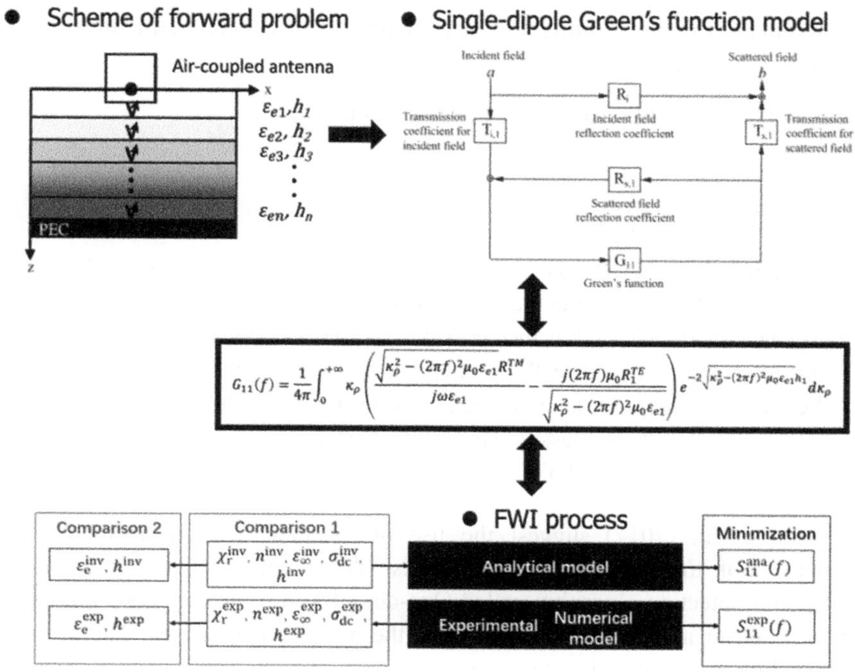

Figure 6.12. *Global radar full waveform inversion process for estimating dielectric and geometric parameters of planar layered media (Guan 2017). For a color version of this figure, see www.iste.co.uk/serhir/radar.zip*

Nowadays, research and expertise tasks related to processing by EM full waveform inversion allow us to respond by radar imaging to several unresolved or partially resolved objectives in the operational field of the evaluation of civil engineering structures:

– thickness estimation of thin and very thin continuous pavement layers;

– location and characterization of pavement layer debonding;

– estimation of the tack coat quality in pavement layers;

– continuous estimation of the wearing course compaction (new pavement or recent surfacing);

– assessment of the pavement layer roughness;

– estimation of water content gradients in concrete and building materials;

– automatic trouble spotting and water mapping of concrete structures under a waterproofing screed;

– underground utility network characterization.

6.5.4. *Machine learning processing*

The interpretation of GPR data is often complex and needs to be handled by experts. The application of artificial intelligence on radar data can provide an automatic solution to realize data interpretation. It is then possible to carry out anomaly detection, and even go as far as its classification or the estimation of geometric parameters by regression (Rasol et al. 2022; Cano-Ortiz et al. 2022). For example, machine learning techniques can statistically detect or classify surveyed areas and be a good complement to high-resolution processing techniques for crack and debonding.

Todkar et al. (2017) suggest the use of machine learning methods as an automatic tool to detect horizontal cracks (debonding). Todkar et al. (2021) developed a detection algorithm called One-class SVM to detect debonding, as the training is faster and only needs data from the non-debonding area. The algorithm was tested on modeled and experimental data sets, and the debonding detection rate remained over 0.95. Such an algorithm can be applied on different pavement structures and with different types of radar data. The constraint is to know a healthy area for learning, otherwise unknown cracks could be assumed to be healthy.

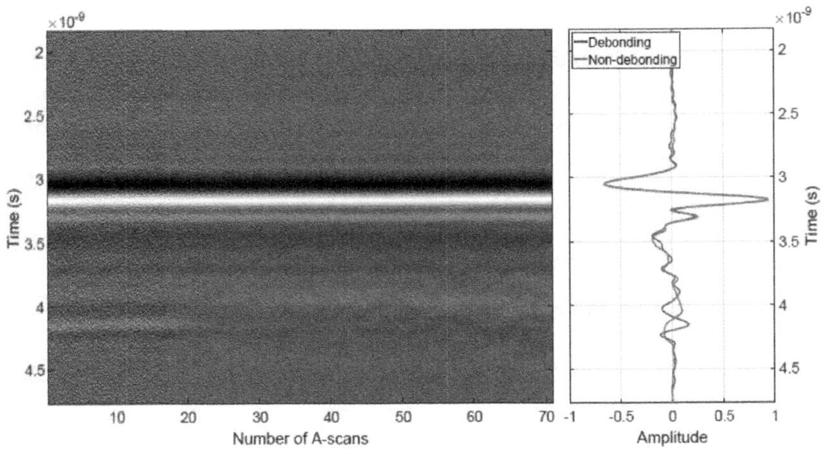

Figure 6.13. *Radar image with (on the left) and without (on the right) debonding. For a color version of this figure, see www.iste.co.uk/serhir/radar.zip*

This algorithm is very good for detecting defects such as cracks, but it does not allow us to classify different kinds of defects. It is therefore very useful for identifying such anomalies with complementary techniques, such as high-resolution methods or classification using artificial intelligence, and then complete the diagnosis.

Another more complex example concerns the characterization of pavement tack coat. The physical and geometrical characteristics of the problem are complex to apprehend according to Andreoli et al. (2022). As for debonding, machine learning can be used. The problem is to determine whether the emulsion has been applied, and moreover if there is an under, correct or over dosage. A two class SVM algorithm was used to determine the absence or not of emulsion on modeled and experimental data. The algorithm shows performances around 0.9, which is acceptable and demonstrates the electromagnetic sensitivity to the presence or not of emulsion (Figure 6.14).

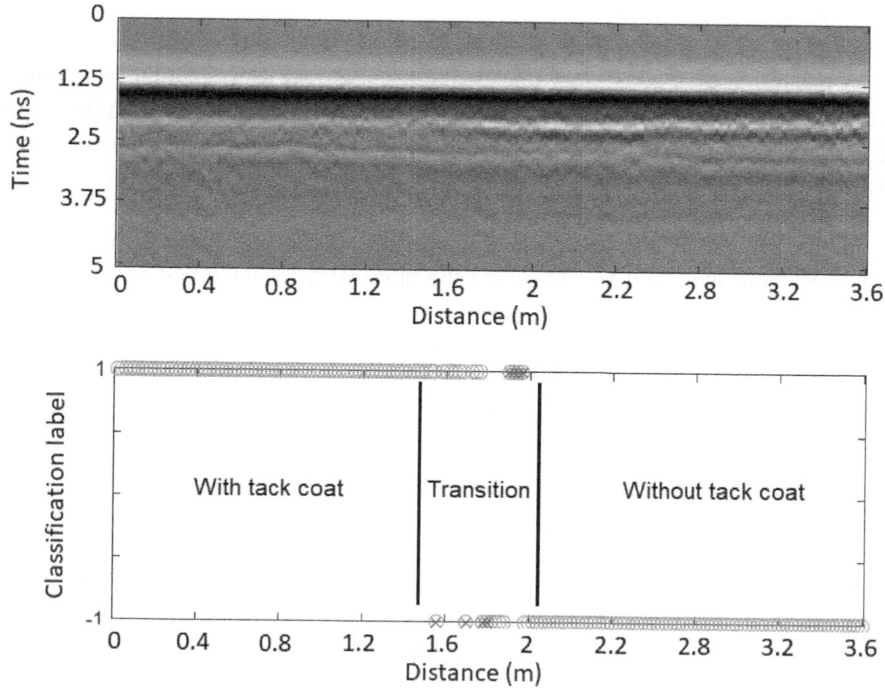

Figure 6.14. *Pavement tack coat detection using two-class SVM algorithm. For a color version of this figure, see www.iste.co.uk/serhir/radar.zip*

The control of the presence or not of emulsion, and the estimation of the quantity of emulsion are often linked to the acceptance of works on new pavements or following their rehabilitation, which does not allow the approach by detection of an anomaly like before (the assumption of a healthy zone cannot be considered). After demonstrating the interest of this method, it remains to be able to generalize the approach to become independent of the acquisition equipment, as well as of the pavement structure.

Methods based on artificial intelligence are mainly based on complete and representative data sets which are complicated to obtain, especially with ground truth. Those data are also dependent of sensors used. Therefore, it is important to integrate the physics behind the data to have an artificially intelligent algorithm, which does not depend on the sensors or materials, as was done in Todkar et al. (2021).

6.5.5. *New orientations*

In the frame of less advanced academic research, we can mention two studies from the French ANR project, named ACIMP. The first one is focused on the effect of the layer interface rugosity on the reflection of radar waves, and the second is focused on the use of GB-SAR for road investigation.

In the context of the detection of embedded cracks, in order to help in the detection, there is a need to understand the physical mechanisms of scattering and propagation that occur. Knowing that the frequency range of analysis is wide (typically, 0.5–20 GHz), the interfaces separating the different layers cannot always be considered as perfectly flat. Indeed, as soon as the height variations of the considered interfaces are greater than about one-tenth of the electromagnetic wavelength, they can no longer be neglected: for a maximum frequency of about 20 GHz, height variations of the upper interface about 2 mm cannot be neglected. This fact is reinforced for the inner and lower interfaces, where the electromagnetic wavelength to be considered is that of the considered overlying medium, being less than that in the vacuum. Furthermore, the scattering in transmission through the different rough interfaces must also be considered in the roughness evaluation (Pinel et al. 2011), which again increases the electromagnetic roughness to be considered.

This configuration greatly complicates the resolution of the problem, not only for the electromagnetic modeling, but also for constructing the signal model. Indeed, once the electromagnetic problem is resolved, it is necessary to incorporate the electromagnetic roughness of the interfaces in the signal model in order to enable an accurate signal processing (detection, estimation, etc.). In Songolo et al. (2022), the generalized pile method has been successfully implemented to rigorously calculate the frequency and time responses of the scattered field for an object buried between two random rough surfaces. This method can be used to simulate the GRP signal of a debonding.

For the second study, compared to typical GPR acquisitions delivering 1D vertical reflectivity profiles along a transect, side-looking configurations permit the measurement of wide areas with a diversity of incidence angles, and advanced 2D or 3D imaging modes may be used to estimate surface roughness and texture, volumetric heterogeneities, humidity and dielectric properties, as well as the vertical structure of the scene. Synthetic on its principle, synthetic aperture radar (SAR) imaging is a coherent electromagnetic signal processing technique that is able to map the reflectivity of objects or environments using spatial diversity, that is, emitting and/or receiving electromagnetic waves from different positions. An equivalent array or aperture larger than the actual antenna system may be synthesized, whose narrower beam leads to an improved spatial resolution in the direction of motion. SAR systems operated in side-looking configurations generally combine the synthetic aperture principle with classical spectral diversity, that is, measuring time delays, in order to produce 2D (azimuth-range) maps of environments. Tomographic SAR (TomoSAR) is an extension of SAR imaging from 2D to 3D. In Wu et al. (2021), the authors obtained theoretical and experimental results for different radar configurations on several asphalt layers slabs. The first results have shown that forward-scattering SAR made with a ground-based SAR system seems to deliver high accuracy and efficiency results for road investigations and deserves further investigation.

6.6. Conclusion

For the last 20 years, important academic and innovative studies on radar methodologies have been developed to tend toward a reliable assessment of flexible pavements, in which the French RST has taken a notable place. These actions have combined impulse and SFR technologies with specific processing or imaging for poorly solved problems, such as the measurement of the thickness of very-thin asphalt layers, or the evaluation of the compactness and bonding of asphalt layers.

These promising results confirm that GPR has become an essential ND technique in road structure assessment and maintenance, and that it is important to continue to improve them.

6.7. Acknowledgements

The authors would like to gratefully acknowledge the different projects ANR-ACIMP (18-CE22-0020), COST Action TU1208, FISSURE and ECODEM research actions for their support.

6.8. References

AASHTO (2013). Designation R37-04. Standard practice for application of ground penetrating radar (GPR) to highways. American Association of State Highway and Transportation Officials, Washington, DC.

AFNOR (1997). Standard NF P94-062. Sols : reconnaissance et essais – Mesure de la masse volumique en place – Diagraphie à double sonde gamma. Association Française de Normalisation, Paris.

AFNOR (2012). Standard NF EN 12697-6. Mélanges bitumineux – Méthodes d'essai – Partie 6 : Détermination de la masse volumique apparente des éprouvettes bitumineuses. Association Française de Normalisation, Paris.

Al-Qadi, I.L. and Lahouar, S. (2005). Measuring layer thicknesses with GPR – Theory to practice. *Construction and Building Materials*, 19, 763–772.

Andreoli, G., Ihamouten, A., Buliuk, V., Fauchard, C., Jaufer, R., Dérobert, X. (2022a). Numerical parametric study to classify and estimate pavement characteristics using GPR and machine learning methods. *Data Science in Pavement Symposium Proceedings*, McLean, VA.

Andreoli, G., Ihamouten, A., Jaufer, R., Todkar, S., Guibert, D., Nguyen, M.L., Simonin, J.M., Tissier, J., Dérobert, X. (2022b). The use of machine learning method to classify electromagnetic characteristics of pavement tack coat. *GPR2022 Conference Proceedings*, Golden, CO.

Araujo, S., Delbreilh, L., Laguerre, L., Dumont, H., Dargent, E., Fauchard, C. (2016). Rock permittivity characterization and application of electromagnetic mixing models for density/compactness assessment of HMA by means of step-frequency radar. *Near Surface Geophysics*, 14(6), 551–562.

Araujo, S., Beaucamp, B., Delbreilh, L., Dargent, E., Fauchard, C. (2017). Compactness/density assessment of newly-paved highway containing recycled asphalt pavement by means of non-nuclear methods. *Construction and Building Materials*, 154, 1151–1163.

ASTM International (2007). D1188-07e1. Standard test method for bulk specific gravity and density of compacted bituminous mixtures using coated samples. American Society for Testing and Materials International, West Conshohocken, PA.

ASTM International (2011a). D2950/D2950M. Standard test method for density of bituminous concrete in place by nuclear methods. American Society for Testing and Materials International, West Conshohocken, PA.

ASTM International (2011b). D6432-11. Standard guide for using the surface ground penetrating radar method for subsurface investigation. American Society for Testing and Materials International, West Conshohocken, PA.

ASTM International (2015a). D6087-08. Standard test method for evaluating asphalt-covered concrete bridge decks using ground penetrating radar. American Society for Testing and Materials International, West Conshohocken, PA.

ASTM International (2015b). D4748-10. Standard test method for determining the thickness of bound pavement layers using short-pulse radar. American Society for Testing and Materials International, West Conshohocken, PA.

Baltzer, S., Pratt, D., Weligamage, J., Adamsen, J., Hildebrand, G. (2010). Continuous bearing capacity profile of 18,000 km Australian road network in five months. *TRB Cong.*, Washington, DC.

Benedetto, A. and Pajewski, L. (2015). *Civil Engineering Applications of Ground Penetrating Radar*. Springer, Berlin.

BRRC (2016). Methodologies for the use of ground-penetrating radar in pavement condition surveys. Belgian Road Research Centre, Method of Measurement, ME91/16.

Cano-Ortiz, S., Pascual-Muñoz, P., Castro-Fresno, D. (2022). Machine learning algorithms for monitoring pavement performance. *Automation in Construction*, 139, 104309.

Carrik Utsi, E. (2017). *Ground Penetrating Radar: Theory and Practice.* Elsevier Butterworth-Heinemann, Oxford.

Chahine, K., Baltazart, V., Wang, Y. (2010). Interpolation-based matrix pencil method for parameter estimation of dispersive media in civil engineering. *Signal Processing*, 90(8), 2567–2580.

Chahine, K., Baltazart, V., Wang, Y. (2011). Parameter estimation of damped power-law phase signals via a recursive and alternately projected matrix pencil method. *IEEE Transactions on Antennas and Propagation*, 59(4), 1207–1216.

COST Action TU1208 (2013–2018). Action TU1208: Civil Engineering Applications of Ground Penetrating Radar: COST Success Story [Online]. Available at: www.GPRadar.eu.

Daniels, D.J. (1996). *Ground Penetrating Radar*. IEE Radar, Sonar & Navigation, London.

Daniels, D.J. (2004). *Ground Penetrating Radar*, 2nd edition. IEE Radar, Sonar & Navigation, London.

Dérobert, X. (2003). Techniques radar appliquées au génie civil. Laboratoire central des ponts et chaussées, Paris.

Dérobert, X., Fauchard, C., Côte, P., Le Brusq, E., Guillanton, E., Dauvignac, J.Y., Pichot, C. (2001). Step frequency radar applied on thin road layers. *Journal of Applied Geophysics*, 47, 317–325.

Dérobert, X. et al. (2019). Guideline for flexible pavement investigation with ground penetrating radar (GPR). COST Action TU1208.

Dérobert, X., Baltazart, V., Simonin, J.-M., Durand, O., Norgeot, C., Todkar, S. (2020a). Radar database over large debonded areas into pavement structure. Gustave Eiffel University. doi: 10.25578/N0RXSQ.

Dérobert, X., Baltazart, V., Simonin, J.-M., Hui, H.Y., Ching, P.H. (2020b). Radar database over narrow debonded areas into pavement structure. Gustave Eiffel University. doi: 10. 25578/MM9Q0R.

Dérobert, X., Baltazart, V., Simonin, J.-M., Todkar, S.S., Norgeot, C., Hui, H.Y. (2021). GPR monitoring of artificial debonded pavement structures throughout its life cycle during accelerated pavement testing. *Remote Sensing*, 13(8).

Diamanti, N. and Redman, D. (2012). Field observations and numerical models of GPR response from vertical pavement cracks. *Applied Geophysics*, 81, 106–116.

Elseicy, A., Alonso-Diaz, A., Solla, M., Rasol, M., Santos-Assunçao, S. (2022). Combined use of GPR and other NDTs for road pavement assessment: An overview. *Remote Sensing*, 14, 4336.

EuroGPR Guideline (2016). Guidelines for pavement structural surveys. European GPR Association, GS1601.

Fauchard, C., Li, B., Ferrari, E., Laguerre, L., Derobert, X., Benjelloun, N., Mazari, B. (2009). Estimation of compaction of bituminous mixtures at microwave frequencies. *NDTCE'2009 Proceedings*, Nantes.

Fauchard, C., Beaucamp, B., Laguerre, L. (2015). Non-destructive assessment of hot mix asphalt compaction/density with a step-frequency radar: Case study on a newly paved road. *Near Surface Geophysics*, 13, 289–297.

Glover, F. (1986). Future paths for integer programming and links to artificial intelligence. *Computers and Operations Research*, 13(5), 533–549.

Guan, B. (2017). Caractérisation des matériaux du bâtiment par méthodes non destructives électromagnétiques : application au tuffeau. PhD thesis, Cerema/IFSTTAR – University of Nantes.

Guan, B., Ihamouten, A., Dérobert, X., Lambot, S., Villain, G. (2017). Near-field full-waveform inversion of ground-penetrating radar data to monitor water front in limestone. *IEEE Journal Selected Topics in Applied Earth Observations and Remote Sensing*, 10(10), 4328–4336.

Ihamouten, A., Villain, G., Dérobert, X. (2012). Complex permittivity frequency variations from multi-offset GPR data: Hydraulic concrete characterization. *IEEE Transactions on Instrumentation and Measurement*, 61, 1636–1648.

Ihamouten, A., Bosc, F., Guan, B., Le Bastard, C., Fauchard, C., Lambot, S., Dérobert, X. (2018). Full-waveform inversion using the step frequency radar in the GPR band to characterize the tack coat in hot-mix asphalt (HMA) layers of flexible pavements. *NDT & E International*, 95, 17–25.

Jol, H.M. (2009). *Ground Penetrating Radar: Theory and Applications*. Elsevier, Amsterdam and Boston, MA.

Kirkpatrick, S., Gelatt, C., Vecchi, M. (1983). Optimization by simulated annealing. *Science*, 220(4598), 671–680.

Lai, W.L., Dérobert, X., Annan, A.P. (2018). A review of ground penetrating radar application in civil engineering: A 30-year journey from locating, testing and evaluation to imaging and diagnosis. *NDT & E International*, 96, 58–78.

Lambot, S. and André, F. (2014). Full-wave modeling of near-field radar data for planar layered media reconstruction. *IEEE Transactions on Geoscience and Remote Sensing*, 52(5), 2295–2303.

Lambot, S., Slob, E., den Bosch, I.V., Stockbroeckx, B., Vanclooster, M. (2004). Modeling of ground-penetrating radar for accurate characterization of subsurface electric properties. *IEEE Transactions on Geoscience and Remote Sensing*, 42(11), 2555–2568.

Lambot, S, Giannopoulos, A., Pajewski, L., Sato, M., Slob, E. (2016). Advances in ground penetrating radar research and applications. *IEEE Journal of Selected Topics in Applied Earth Observations and Remote Sensing*, Special Issue, 9(1), 5–8, Wuhan.

Le, C., Wang, Y., Baltazart, V. (2008). Algorithmes haute résolution linéaires pour l'estimation des temps de retard dans un contexte de bruit quelconque. *Revue de Traitement de Signal*, 25(4–5), hal-00435932.

Le Bastard, C., Baltazart, V., Wang, Y., Saillard, J. (2007). Thin-pavement thickness estimation using GPR with high-resolution and super-resolution methods. *IEEE Transactions on Geoscience and Remote Sensing*, 45(8), 2511–2519.

Le Bastard, C., Wang, Y., Baltazart, V., Dérobert, X. (2014). Time delay and permittivity estimation by ground penetrating radar with support vector regression. *IEEE Transactions on Remote Sensing*, 11(4), 873–877.

Le Bastard, C., Pan, J., Wang, Y., Todkar, S.S., Baltazart, V., Pinel, N., Ihamouten, A., Dérobert, X., Bourlier, C. (2019). A linear prediction and support vector regression based debonding detection method using step-frequency ground penetrating radar. *Geoscience and Remote Sensing Letters*, 16(3), 367–371.

Liu, T., Klotzsche, A., Pondkule, M., Vereecken, H., Su, Y., van der Kruk, J. (2018). Radius estimation of subsurface cylindrical objects from ground-penetrating-radar data using full-waveform inversion. *Geophysics*, 83(6), H43–H54.

Loizos, A. and Plati, C. (2007). Accuracy of pavement thicknesses estimation using different ground penetrating radar analysis approaches. *NDT & E International*, 40, 147–157.

Mara Nord Project Recommendations (2012). The use of GPR in road rehabilitation projects. Interreg-Project IV/A/Nord [Online]. Available at: https://www.yumpu.com/user/maranord.ramk.fi.

Pinel, N., Le Bastard, C., Baltazart, V., Bourlier, C., Wang, Y. (2011). Influence of layer roughness for road survey by ground penetrating radar at nadir: Theoretical study. *IET Radar, Sonar & Navigation*, 5, 650–656.

Rasol, M., Pais, J.C., Pérez-Gracia, V., Solla, M., Fernandes, F.M., Fontul, S., Ayala-Cabrera, D., Schmidt, F., Assadollahi, H. (2022). GPR monitoring for road transport infrastructure: A systematic review and machine learning insights. *Construction and Building Materials*, 324, 126686.

Saarenketo, T. and Scullion, T. (2000). Road evaluation with ground penetrating radar. *Journal of Applied Geophysics*, 43, 119–138.

Sihvola, A.H. and Kong, J.A. (1988). Effective permittivity of dielectric mixtures. *IEEE Transactions on Geoscience and Remote Sensing*, 26, 420–429.

Simonin, J.M., Baltazart, V., Hornych, P., Kerzrého, J.P., Dérobert, X., Trichet, T., Durand, O., Alexandre, J., Joubert, A. (2012). Detection of debonding and vertical cracks with ND techniques during accelerated tests. *Proceedings of the 4th International Conference on Accelerated Pavement Testing*, Davis, CA.

Simonin, J.M., Kerzreho, J.P., Hornych, P., Gouy, T. (2013). Comparison of NDT methods to detect debonded interfaces. *Bearing Capacity of Roads, Railways & Airfields Conference Proceedings*, Trondheim.

Simonin, J.M., Hornych, P., Baltazart, V., Dérobert, X., Thibaut, E., Sala, J., Utsi, V. (2014). Case study of detection of artificial defects in an experimental pavement structure using 3D GPR. *GPR'2014*, Brussels.

Solla, M. and Lagüela, S. (2019). Recent advances in GPR imaging. *Remote Sensing*, Special Issue, Basel.

Solla, M., Lagüela, S., Gonzáles-Jorge, H., Arias, P. (2014). Approach to identify cracking in asphalt pavement using GPR and infrared thermographic methods: Preliminary findings. *NDT & E International*, 62, 55–65.

Solla, M., Pérz-Gracia, V., Fontul, S. (2021). A review of GPR application on transport infrastructures: Troubleshooting and best practices. *Remote Sensing*, 13, 672.

Songolo, M., Pinel, N., Bourlier, C. (2022). Full wave modeling of electromagnetic scattering by an object buried between two rough surfaces: Application to GPR. *Progress in Electromagnetics Research B*, 96, 133–152.

Sun, M., Wang, Y., Le Bastard, C., Pan, J., Ding, Y. (2017). Signal subspace smoothing technique for time delay estimation using MUSIC algorithm. *Sensors*, 17, 2868.

Sun, M., Le Bastard, C., Wang, Y., Pinel, N., Pan, J., Baltazart, V., Simonin, J.M., Dérobert, X. (2018). Time delay and interface roughness estimation using modified ESPRIT with interpolated spatial smoothing technique. *IEEE Transactions on Remote Sensing*, 56, 1475–1484.

Tankeu, B. (2023). High resolution time delay estimation methods for processing GPR data collected over debonded pavement structures. PhD dissertation, University of Nantes.

Tankeu, B., Baltazart, V., Wang, Y., Guilbert, D. (2021). Application of PUMA algorithm to time delay estimation for pavement survey by ground penetrating radar. *Remote Sensing*, 13, 3456.

Todkar, S., Le Bastard, C., Ihamouten, A., Baltazart, V., Dérobert, X., Fauchard, C., Guilbert, D., Bosc, F. (2017). Detection of debondings with ground penetrating radar using a machine learning method. *IWAGPR'2017*, Edinburgh.

Todkar, S., Le Bastard, C., Baltazart, V., Ihamouten, A., Dérobert, X. (2018). Comparative study of classification algorithms to detect interlayer debondings within pavement structures from step-frequency radar data. *IGARSS Proceedings*, Valencia.

Todkar, S., Le Bastard, C., Baltazart, V., Ihamouten, A., Dérobert, X. (2019). Performance assessment of SVM-based classification techniques for the detection of artificial debondings within pavement structures from step-frequency A-scan radar data. *NDT & E International*, 107, 102128.

Todkar, S., Baltazart, V., Ihamouten, A., Dérobert, X., Guilbert, D. (2021). One-class SVM based outlier detection strategy to detect thin interlayer debondings within pavement structures using ground penetrating radar data. *Journal of Applied Geophysics*, 192, 104392.

Van der Kruk, J.V., Kalogeropoulos, A., Hugenschmidt, J., Klotzsche, A., Busch, S., Vereecken, H. (2014). Full-waveform inversion of GPR data for civil engineering applications. *EGU General Assembly, Geophysical Research Abstracts*, Vienna.

Wu, R., Li, X., Li, J. (2002). Continuous pavement profiling with ground-penetrating radar. *IEE Proceedings – Radar, Sonar and Navigation (IET)*, 149, 183–193.

Wu, M., Ferro-Famil, L., Wang, Y. (2021). Comparison of radar imaging configurations for the characterization and diagnosis of roadways. *IEEE International Geoscience and Remote Sensing Symposium*, Brussels.

Yamada, H., Ohmiya, M., Ogawa, Y., Itoh, K. (1991). Superresolution techniques for time-domain measurements with a network analyzer. *IEEE Transactions on Antennas and Propagation*, 39, 177–183.

7

GPR for Tree Roots Reconstruction under Heterogeneous Soil Conditions

Abderrahmane ABOUDOURIB[1], Mohammed SERHIR[2] and Dominique LESSELIER[3]

[1] *Université Paris-Saclay, CNRS, CentraleSupélec, Laboratoire de Génie Electrique et Electronique de Paris, Gif-sur-Yvette, France*

[2] *Université Paris-Saclay, CentraleSupélec, SONDRA, Gif-sur-Yvette, France*

[3] *Université Paris-Saclay, CNRS, CentraleSupélec, Laboratoire des Signaux et Systèmes, Gif-sur-Yvette, France*

7.1. Introduction

Identified in 2007 in New York State, the oldest fossil tree is over 385 million years old. Trees and forests have an undeniable contribution to the appearance of life on the planet Earth. Nowadays, there is a growing awareness of the need to protect the environment by preventing the destruction of forest areas. Appropriate management of this natural heritage is more necessary than ever.

Of all organs of the tree, the roots are vitally important because they perform functions that are crucial to trees and ecosystems: they anchor the tree, absorb water and nutrients, store carbon and change the properties of the surrounding soil.

For a color version of all the figures in this chapter, see www.iste.co.uk/serhir/radar.zip.

Ground Penetrating Radar,
coordinated by Mohammed SERHIR and Dominique LESSELIER. © ISTE Ltd 2024.

However, even though roots account for between about 10% and 65% of the tree's total biomass, they are generally found below the soil surface, limiting our understanding of the development of the root system and its interaction with the soil. Site conditions, soil composition, water, mineral content and other factors are responsible for the development of the tree's root structure. The functioning of the tree root system (growth, architecture, interactions with the soil, etc.) and the estimation of root-related parameters (size, biomass, density, etc.) are of paramount importance and call for strong research.

In civil engineering, tree roots destructively interact with man-made infrastructures, causing mechanical damage to sidewalks, buildings, sewers, or enlarging pipe fissures (Randrup et al. 2001; Daniels 2005). Roots are capable of applying significant radial pressures (up to 8×105 N/m^2) damaging sidewalks, pipes and walls of nearby buildings.

In the global warming context, elevated CO_2 concentration combined with accelerated land exploitation significantly alters the sustainability of the Earth's ecosystem. Understanding the root biomass behavior and the factors that regulate it is part of the solution to modeling environmental dynamics. Tree roots are a key structural component of trees significantly contributing to the ecosystem carbon budget, accounting for 20–40% of the total forest carbon storage (Brunner and Godbold 2007). Coarse root biomass correlates positively with elevated CO_2 concentration (Miller et al. 2006), indicating that coarse roots can absorb and store excessive CO_2 from the atmosphere (Stover et al. 2007).

Roots are inherently subterranean (with the exception of some species) and respond to local environmental changes (Day et al. 2010). They are opportunistic and grow mostly or at least better where favorable conditions allow. Their architecture also changes with growth over time. Usual methods for studying root systems, like excavation, uprooting and the air-spade method, are destructive, which is making repeatability impossible (Guo et al. 2013).

In contrast, ground-penetrating radar (GPR) as an intrinsically non-destructive method can locate underground objects (e.g. landmines, utilities, and cables) (Daniels 2005), and it is increasingly utilized for imaging tree root systems (Guo et al. 2013), root detection (Dannoura et al. 2008), estimation of root orientation (Tanikawa et al. 2013; Qixin et al. 2018) and biomass quantification (Cui et al. 2013). The limiting factors of root detection (Hirano et al. 2009) are then especially important, namely root diameter and water content, depth, orientation, root distribution and soil properties (composition and moisture).

Another aspect of this research field is three-dimensional root systems mapping. The topological structure of tree roots and their growth dynamics are indeed an indicator of environmental stresses (Miller 1986). As an example, in Ohashi et al.

(2019), such a root system architecture (RSA) is analyzed using GPR and an algorithm based on root point coordinates. In Wu et al. (2014), a reconstruction based on root points and employing a series of concentric circular survey lines is proposed, a key limitation being that root points are picked out from visual identification, which is both laborious and time consuming.

In the present chapter, a processing framework for tree-root reconstruction under heterogeneous conditions (Aboudourib et al. 2021) using GPR is proposed and investigated. The core of this GPR data processing is based on the matched filter (Leuschen and Plumb 2001) as a wave migration technique. Heterogeneous soil conditions then make GPR processing challenging. The framework has been tested considering a heterogeneous medium based on an FDTD electromagnetic simulation scenario. The SVD-based clutter removal approach has been chosen since it clearly outperforms other techniques and it offers flexibility in clutter reduction. There exist several solutions to reduce the effect of background noise, mean/median subtraction, high-pass filtering, shifted and scaled background removal and principal component analysis (PCA) approaches (Persico and Soldovieri 2008; De Coster and Lambot 2019; Chen and Qian 2011).

The relative permittivity of the heterogeneous soil itself is inaccessible. But GPR is a delay-based detection technique where root depth is to be decided for a known relative permittivity of the media of propagation. So, during in situ GPR surveys, operators perform additional measurements in order to assess the soil relative permittivity by empirical methods like borehole GPR, average envelope amplitude, frequency shift methods or full-wave inversion techniques (Lambot et al. 2006; Liu et al. 2017). Otherwise, methods using dedicated sensors are based on a soil moisture analysis. Indeed, using models such as Topp's one (Topp et al. 1980), the measured soil moisture can be translated into a relative dielectric constant.

However, alternatives do exist. Hyperbola fitting is a processing method that can be employed to extract information about the propagation medium (Mertens et al. 2016; Dou et al. 2017). Yielding the relative permittivity of the heterogeneous soil based on GPR hyperbola parameters, the so-called randomized hough transform (RHT) (Li et al. 2016) is proposed. This method is efficient and semi-automatic. Tested with homogeneous soils (the soil permittivity being known), a good correlation between actual and estimated values is reached, which does substantiate its use.

The GPR processing steps are illustrated by electromagnetic (EM) simulations in complex scenarios. A realistic dichotomous root prototype is tested using the FDTD open-source simulations environment gprMax software (Warren et al. 2016). Heterogeneous soils can be investigated by examining the effect of increasing soil moisture with parametric compositions (sandy and clay soils). To provide experimental insight, controlled laboratory experiments using an in-house bi-static

GPR system are then run for a simplified 3D root structure reconstruction from GPR data.

7.2. Materials, definitions and methods

7.2.1. *Root description and modeling*

This section is devoted to detailing the models/hypothesis adopted for the FDTD EM simulations in gprMax (Warren et al. 2016) to model the wave propagation for GPR purposes.

Roots exhibit complex morphological structures composed of the stele, cortex and epidermis. Yet, to the best of our knowledge, no study appears to have addressed the radar signal reflected from the interfaces between the different parts of the internal root. Consequently, root modeling is simplified by considering roots as a complex mixture of homogeneous cellular material, wood, water and air. Specifically, hydrogen (H) bonded water to the hydroxide (OH) cell wall is defined as *bond water*. The remaining water held in the cell wall cavity with the ability to move freely is defined as *free water*. This is sketched in Figure 7.1.

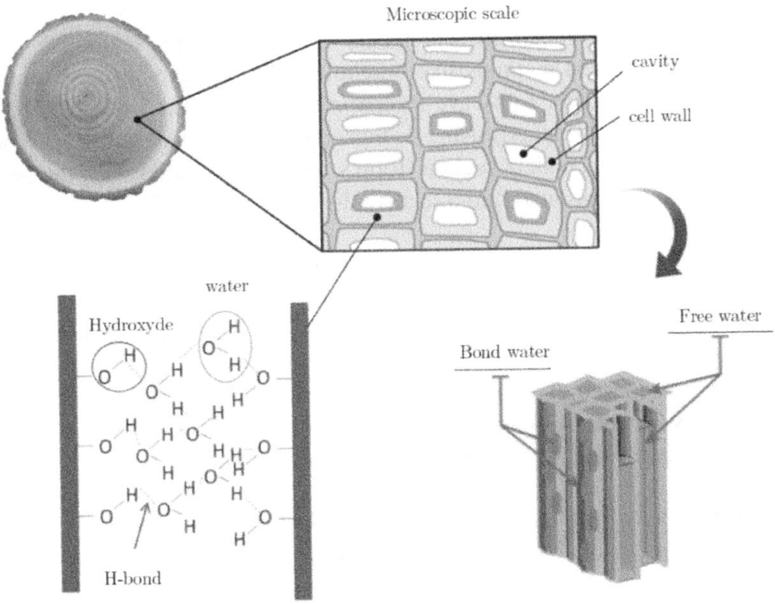

Figure 7.1. *Bond and free water illustration in the wood structure*

The fiber saturation point (FSP) is defined in the literature as the stage in the wood drying process at which the moisture content is 30%. Only bound water remains in the cell walls by absorption. Free water has been removed from the cell cavities (Berry and Roderick 2005). The water content around the FSP greatly affects the physical and mechanical properties of the wood. Thus, depending upon percentages of free and bound water, the gravimetric water content of the roots θ_g could be divided into two regions, each defining a different composition of the root.

$$\begin{aligned}\theta_g &= \frac{m_{water}}{m_{wood}} \times 100(\%)\\ \theta_g &= \frac{m_{wet} - m_{dry}}{m_{dry}} \times 100(\%)\end{aligned} \quad [7.1]$$

In the above, θ_g is the gravimetric water content of the root, m_{water} is the weight of the total water and m_{wood} is the weight of the dry root. In practice, measuring the wood gravimetric water content requires drying the sample under study to a constant mass, and then comparing the mass of the sample before (m_{wet}) and after (m_{dry}) drying.

The water content is related to the root dielectric constant (Al Hagrey 2007). The correlation between these two quantities is not the same below and above FSP. For roots above, the Maxwell–Garnett model can be applied to compute the root dielectric constant (Paz et al. 2011). This model is based on the mixing theory, assuming that the inclusion of a given shape is distributed in a host environment. In a root structure, the air is the host, while free water and solid solution (mixture of wood and bound water) are inclusions. The air is an electrodynamically isotropic host and the two inclusions are randomly oriented ellipsoids. The Maxwell–Garnett model equation can be derived as (Koledintseva et al. 2004):

$$\varepsilon_m = \varepsilon_a + \frac{\frac{1}{3}\sum_{i=0}^{2}\theta_i(\varepsilon_i - \varepsilon_a)\sum_{j=0}^{3}\frac{\varepsilon_a}{\varepsilon_a + N_{i,j}(\varepsilon_i - \varepsilon_a)}}{1 - \frac{1}{3}\sum_{i=0}^{2}\theta_i(\varepsilon_i - \varepsilon_a)\sum_{j=0}^{3}\frac{N_{i,j}}{\varepsilon_a + N_{i,j}(\varepsilon_i - \varepsilon_a)}} \quad [7.2]$$

where ε_m is the dielectric constant of the mixture, ε_a is the dielectric constant of the host medium, ε_i is the dielectric constant of the ith inclusion ($i = 1, 2$ in this case, for free water and solid solution, respectively), θ_i is the volumetric fraction of the ith inclusion and $N_{i,j}$ is the depolarization factor of the i inclusion in the direction of the j semi-axis ($j = 1, 2, 3$ for semi-axes in the x, y, z directions).

The values to be considered for the root model are $\varepsilon_a = 1$ for air, $\varepsilon_1 = 81$ for free water and $\varepsilon_2 = 10$ for the solid solution (from Paz et al. (2011)). Roots are considered as being circular cylinders. The values of depolarization factors of thin cylinders as reported, for example, in Koledintseva et al. (2004) are $N_1 = 0.5$ (x axis), $N_1 = 0$ (y axis) and $N_1 = 0.5$ (z axis). Regarding the volumetric fractions of the inclusions (θ_1 for free water and θ_2 for the solid solution), they can be calculated on the basis of the root composition model (RCM) proposed in Li et al. (2013). The RCM requires as input the dry weight of the root.

Figure 7.2 shows two different empirical models that have been identified in the literature describing the relationship between the diameter (d) and the dry root weight (m_{wood}). In the Li et al. (2013) model, it is a regression based on the measurement involving 254 roots belonging to 26 trees of five different species. The diameters of the studied roots evolve from 0.2 to 3.9 cm. However, the regression model presented in Hirano et al. (2012) is based on the measurement of 56 root samples belonging to the same species (pine). From the comparison provided by Figure 7.2, these two empirical models however seem to be in very strong agreement.

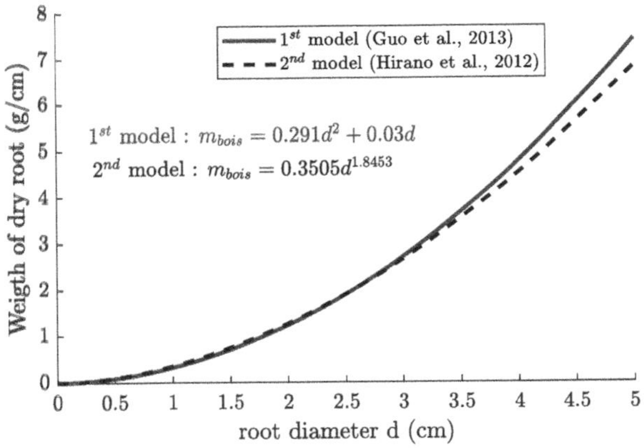

Figure 7.2. *Comparison of the empirical models proposed in Li et al. (2013) and Hirano et al. (2012)*

Considering the model presented in Li et al. (2013), the calculation of the volumetric fractions of free water and a solid solution is given by

$$\begin{aligned}\theta_1 &= m_{bois}\frac{4}{\pi d^2}(\theta_g - 0.3) = \left(0.370 + \frac{0.038}{d}\right)(\theta_g - 0.3)\\ \theta_2 &= 0.95\, m_{bois}\frac{4}{\pi d^2} = 0.352 + \frac{0.036}{d}\end{aligned} \qquad [7.3]$$

The resulting relative permittivity of the root for a given diameter and gravimetric water content is

$$\varepsilon_m = 1 + \frac{27.63\theta_1 + 4.09\theta_2}{1 - 0.65\theta_1 - 0.54\theta_2} \qquad [7.4]$$

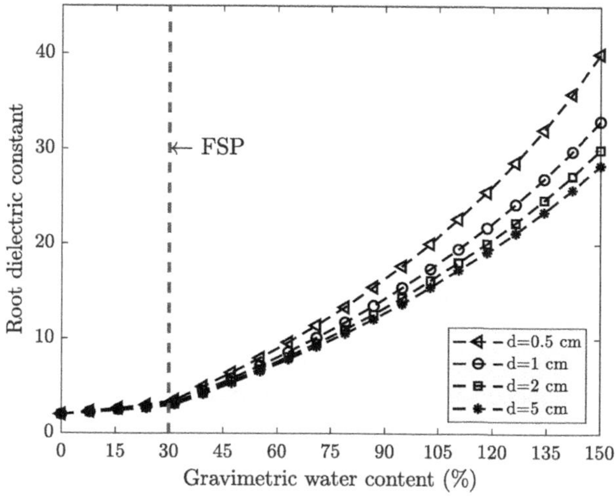

Figure 7.3. *Dielectric permittivity of roots as a function of their gravimetric water content for different diameters*

Figure 7.3 shows the dielectric constant of the root as a function of its gravimetric water content. The relative permittivity of the root decreases with increasing diameters. As the roots develop, their diameters expand and their porosity increases. Note that the root water content is sometimes expressed as a volumetric fraction (θ_v) instead of a gravimetric one (θ_g). These two quantities are related by the specific gravity (SG) of the root wood as expressed as $\theta_v = \theta_g \times SG$. Passing from volumetric to gravimetric water content requires the specific gravity of the root wood under test. A list of specific values according to tree species is particularly found on the Wagner Meters website (see: https://www.wagnermeters.com/specific-gravity/). For roots below FSP, the Maxwell–Garnett model cannot be used directly since the dielectric permittivity of bound water is governed by more complex laws. Dry roots are composed exclusively of dry woody material and air. For dry roots below FSP, the dielectric constant follows linear interpolation.

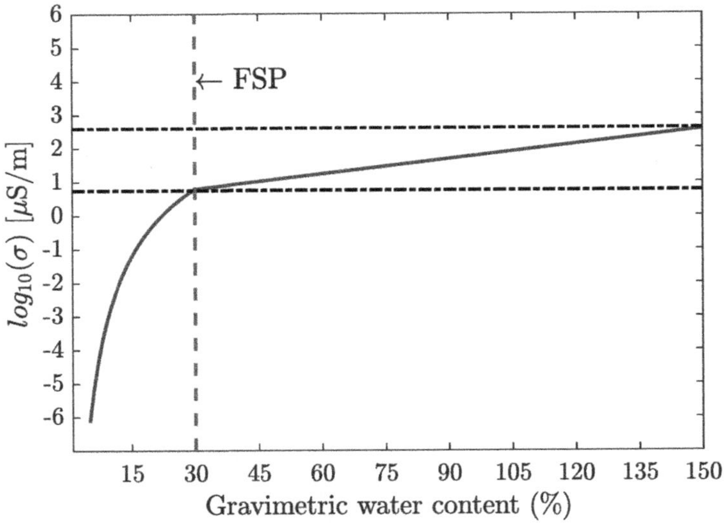

Figure 7.4. *Root electrical conductivity as a function of the root gravimetric water content*

In Glass and Zelinka (2010), the electric conductivity of wood depends on its moisture content. For roots below FSP, a relationship established by Straube et al. (2002) is used to estimate the dielectric conductivity of roots:

$$\log_{10}(\theta_g) = 2.99 - 2.113\left(\log_{10}\left(\log_{10}\left(\frac{1}{\sigma}\right)\right)\right) \quad [7.5]$$

Figure 7.4 shows that below FSP, the electrical conductivity decreases with increasing moisture content. When the FSP is approached, the electric conductivity increases almost linearly with water content. Therefore, for roots above FSP, the electrical conductivity is assumed to increase linearly with the increasing water content, and this assumption is confirmed by the results in Glass and Zelinka (2010).

In order to achieve realistic electromagnetic modeling of tree root reconstruction under heterogeneous conditions, root and soil models are based on established definitions found in the literature. Indeed, several parameters are taken into account to sketch the root model presented in Figure 7.5. A dichotomous branching pattern is considered as one of the most common types of branching in tree roots (Oppelt et al. 2001). The geometrical structure and the roots positioning in their system (topology) have been widely studied over the last decades. The architecture of the roots is set up according to a more or less complex branching process. The roots are not connected to each other randomly but according to a hierarchical pattern. Two main branching patterns have been observed in trees: dichotomous and chevron. In a branching diagram, the position of a root is often defined by its "branching order". The first order corresponds to the first emitted root and the last one to the most distant in the structure. In chevron architectures, branching is confined to the main axis, whereas a dichotomous system has equal branching on all links.

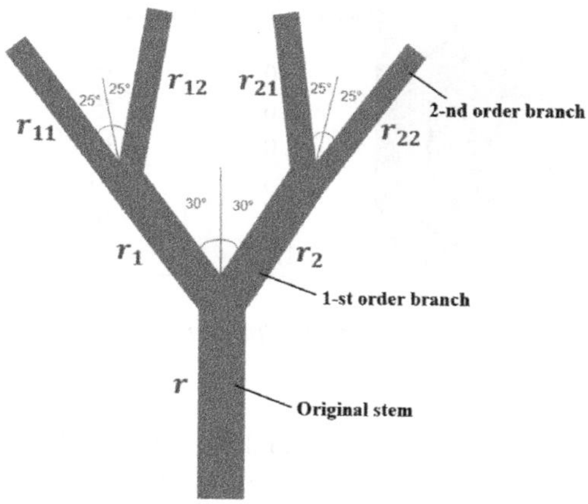

Figure 7.5. *The 3D dichotomous root model used for numerical simulation*

We have adopted the dichotomous hypothesis in this analysis. The original stem divides into two, each forming a branch. For the chosen model, only the two first branching orders are considered for simplicity. With regard to branch diameters, those follow a model that goes back to Leonardo da Vinci (Oppelt et al. 2001) that describes the relationship between parents and daughters. Assuming roots as concentric cylinders, the cross-sectional area of every parent branch is equal to the sum of the cross-sectional areas of the daughter segments. This can be expressed in terms of diameters as

$$d^2 = \sum_{i=1}^{n} d_i^2 \qquad [7.6]$$

where d stands for the parent branch diameter, d_i are the diameters of the daughter segments and n is the number of branches. The root model starts from an original stem with a diameter of 4 cm, and the daughter segment diameters follow equation [7.6]. The daughter diameters are listed in Table 7.1.

An important factor to be considered for root modeling is the relative angle between the roots and the GPR probing direction. The root orientation indeed influences the detection accuracy (Qixin et al. 2018). In Tanikawa et al. (2013), it is reported that when the cross angle between root orientation and probing line is beyond the detection range (45°–135°), it is difficult to identify roots on GPR radargrams. In light of this analysis, and taking into consideration the probing scenario that will be used in the simulation, the maximum cross angle has been set to 60°.

Root segment	Diameter (cm)	Length (cm)
r	4	40
r_1	3.0	35
r_2	2.6	35
r_{11}	2.0	30
r_{12}	2.3	30
r_{21}	1.9	30
r_{22}	1.6	30

Table 7.1. *Diameters and lengths of root model segments*

Here, the relative permittivity ε_{root} and the electrical conductivity σ_{root} are defined based on the model presented above. In Aboudourib et al. (2019) the root detection is linked with its volumetric water content. In the following, the proposed root model is defined with a volumetric water content of 50%, which corresponds to a relative permittivity $\varepsilon_{root} = 24$ and an electrical conductivity $\sigma_{root} = 0.63$ mS/m. The magnetic permeability and magnetic conductivity are set to 1 and 0, respectively.

7.2.2. *Dielectric models for heterogeneous soil*

The tree roots' dielectric properties modeling has been presented above. We are interested in the present section about the dielectric properties of natural soils as encountered in the literature. Two families of models are usually considered, the ones which are dependent on the volumetric content of the components (independent of the frequency) and the ones which take into account the relaxation phenomenon (frequency dependent).

The Topp et al. (1980) model is the first one. It is widely used for GPR purposes that describe the dielectric response of a quasi-homogeneous soil as a function of the soil moisture. The model is inappropriate for clays and soils rich in organic matter, but is quite suitable for sandy and silty soils over a wide range of moisture content (up to 60%) in the frequency band (1 MHz–1 GHz). The second model, the "Complex Refractive Index Model" (CRIM), exploits the volume mixing law and considers that the propagation time of an electromagnetic pulse in a medium is the sum of the propagation time within each of its components. The CRIM works well in the frequency range (10 MHz–1 GHz). It has the advantage of being a volumetric model which requires the knowledge of each material permittivity and its fractional volume contribution. Its general formula is proposed in Wharton et al. (1980). Overall, these two models lead to comparable results as is shown in Figure 7.6.

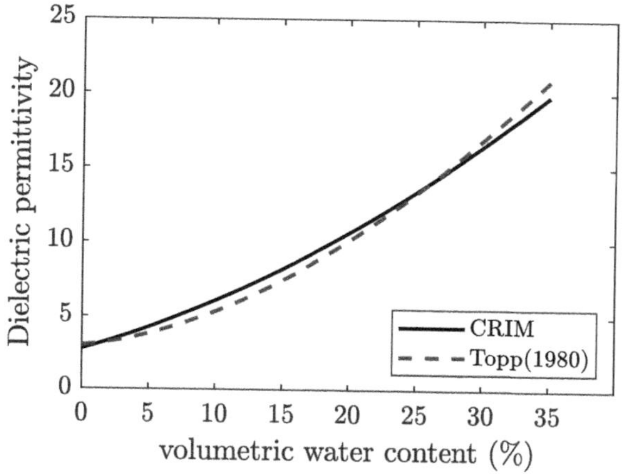

Figure 7.6. *Relative soil permittivity modeled by Topp et al. (1980) and CRIM for a sandy soil*

For tree root reconstruction under the heterogeneous soil hypothesis, we have used here the semi-empirical model of Peplinski (Dobson et al. 1985; Peplinski et al. 1995) combined with a fractal description of the soil heterogeneity (Turcotte 1997), which are both being implemented in the gprMax software and readily used.

In Dobson et al. (1985), a model is proposed that relates the soil dielectric value $\varepsilon = \varepsilon' + i\varepsilon''$ to the clay mass fraction (C), the sand mass fraction (S), the moisture content f_w, the soil particle density ρ_s and the bulk density ρ_a. This approach is accurate in the frequency range typically of (1.4 – –18 GHz). It underestimates ε' and overestimates ε'' for frequencies below 1.4 GHz. To overcome this, Peplinski et al. (1995) proposed a correction to the semi-empirical model originally suggested by Dobson et al. (1985). The updated model is now valid for a frequency range of (0.3––1.3 GHz). All of these models are based on the volume mixing law considering a three-phase soil composed of soil particles, water inclusions and air. In contrast to Topp's model, Peplinski's semi-empirical model has the advantage also to yield estimates of soil conductivity.

Figure 7.7 illustrates a comparison (real part) between relative permittivities provided by Peplinski's and Topp's models. The soil properties considered in Peplinski's model are $\rho_a = 1\ \text{g/cm}^3$ and $\rho_s = 2.66\ \text{g/cm}^3$. The black curve is for a clayey sand ($S = 0.1$ and $C = 0.9$). The permittivity curve appears in strong agreement with Topp's results for this soil type. However, as the sand fraction (S) increases, the gap between the two models becomes more important (check on the green and blue curves). For sandy soils, the semi-empirical model in effect overestimates the real permittivity compared with the Topp model.

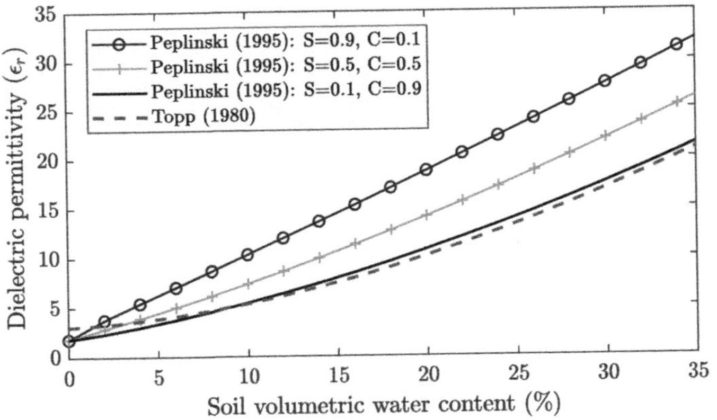

Figure 7.7. *Relative dielectric permittivity calculated at 900 MHz using the models of Peplinski et al. (1995) and Topp et al. (1980). The soil properties are defined by $\rho_a = 1$ g/cm^3 and $\rho_s = 2.66$ g/cm^3*

In the following simulations, a sandy soil with $S = 0.7$, $C = 0.3$, $\rho_s = 2.66$ g/cm^3, $\rho_b = 2$ g/cm^3, and $f_w = 1 - -10\%$ is assumed. The corresponding soil including the root model is presented in Figure 7.8. Regarding the heterogeneous distribution of the soil, it is modeled using the fractal correlated noise (Turcotte 1997). As it is stated in Giannakis et al. (2016), soil-related environmental properties obey fractal laws. Fractals come from the convolution of Gaussian noise with the inverse Fourier transform of $1/k^b$, where k is the wavenumber and b is a constant related to the fractal dimension (Giannakis et al. 2015). Here, we consider a soil composed of 20 elements that are stochastically distributed using a fractal dimension β of 1.5 (β between 0 and 3). The heterogeneities are distributed with the same weight along the three directions Ox, Oy, and Oz. Figure 7.8 shows the stochastic distribution of the heterogeneous soil used in the simulations. More details about the semi-empirical model of Peplinski's soil and the fractal model can be found in Giannakis et al. (2016).

7.2.3. *Forward electromagnetic simulation*

The scenario considered in gprMax is sketched in Figure 7.8. The 3D simulation is run in a parallelepipedic domain sized 2 m × 1 m × 0.6 m considering a bistatic configuration involving y-polarized dipoles in a common-offset GPR acquisition mode. The separation between transmitter and receiver dipoles and the distance between antennas and the soil surface is set to 0.10 m and 0.05 m, respectively. The excitation waveform $s(t)$ is assumed to be a Ricker pulse with a central frequency of 1 GHz:

$$s(t) = -2\sqrt{\frac{e}{2\xi}}\xi(t-\tau)\exp(-\xi(t-\tau)^2) \qquad [7.7]$$

where $\xi = 2\pi^2 {f_c}^2$, $\tau = 1/f_c$ and f_c is the central frequency.

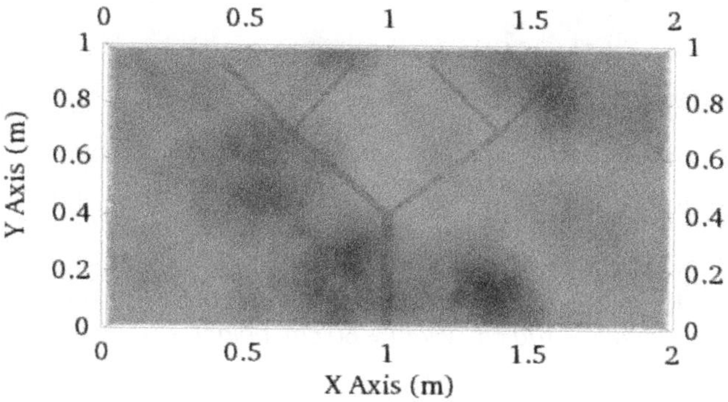

Figure 7.8. *Top view of the 3D chosen scenario in gprMax. The root model is buried at a* 0.15 *m depth in the heterogeneous medium*

The sketched root shown in Figure 7.5 is buried along the y-direction at a depth of 0.15 m. The simulation is run for different Tx-Rx positions, with a displacement step of 0.01 m along the x-direction to yield B-scans, then along the y-direction with a step of 0.01 m to generate the C-scan. The raw data are processed using MATLAB.

7.2.4. *Laboratory experiments*

The laboratory bistatic GPR system comprises a sandbox, an Antenna Positioning System (APS), a Keysight N9913A Vector Network Analyzer (VNA) and two folded complementary bowtie antennas. Antennas are placed at a distance of 5 cm in front of the sandbox and the antenna positioning system allows their displacements into x and y directions so as to perform the C-scan (Figure 7.9).

The S_{21} measured data are collected using the Keysight-VNA in the frequency band [0.3 – 3.3] GHz using the stepped frequency continuous wave (SFCW) strategy. The configuration parameters used for the VNA during this experiment are listed in

Table 7.2. The B-scan radargram results from the inverse Fourier transformation of the S_{21} measured over the considered bandwidth with a spatial step of 0.015m along the x-direction.

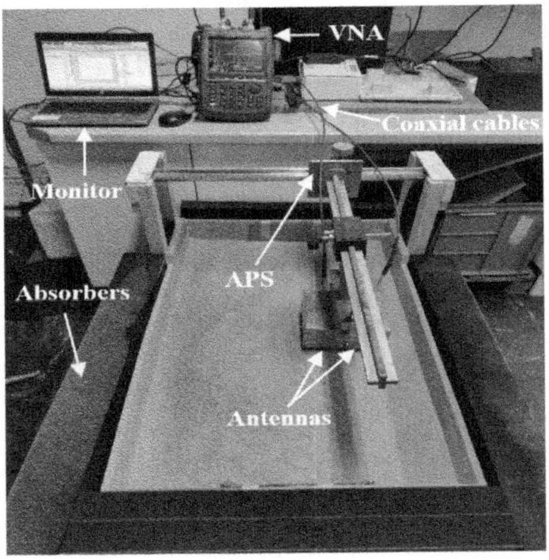

Figure 7.9. *Laboratory GPR system (box dimensions: length = 100 cm, width = 60 cm, height = 40 cm)*

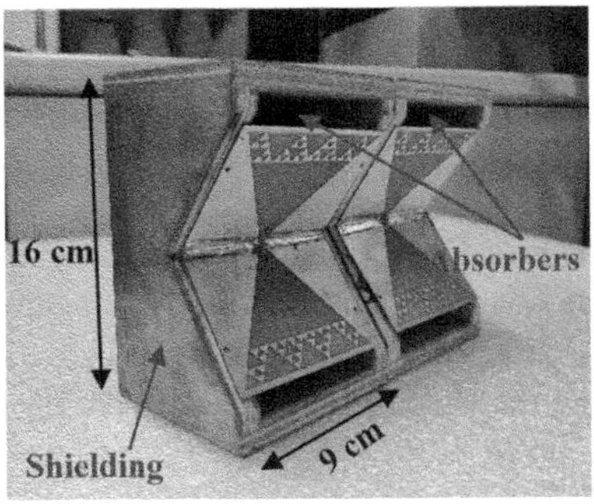

Figure 7.10. *Fractal folded complementary bowtie antennas used in laboratory experiments*

Fractal-folded complementary bowtie antennas (Liu et al. 2018) are used to perform the measurements. The antennas are described in Figure 7.10. Their design is optimized with fractals in order to enlarge the covered bandwidth. The antenna is enclosed in a copper-shielded cavity to prevent backward radiation. Absorbers are placed inside the cavity to minimize the coupling between the antenna and the shielding.

VNA parameters	Value
Number of frequency points	201
Frequency step size	15 MHz
Start frequency	300 MHz
Stop frequency	3300 MHz
I.F. bandwidth	300 Hz
Tx power	-10 dBm

Table 7.2. *Summary of the VNA configuration parameters used for this experiment*

Laboratory experiments are carried out using a root prototype with dielectric properties that have been measured using an open-ended coaxial probe 85070E. The measured dielectric permittivity and conductivity are presented in Figure 7.11. The root prototype dielectric properties slightly vary over the frequency range of measurements, with mean values of $\varepsilon_r = 25.6$ and $\sigma = 0.54$ mS/m. The GPR measurement box is of limited size. The root prototype is limited to the first dichotomous branching pattern as shown in Figure 7.11(a). The cylindrical wooden segments are 0.25 m in length. Diameters respect Leonardo Da Vinci's model, as previously cited.

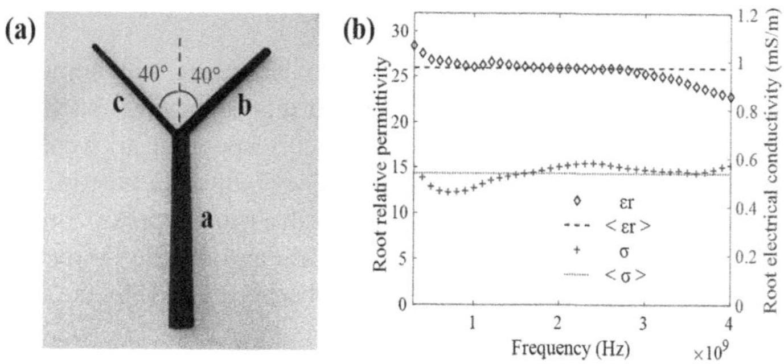

Figure 7.11. *(a) Root prototype used for laboratory experiments. Diameters of segments: $D_a = 1.9$ cm, $D_b = 1.5$ cm, $D_c = 1$ cm. (b) Root prototype dielectric properties measurements using the 85070E dielectric probe*

7.3. Data-processing framework

The processing framework has been developed in MATLAB. Applied to both numerical and experimental GPR data, the description of the processing flowchart is shown in Figure 7.12. To operate the matched filtered migration technique, the soil relative permittivity is supposed to be known as well as to determine the depth of the buried object. Time-zero correction corresponds to adjusting/identifying the spatial origin of the echoes received by the antenna resulting from the soil interface reflection.

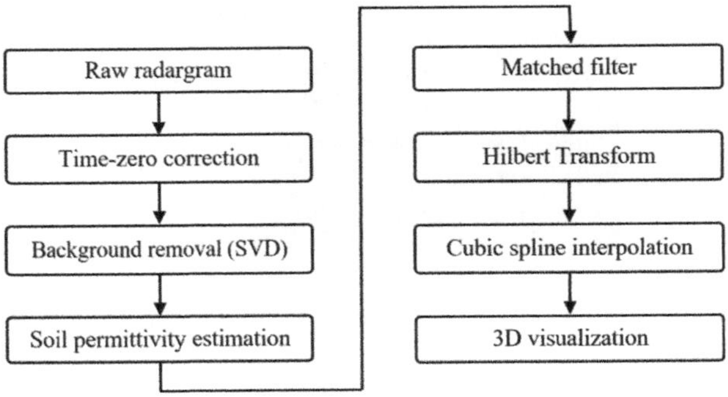

Figure 7.12. *Flowchart of post-processing techniques used for both numerical and experimental GPR data*

7.3.1. *Background removal*

Background removal is widely commented on in the literature. Here, a singular value decomposition (SVD) approach (Cai et al. 2017) is employed. SVD helps to filter out the antennas' coupling and the air–soil interface reflection. It also reduces the clutter due to soil heterogeneity.

The SVD approach reconstructs the measured data using only some eigenvalues, the large ones being associated with strongly correlated features such as antennas' coupling and ground bounce, and the low ones corresponding to weakly correlated features, namely target responses and high-frequency clutter. In practice, an empirical strategy is followed while identifying the eigenvalues to suppress. Since the clutter is associated with the first p singular values, the radargram can be reconstructed by projecting the measured data on the orthogonal subspace defined by the p first singular values.

7.3.2. *Soil dielectric permittivity estimation*

The soil relative permittivity is determined using the RHT. Based on hyperbola parameters fitting, this technique is used for the soil relative permittivity identification and consequently the root position assessment (Windsor et al. 2013; Li et al. 2016; Liu et al. 2019). Indeed, when GPR antennas are moved along the survey direction above the root, the hyperbolic reflection generated by the buried root can be expressed using three parameters in the hyperbola equation

$$\frac{t^2}{a^2} - \frac{(x - x_0)^2}{b^2} = 1 \qquad [7.8]$$

with

$$a = t_0 \quad \text{and} \quad b = \frac{t_0}{2} \frac{c}{\sqrt{\varepsilon_r}} \qquad [7.9]$$

x expresses the antenna position, t stands for the two-way travel time of the wave, (x_0, t_0) are the coordinates of the hyperbola apex, c is the speed of light in free space and ε_r is the soil relative permittivity. In equation [7.8], the root diameter is ignored. This is consistent since the root diameters in this study are small compared to the working wavelength (Liu et al. 2019).

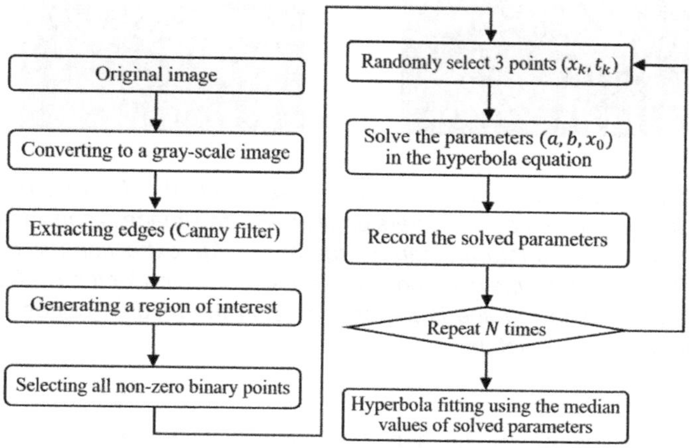

Figure 7.13. *Flowchart of the RHT-based algorithm for hyperbola parameters estimation*

The flowchart of the RHT method is sketched in Figure 7.13. An example of a radargram (deriving from laboratory experiments) processed using this algorithm is shown in Figure 7.14. After time-zeroing and background removal operations, the B-scan profile is transformed into a gray-scale image as shown in Figure 7.14(a). The edge of the hyperbola of interest is extracted using a Canny edge detector (Figure 7.14(b)), as classically used in image analysis applications (Canny 1986). The image area is then limited to a specific region of interest (ROI), which considerably reduces the computational time (Figure 7.14(c)). The ROI definition can be achieved using one of the following methods:

1) manually selecting the hyperbola of interest;

2) adjusting the Canny filter threshold and the standard deviation of the Gaussian filter on which the Canny filter is based.

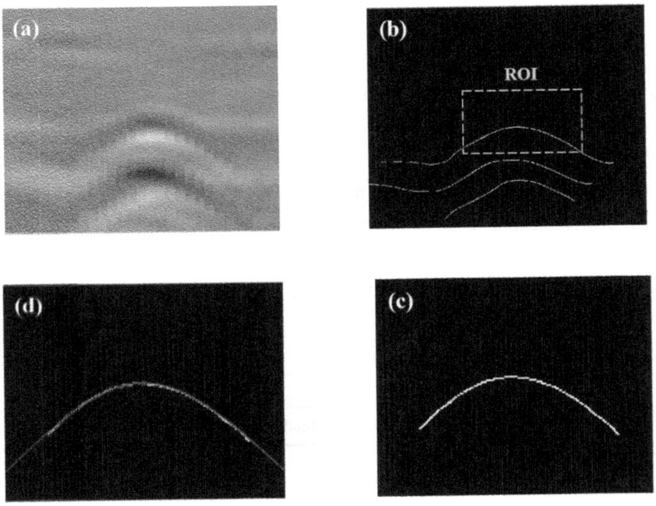

Figure 7.14. *Example of a radargram processed with the RHT-based algorithm. (a) After preprocessing, a gray-scale version of the B-scan image is considered. (b) The GPR B-scan is converted into a binary image after edge extraction using the Canny filter. (c) An ROI containing the hyperbola of interest is selected for the RHT processing. (d) The hyperbola fitting red line is performed using the median values of the three estimated hyperbola parameters*

The points belonging to the hyperbola represent all non-zero points of the resulting image. To fit the three hyperbola parameters of equation [7.8], three points (x_k, t_k) are randomly selected among the recovered non-zero points (belonging to the hyperbola and resulting from the edge detection) at each iteration of the RHT loop. The assessed parameters are then recorded and the process with three new points coordinates is repeated. Finally, as shown in Figure 7.14(d), the hyperbola fitting is achieved using the median values of the solutions (a, b, x_0) that are Hough transformed. The RHT technique for soil permittivity retrieval is tested based on EM simulations using gprMax involving homogeneous soils with different known values of relative permittivity. Besides evaluating the accuracy of the proposed method, this test yielded the number of iterations required in the RHT loop. Less than 100 iterations appear needed to converge to stable accuracy. The results are presented later on.

7.3.3. *Matched filter technique*

The matched filter technique is the basic principle of synthetic aperture radar (SAR) imaging (Leuschen and Plumb 2001; Chan and Koo 2008). This migration technique aims at maximizing the signal-to-noise ratio (SNR) at the receiver level.

Actually, when a matched filter is applied to a received signal, the SNR is maximized at a single point in time corresponding to the delay to the target (under the hypothesis that the target is small). From the geometry sketched in Figure 7.15, where (x, y) are coordinates of the root position, the delay T_d can be computed as follows:

$$\left(\frac{cT_d}{2\sqrt{\varepsilon_r}}\right)^2 = (r-x)^2 + y^2 \quad [7.10]$$

which can be rewritten as

$$T_d = \frac{2\sqrt{\varepsilon_r}}{c}\sqrt{(r-x)^2 + y^2}. \quad [7.11]$$

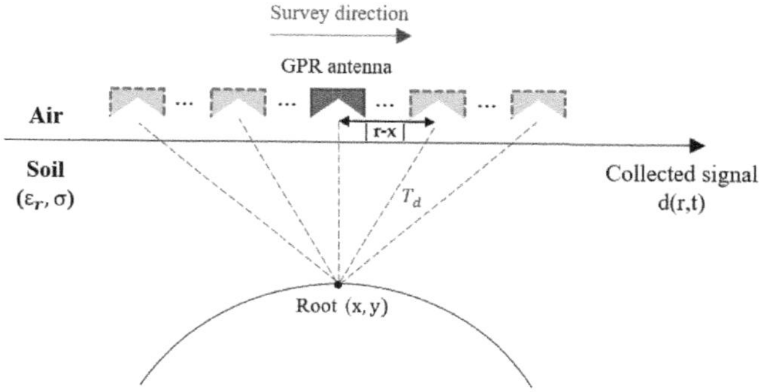

Figure 7.15. *Linear SAR data collection*

As expressed in equation [7.12], the impulse response $h(t)$ of the matched filter which maximizes the SNR is a scaled, conjugated, time-reversed and delayed version of the transmitted signal (antenna excitation signal)

$$h(t) = as^*(T_d - t) \quad [7.12]$$

a is a normalization factor (generally $a = 1$). The excitation waveform expression $s(t)$ is given in equation [7.7], so the matched filter impulse response expression is

$$h(t) = 2\sqrt{\frac{e}{2\xi}}\xi(t+\tau-T_d)\exp(-\xi(t+\tau-T_d)^2). \quad [7.13]$$

The matched filter impulse response $h(t)$ is convolved with the received signal $d(t)$ to yield the output $y(t)$

$$y(t) = d(t) * h(t) = \int_{-\infty}^{+\infty} d(\alpha)h(t-\alpha)\,\mathrm{d}\alpha. \quad [7.14]$$

$y(t)$ results in an image where pixels at the location of the root have a higher SNR compared with the surrounding pixels.

7.4. Results

7.4.1. *Results of simulation*

In this section, we discuss the results of the proposed reconstruction framework for tree root reconstruction. Firstly, the RHT method is applied to estimate the soil's relative permittivity in homogeneous soil conditions. Nine scenarios with different soil dielectric values ranging from 2 (for dry soils) to 10 (for wet soils) are simulated in gprMax. The accuracy of the retrieved relative permittivity is seen in Figure 7.16.

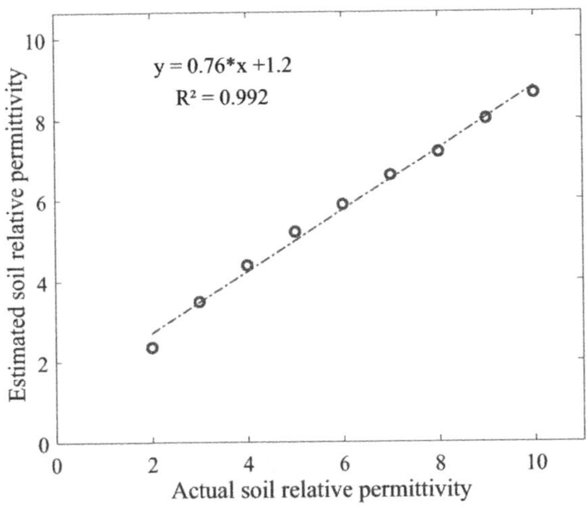

Figure 7.16. *Results of soil relative permittivity estimation using the RHT method. Each blue circle corresponds to a performed simulation*

Quantitatively speaking, the permittivity extraction error uses the root mean square relative error (RMSRE) expressed as

$$RMSRE = \sqrt{\frac{1}{n}\sum_{i=1}^{n}\left(\frac{\varepsilon_{\text{soil},i}^{\text{act}} - \varepsilon_{\text{soil},i}^{\text{est}}}{\varepsilon_{\text{soil},i}^{\text{act}}}\right)^2} \quad [7.15]$$

where $\varepsilon_{\text{soil},i}^{\text{act}}$ is the actual soil relative permittivity for simulation i, $\varepsilon_{\text{soil},i}^{\text{est}}$ is the estimated soil relative permittivity for simulation i and n is the number of performed simulations. Here, a RMSRE value of 7.89% is achieved.

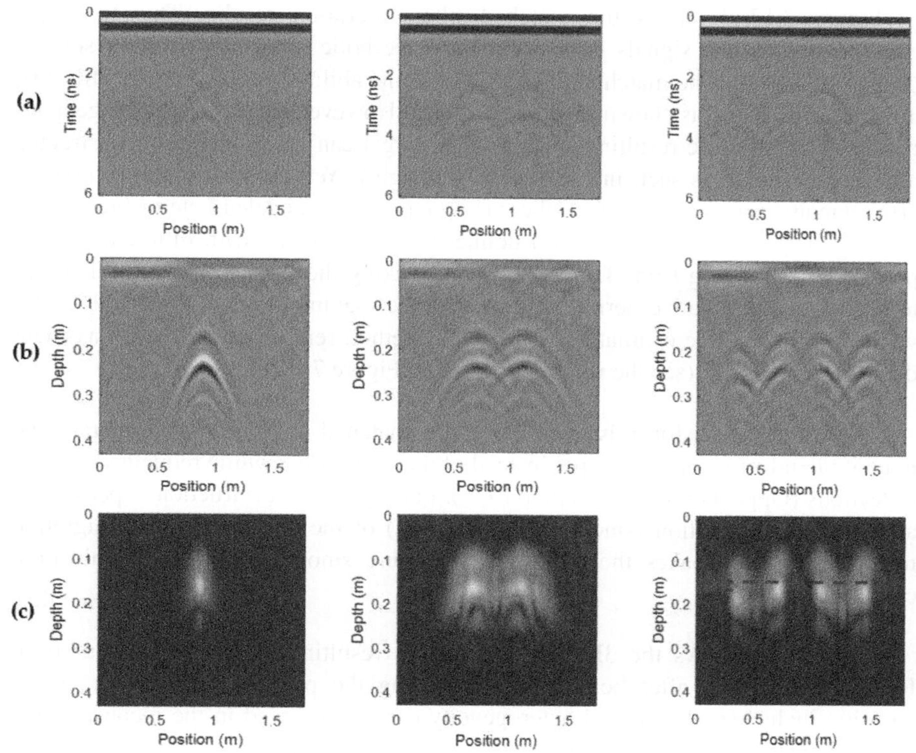

Figure 7.17. *Effects of processing procedures on numerical data. From left to right: GPR profile from the bottom part, the middle part, and the top part of the root prototype. (a) Raw radargrams time-zero corrected. (b) The time axis is converted into the depth axis and background subtraction is performed using the SVD approach. (c) Matched filter, Hilbert transform and cubic spline interpolation are performed to focus the root signature, correct multiple reflections and smooth the image, respectively. Red dashed lines indicate the error in depth determination*

Figure 7.17 shows the results of the processing framework for three different root cross-sections resulting from three B-scan along the axis y = 0.10 m, y = 0.50 m and y = 0.90 m as depicted in Figure 7.8. From Figure 7.17(a), the root responses in the radargram are not clearly visible because of antenna–soil interactions. The RHT method to estimate the relative permittivity of the considered heterogeneous soil results in a value of 4.5. In Figure 7.17(b), besides the attenuation of the air–soil interface reflection and the antennas' coupling, the clutter due to the soil heterogeneity is significantly mitigated because of the SVD-based clutter removal approach. In this case study, excluding the three largest singular values helps us to obtain images with reduced clutter.

Figure 7.17(c) shows the matched filter migration results. This technique back-propagates the signals composing the hyperbolic reflections to their sources (scattering target). The matched filter has shown its ability to separately identify two adjacent reflections as shown in Figure 7.17(c). However, due to the heterogeneous nature of the soil, the resulting spot from the target can eventually be reconstructed after a time delay as seen in Figure 7.17(c) (right). Yet, the root depth estimation error remains acceptable. In fact, the parameters of the modeled heterogeneous soil ($\beta = 1.5$, $f_w = [0 - 10\%]$, ...) define 20 soil elements with different relative permittivities ranging from 3.7 to 7.7. Simply taking the averaged value (calculated to be 6.1) as the relative permittivity of soil, the maximum depth error is 13.33%, while using the value estimated by the RHT method (employed here), a maximum depth error is 6.67% (see the red dashed lines in Figure 7.17(c) (right)).

The Hilbert transform is used after the matched filter step to generate the magnitude information (envelope) instead of their amplitude while reducing multiple reflections (Oppenheim and Schaffer 1975). Finally, 3D reconstruction is performed after cubic interpolation (smoothing the images) of the different B-scan migration results. This step makes the entire root structure smooth for a 3D visualization enhancement.

Figure 7.18 shows the 3D root architecture resulting from C-scan data (from FDTD simulations) after being processed using the proposed framework. Some parasitic inclusions due to soil heterogeneity can be observed in the reconstruction result. By setting a convenient threshold, we can filter out the pixels corresponding to the soil heterogeneity. The risks are to disregard parts belonging to the root architecture. Hence, a trade-off must be identified between the pixels corresponding to the root and the ones due to soil inclusions. Only pixels with normalized intensity higher than 0.7 are considered for the reconstruction shown in Figure 7.18. The reconstructed root structure is altered by the soil clutter. However, the shape of the initially simulated root prototype can be certainly recognized. The migration results exhibit strong reflections between adjacent root segments (daughters) of the second-order branching when this distance is less than 11.7 cm (the GPR horizontal resolution). Indeed, below GPR resolution, the daughter segments are not sufficiently spaced to be separated, which leads to overlapping of hyperbolas and yielding an ambiguous reflection after matched filter migration.

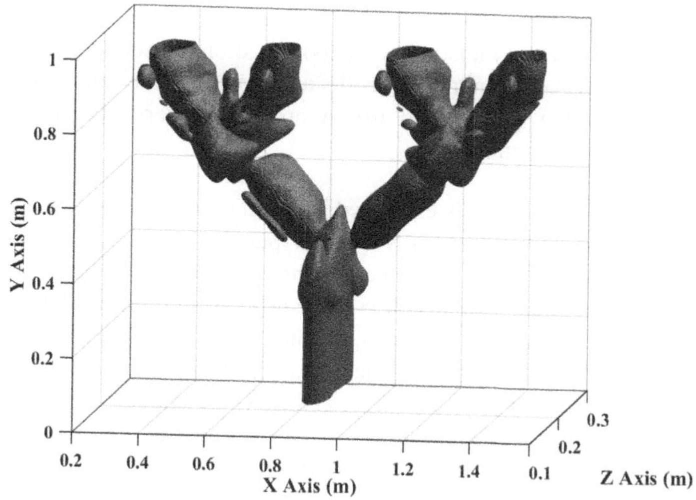

Figure 7.18. *3D representation of the simulated root model architecture after reconstruction using the proposed processing framework*

Soil moisture / Soil composition	Moderately wet f_u = [0 – 10%]	Wet f_u = [0 – 20%]	Highly wet f_u = [0 – 35%]
Sandy soil (S = 0.7 ; C = 0.3)	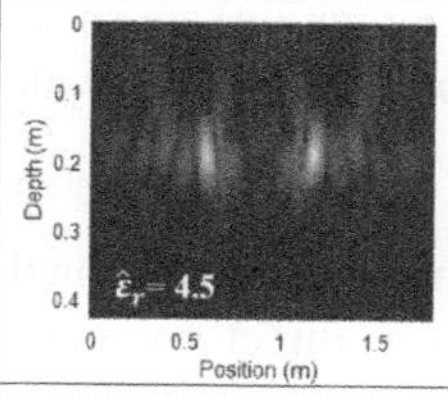	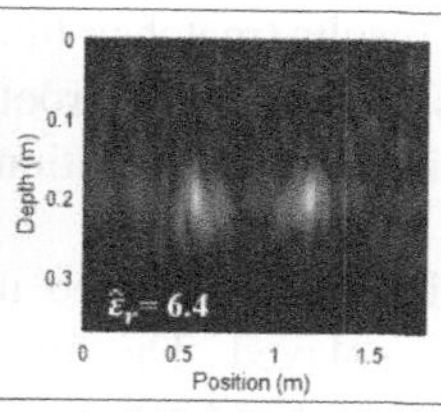	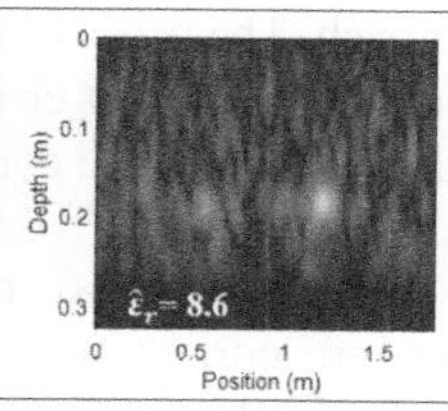
Clay soil (S = 0.3 ; C = 0.7)	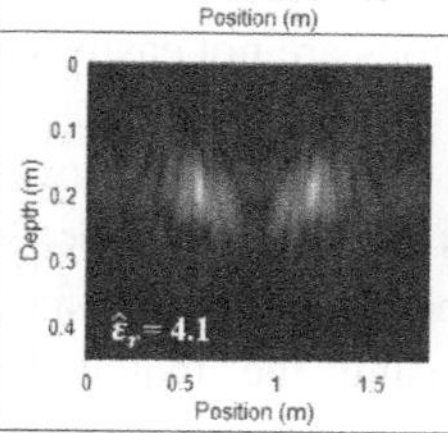	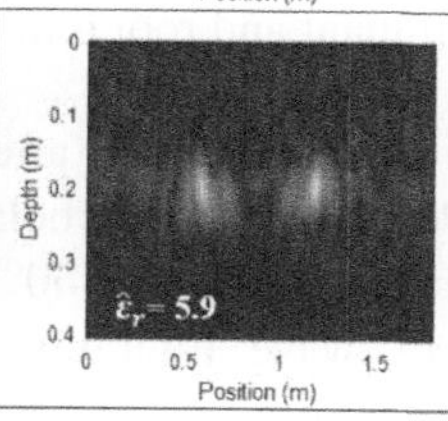	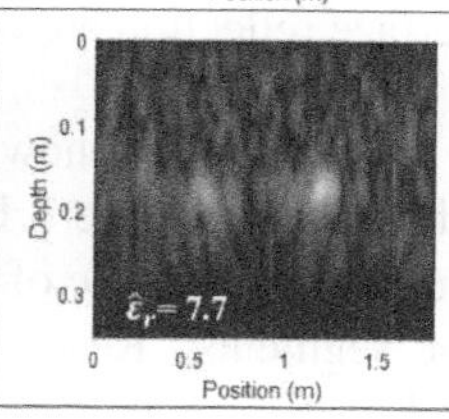

Table 7.3. *Numerical simulation results on the root model cross-section* y = 0.60 m *involving six different soil types.* $\hat{\varepsilon}_r$ *stands for the estimated soil dielectric permittivity using the RHT approach,* f_u *is the soil water content fraction,* S *and* C *are the sand and clay fractions, respectively*

The effect of soil heterogeneity on the proposed framework is tested through supplementary simulations under gprMax where we have used six different soils (variable compositions and moisture contents) based on the semi-empirical model of Peplinski. In Table 7.3, the migration results are shown for the same cross-section of the root (survey along $y = 0.60$ m) in order to observe the electromagnetic interaction with the different soils.

The soil dielectric constant is estimated for each scenario using the RHT method. As expected, the dielectric permittivity (real part) of clay soils is often smaller compared with sandy ones. Due to their higher porosity, clay environments could have a large fraction of air (Daniels 2005). The processing framework has been tested for the six configurations and the results are shown in Table 7.3. The SNR is getting lower with increasing soil moisture. The signatures from the two symmetrical parts of the root model are not perfectly similar due to the spatial variation of the soil moisture in a heterogeneous environment. In sandy soil with high water content, this phenomenon made one of the root signatures very attenuated.

7.4.2. *Experimental results*

The measurement data are acquired following a step frequency strategy (VNA) and the data are inverse Fourier transformed to get transient domain matrices (B-scan and C-scan). The processing framework is applied. The soil permittivity in these laboratory-controlled conditions has been estimated at 2.5 using the RHT-based approach. The processing results (matched filter migration) are shown in Figure 7.19, where two different cross-sections of the root prototype (survey along $y = 0.10$ m and $y = 0.35$ m) are considered for illustration.

In Figure 7.19(a), radargrams collected using the fractal bowtie antennas after time-zero correction are displayed. The antennas' direct coupling and the air/soil interface reflection are dominant and root reflections are not easily visible.

Figure 7.19(b) shows clear hyperbolas after background removal using the SVD technique. In Figure 7.19(b) (right), hyperbolas from the two root segments are not symmetrical because of the cross-angle of 50° between the antenna scan line and the root segments. Results of wave migration using the matched filter technique combined with Hilbert transform are presented in Figure 7.19(c), where a cubic spline interpolation has been performed to smooth the 3D representation. As it can be noted from Figure 7.19(c) (right), root signatures are well focused, but not located at exactly the same depth since root segments have been placed manually in the soil of the experiments, which causes a slight difference in the predicted depth.

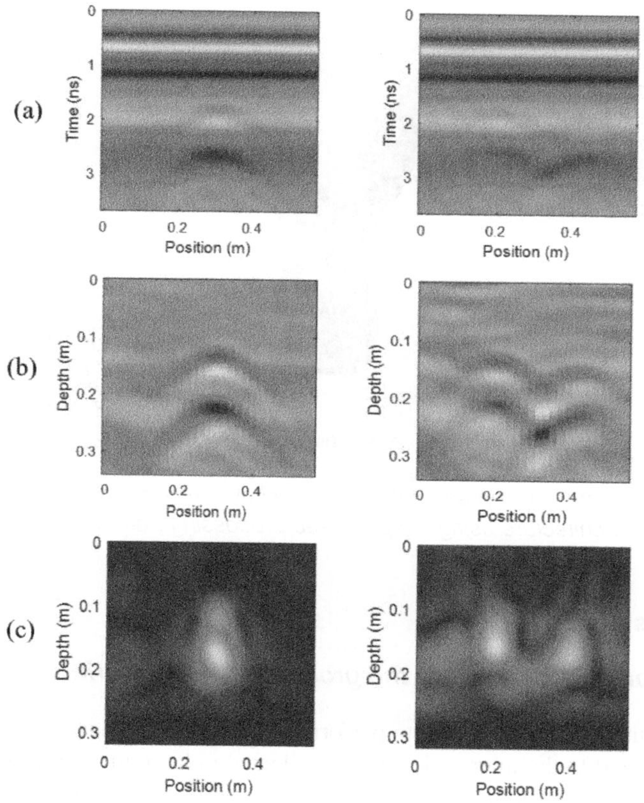

Figure 7.19. *Effects of processing procedures on the bottom part of the root prototype (left) and the top part of the root prototype (right). (a) Raw radargrams time-zero corrected. (b) The time axis is converted into the depth axis and background subtraction is performed using the SVD approach. (c) Matched filter, Hilbert transform, and cubic spline interpolation are performed to focus the root signature, correct multiple reflections and smooth the image, respectively*

Figure 7.20 shows the results of the root architecture 3D reconstruction based on C-scan data issued from the laboratory-controlled experimental measurement set. The reflection from the bottom part of the root prototype exhibits a large reflection in comparison with other root segments since this one has the largest diameter and has been placed perpendicularly to the GPR antenna line scan.

It should be noted that the soil used in the experimental setup is homogeneous unlike that modeled in the gprMax simulations, resulting in less clutter compared with the numerical results. The 3D reconstruction of the root based on the experimental data does not suffer from spurious inclusions around the root architecture.

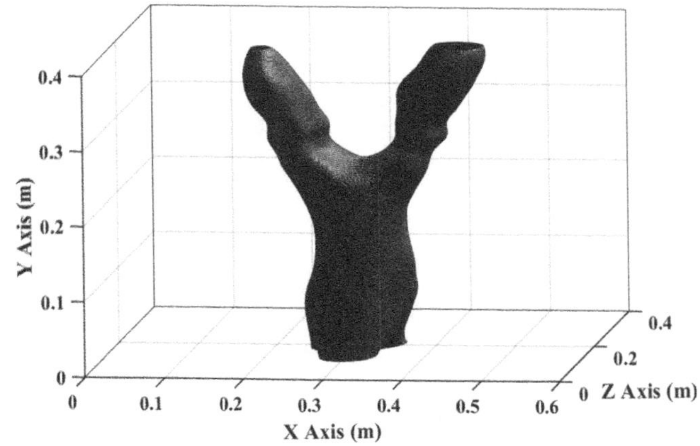

Figure 7.20. *3D representation of the reconstructed root prototype architecture using the proposed processing framework*

7.5. Discussion

7.5.1. *Computational times of the processing framework*

The computation time required to perform the proposed processing framework is directly correlated to the matrix size of the collected radargram data. In this study, a B-scan from gprMax is made of 90 A-scans, each with 1213 time-domain samples. The B-scan from the laboratory-controlled measurement is made of 20 A-scans, each with 111 time-domain samples. In Table 7.4, computational times required to process numerical and experimental data are given, being noticed that the estimation of the relative soil permittivity by the RHT method is performed on only one radargram.

Procedure	Computational times	
	Simulation	Measurements
RHT method for ε_{soil} estimation	13.58 s	13.23 s
Matched filter application	45.46 s	1.16 s
Total processing framework	59.54 s	14.47 s

Table 7.4. *Computational times of the proposed framework per radargram using a PC with Intel Xeon 3.80 GHz processor*

7.5.2. *Potentialities of the proposed method*

This processing framework proposes a set of tools (chosen among many others) for tree root reconstruction under heterogeneous soil conditions. Firstly, the use of the

SVD approach does not only make the root signature detectable. Secondly, for the soil (effective) relative permittivity estimation process, the Canny operator uses a filter that reduces the noise around the hyperbola of interest. This offers flexibility by adjusting the Canny filter thresholds. Lastly, the matched filter is proposed as a noise-resistant wave migration technique. The combination of these tools with the Hilbert transform results in good target localization. Some of the signal processing procedures and the corresponding effects have been already introduced by the authors (Aboudourib et al. 2019). However, unlike in the existing studies where the heterogeneous soil relative permittivity is computed by simply considering the average value of the soil elements, the RHT method is applied to estimate this value more accurately. The soil permittivity is then used within the matched filter process, where it is an important parameter to estimate the wave speed in the soil media. As a result, root segments are located which yield the 3D reconstruction of the root architecture.

7.5.3. *Limitations of the proposed method*

The proposed processing framework shows interesting capabilities to reconstruct the 3D buried root architecture. However, limitations still exist in its practical applications. In fact, high frequencies are often employed in GPR for highly resolved radargrams. However, increasing the frequency of operation, the root diameter cannot be neglected in the processing framework, GPR signals become more sensitive to the natural heterogeneity of the soil, which results in unwanted scattering (clutter) that degrades the SNR of measured data (Takahashi et al. 2013).

For field measurements, where the soil is highly heterogeneous and the soil moisture is spatially varying, the root hyperbolic reflections are accompanied by a reduced SNR ratio and the soil dielectric permittivity estimation is less accurate. This alters the results of the processing framework and necessitates the exploration of alternative methods to better estimate soil permittivity. In addition, when the contrast between soil and roots is not significant enough (e.g. a dead root), the root hyperbolic signature is not visible and the detection/processing is difficult. In a previous work (Aboudourib et al. 2019), through numerical simulations and validated through experiments, it has been found that root detection using GPR is inaccessible when the contrast between roots and the hosting medium is low, even with large root diameters.

Finally, a limitation related to the proposed approach is that it still requires human interaction to specify the ROI while estimating the soil permittivity, which is time consuming. Future attention should be paid to the improvement of an automatic ROI detection algorithm to make the process automatic.

7.5.4. *Applicability of the proposed framework to field measurements*

Aiming to be closer to natural conditions in both numerical and experimental configurations, we have chosen one of the most common root branching: a dichotomous root model, including a variety of branch diameters and orientations. In addition, the heterogeneous soil defined using fractal theory is a realistic hypothesis for soil modeling. The variability of this model offers possibilities to study the heterogeneity effect on root architecture processing. The proposed methodology is applicable for field measurements, yet it is worth outlining points related to the application for real-life situations:

1) Tree roots' spatial distribution is influenced by environmental factors, mainly soil moisture and soil type (factors examined here). In real-life situations, especially in urban areas, it is uncommon for trees to have roots deeper than 2 m. Most roots are found close to the soil surface, with 90% or more of all root biomass concentrated in the upper 60 cm (Dobson 1995). Consequently, high frequencies ($\geq$1 GHz) are used more often for root investigation since they reach such depths while allowing good resolution. However, in lossy media, high-frequency signals are rapidly attenuated as they propagate into the ground. To enhance root detectability and allow the application of the processing framework with in situ data, it is proposed to apply an amplitude decay compensation to the field-collected traces (linear or exponential gain).

2) Unlike pipes and other man-made buried objects, tree roots can grow in multiple directions. Thus, root orientation is an important factor that impinges on their detectability. That is, it becomes challenging to identify roots on GPR B-scans when the cross angle between root orientation and survey line is beyond the detection range of about (45°–135°) (Tanikawa et al. 2013). To address such an issue, we could perform GPR scans of the studied area into two perpendicular directions to come across all detectable angles, then combine the two polarization results.

3) In real-case studies, the soil permittivity is determined by measuring the depth to a clear hyperbolic reflection and digging down to check the depth (Mancuso 2011). However, due to the invasive nature of the method, this is difficult or even impossible to monitor root zone soil moisture repeatedly. Moreover, the soil around the tree in real-life is usually inhomogeneous, and the permittivity can vary significantly over different root zones of the same tree. The proposed processing scheme can be efficient in such situations as it can compute the soil permittivity over every single scan through the RHT method.

7.6. Conclusion

In this chapter, a GPR processing framework is proposed so as to reconstruct the 3D architecture of tree root systems under heterogeneous soil conditions. It has been tested with numerical data acquired from the powerful gprMax software involving realistic root and soil models, as well as with experimental data in controlled conditions in the laboratory. Accordingly, the SVD-clutter removal approach has significantly reduced the clutter noise caused by soil heterogeneity. The results of the RHT technique in estimating soil relative permittivity have also been presented. The matched filter method appears as an efficient migration technique to perform 3D reconstructions of root architectures with satisfactory accuracy.

The proposed framework can also be applied to other dielectric and metallic targets as well as with in situ GPR data. Yet, it is important to highlight that the GPR technique is site specific, and comparisons between results can only be achieved under comparable environmental conditions. Herein, numerical simulations and controlled experiments propose a proof of concept of the GPR processing framework, while the ability to detect tree roots is directly related to the site characteristics.

For our laboratory GPR system, the acquisition speed still represents a limitation for field measurements on real tree roots. In this perspective, the development of multistatic configurations involving a number of antenna pairs and employing as a good example linear sampling method (LSM) algorithms (Liu et al. 2016), among others in the same frame of thought, is being considered. Besides the improvement of the acquisition speed, this multistatic system should enhance the achieved resolution as well as the quality of imaging.

7.7. References

Aboudourib, A., Serhir, M., Lesselier, D. (2019). Impact of root diameter and water content on tree roots detection using ground-penetrating radar. In *Proceedings of the 13th European Conference on Antennas and Propagation (EuCAP)*, 1–5, Krakow.

Aboudourib, A., Serhir, M., Lesselier, D. (2021). A processing framework for tree-root reconstruction using ground-penetrating radar under heterogeneous soil conditions. *IEEE Transactions on Geoscience and Remote Sensing*, 59(1), 208–219.

Al Hagrey, S. (2007). Geophysical imaging of root-zone, trunk, and moisture heterogeneity. *Journal of Experimental Botany*, 58(4), 839–854.

Berry, S. and Roderick, M. (2005). Plant-water relations and the fibre saturation point. *The New Phytologist*, 168, 25–37.

Brunner, I. and Godbold, D.L. (2007). Tree roots in a changing world. *Journal of Forest Research*, 12(2), 78–82.

Cai, L., Chao, S., Qi, L. (2017). Random noise de-noising and direct wave eliminating based on SVD method for ground penetrating radar signals. *Journal of Applied Geophysics*, 144, 125–133.

Canny, J. (1986). A computational approach to edge detection. *IEEE Transactions on Pattern Analysis and Machine Intelligence*, 8(6), 679–698.

Chan, Y.K. and Koo, V.C. (2008). An introduction to synthetic aperture radar (SAR). *Progress in Electromagnetics Research*, 2, 27–60.

Chen, G. and Qian, S.-E. (2011). Denoising of hyperspectral imagery using principal component analysis and wavelet shrinkage. *IEEE Transactions on Geoscience and Remote Sensing*, 49, 973–980.

Cui, X., Guo, L., Chen, J., Chen, X., Zhu, X. (2013). Estimating tree-root biomass in different depths using ground-penetrating radar: Evidence from a controlled experiment. *IEEE Transactions on Geoscience and Remote Sensing*, 51, 3410–3423.

Daniels, D.J. (2005). Ground penetrating radar. In *Encyclopedia of RF and Microwave Engineering*, Chang, K. (ed.). Wiley, New York.

Dannoura, M., Hirano, Y., Igarashi, T., Ishii, M., Aono, K., Yamase, K., Kanazawa, Y. (2008). Detection of cryptomeria japonica roots with ground penetrating radar. *Plant Biosystems*, 142, 375–380.

Day, S.D., Wiseman, P.E., Dickinson, S.B., Harris, J.R. (2010). Contemporary concepts of root system architecture of urban trees. *Arboriculture & Urban Forestry*, 36(4), 149–159.

De Coster, A. and Lambot, S. (2019). Full-wave removal of internal antenna effects and antenna-medium interactions for improved ground-penetrating radar imaging. *IEEE Transactions on Geoscience and Remote Sensing*, 57(1), 93–103.

Dobson, M. (1995). Tree root systems. In *Arboricultural Research and Information Note 130/95/ARB*. Arboricultural Advisory and Information Service, Forestry Commission.

Dobson, M., Ulaby, F.T., Hallikainen, M.T., El-Rayes, M.A. (1985). Microwave dielectric behavior of wet soil. Part II: Dielectric mixing models. *IEEE Transactions on Geoscience and Remote Sensing*, 33(1), 35–46.

Dou, Q., Wei, L., Magee, D.R., Cohn, A.G. (2017). Real-time hyperbola recognition and fitting in GPR data. *IEEE Transactions on Geoscience and Remote Sensing*, 55(1), 51–62.

Giannakis, I., Giannopoulos, A., Warren, C. (2015). A realistic FDTD numerical modeling framework of ground penetrating radar for landmine detection. *IEEE Journal of Selected Topics in Applied Earth Observations and Remote Sensing*, 9(1), 37–51.

Giannakis, I., Giannopoulos, A., Warren, C. (2016). A realistic FDTD numerical modeling framework of ground penetrating radar for landmine detection. *IEEE Journal of Selected Topics in Applied Earth Observations and Remote Sensing*, 9(1), 37–51.

Glass, S. and Zelinka, S. (2010). Moisture relations and physical properties of wood. Wood Handbook: Wood as an Engineering Material. Report, U.S. Dept. of Agriculture, 4.1–4.19.

Guo, L., Chen, J., Cui, X., Fan, B., Lin, H. (2013). Application of ground penetrating radar for coarse root detection and quantification: A review. *Plant and Soil*, 362(1–2), 1–23.

Hirano, Y., Dannoura, M., Aono, K., Igarashi, T., Ishii, M., Yamase, K., Makita, N., Kanazawa, Y. (2009). Limiting factors in the detection of tree roots using ground-penetrating radar. *Plant and Soil*, 319(1–2), 15–24.

Hirano, Y., Yamamoto, R., Dannoura, M., Aono, K., Igarashi, T., Ishii, M., Yamase, K., Makita, N., Kanazawa, Y. (2012). Detection frequency of pinus thunbergii roots by ground-penetrating radar is related to root biomass. *Plant and Soil*, 360(1–2), 363–373.

Koledintseva, M., Wu, J., Zhang, H., Drewniak, J., Rozanov, K. (2004). Representation of permittivity for multiphase dielectric mixtures in FDTD modeling. In *2004 International Symposium on Electromagnetic Compatibility (IEEE Cat. No.04CH37559)*, 1, 309–314, Silicon Valley.

Lambot, S., Weihermuller, L., Huisman, J., Vereecken, H., Vanclooster, M., Slob, E. (2006). Analysis of air-launched ground-penetrating radar techniques to measure the soil surface water content. *Water Resources Research*, W11403, 1–12.

Leuschen, C. and Plumb, R. (2001). A matched-filter-based reverse-time migration algorithm for ground-penetrating radar data. *IEEE Transactions on Geoscience and Remote Sensing*, 39, 929–936.

Li, G., Henry, L., Bihang, F., Xihong, C., Jin, C. (2013). Forward simulation of root's ground penetrating radar signal: Simulator development and validation. *Plant and Soil*, 372(1–2), 487–505.

Li, W., Cui, X., Guo, L., Chen, J., Chen, X., Cao, X. (2016). Tree root automatic recognition in ground penetrating radar profiles based on randomized Hough transform. *Remote Sensing*, 8(5), 430, 1–16.

Liu, X., Serhir, M., Kameni, A., Lambert, M., Pichon, L. (2016). Enhanced linear sampling method for GPR imaging. In *International Workshop on New Computational Methods for Inverse Problems (NCMIP 2016)*, Cachan.

Liu, X., Chen, J., Cui, X., Liu, Q., Cao, X., Chen, X. (2017). Measurement of soil water content using ground-penetrating radar: A review of current methods. *International Journal of Digital Earth*, 12, 1–24.

Liu, X., Serhir, M., Lambert, M. (2018). Detectability of junctions of underground electrical cables with a ground penetrating radar: Electromagnetic simulation and experimental measurements. *Construction and Building Materials*, 158, 1099–1110.

Liu, X., Cui, X., Guo, L., Chen, J., Li, W., Yang, D., Cao, X., Chen, X., Liu, Q., Lin, H. (2019). Non-invasive estimation of root zone soil moisture from coarse root reflections in ground-penetrating radar images. *Plant and Soil*, 436(1–2), 623–639.

Mancuso, S. (2011). *Measuring Roots: An Updated Approach*. Springer-Verlag, Berlin/Heidelberg.

Mertens, L., Persico, R., Matera, L., Lambot, S. (2016). Automated detection of reflection hyperbolas in complex GPR images with no a priori knowledge on the medium. *IEEE Transactions on Geoscience and Remote Sensing*, 54(1), 580–596.

Miller, D. (1986). Root systems in relation to stress tolerance. *HortScience*, 21(4), 963–970.

Miller, A., Allen, H., Maier, C. (2006). Quantifying the coarse-root biomass of intensively managed loblolly pine plantations. *Canadian Journal of Forest Research*, 36(1), 12–22.

Ohashi, M., Ikeno, H., Sekihara, K., Tanikawa, T., Dannoura, M., Yamase, K., Todo, C., Tomita, T., Hirano, Y. (2019). Reconstruction of root systems in Cryptomeria japonica using root point coordinates and diameters. *Planta*, 249(2), 445–455.

Oppelt, A.L., Kurth, W., Godbold, D.L. (2001). Topology, scaling relations and Leonardo's rule in root systems from African tree species. *Tree Physiology*, 21(2–3), 117–128.

Oppenheim, A.V. and Schaffer, R.W. (1975). *Digital Signal Processing*, volume 6. Prentice-Hall, Englewood Cliffs.

Paz, A., Thorin, E., Topp, C. (2011). Dielectric mixing models for water content determination in woody biomass. *Wood Science and Technology*, 45, 249–259.

Peplinski, N., Ulaby, F., Dobson, M. (1995). Dielectric properties of soils in the 0.3–1.3-ghz range. *IEEE Transactions on Geoscience and Remote Sensing*, 33(3), 803–807.

Persico, R. and Soldovieri, F. (2008). Effects of background removal in linear inverse scattering. *IEEE Transactions on Geoscience and Remote Sensing*, 46, 1104–1114.

Qixin, L., Xihong, C., Xinbo, L., Jin, C., Xuehong, C., Xin, C. (2018). Detection of root orientation using ground penetrating radar. *IEEE Transactions on Geoscience and Remote Sensing*, 56(1), 93–104.

Randrup, T., Mcpherson, E., Costello, L. (2001). A review of tree root conflicts with sidewalks, curbs, and roads. *Urban Ecosystems*, 5(3), 209–225.

Stover, D.B., Day, F.P., Butnor, J.R., Drake, B.G. (2007). Effect of elevated CO2 on coarse-root biomass in Florida scrub detected by ground-penetrating radar. *Ecology*, 88(5), 1328–1334.

Straube, J., Onysko, D., Schumacher, C. (2002). Methodology and design of field experiments for monitoring the hygrothermal performance of wood frame enclosures. *Journal of Thermal Envelope and Building Science*, 26(2), 123–151.

Takahashi, K., Igel, J., Preetz, H., Sato, M. (2013). Influence of heterogeneous soils and clutter on the performance of ground-penetrating radar for landmine detection. *IEEE Transactions on Geoscience and Remote Sensing*, 52(6), 3464–3472.

Tanikawa, T., Hirano, Y., Dannoura, M., Yamase, K., Aono, K., Ishii, M., Igarashi, T., Ikeno, H., Kanazawa, Y. (2013). Root orientation can affect detection accuracy of ground-penetrating radar. *Plant and Soil*, 373(1–2), 317–327.

Topp, G.C., Davis, J.L., Annan, A.P. (1980). Electromagnetic determination of soil water content: Measurements in coaxial transmission lines. *Water Resources Research*, 16(3), 574–582.

Turcotte, D.L. (1997). *Fractals and Chaos in Geology and Geophysics*. Cambridge University Press.

Warren, C., Giannopoulos, A., Giannakis, I. (2016). gprMax: Open source software to simulate electromagnetic wave propagation for ground penetrating radar. *Computer Physics Communications*, 209(12), 163–170.

Wharton, R.P., Hazen, G.A., Rau, R.N., Best, D.L. (1980). Advancements in electromagnetic propagation logging. In *SPE Rocky Mountain Regional Meeting, Society of Petroleum Engineers*, SPE-9041-MS, Casper.

Windsor, C.G., Capineri, L., Falorni, P. (2013). A data pair-labeled generalized Hough transform for radar location of buried objects. *IEEE Geoscience and Remote Sensing Letters*, 11(1), 124–127.

Wu, Y., Guo, L., Cui, X., Chen, J., Cao, X., Lin, H. (2014). Ground-penetrating radar-based automatic reconstruction of three-dimensional coarse root system architecture. *Plant and Soil*, 383(1–2), 155–172.

8

Sounding Radars and Ground Penetrating Radars Designed for the Exploration of Our Solar System – Focus on Planet Mars

Valérie CIARLETTI
LATMOS/IPSL, UVSQ/Université Paris-Saclay, Sorbonne Université, CNRS, Guyancourt, France

8.1. Introduction

At the beginning of the first space missions in the 1960s, the exploration of our solar system has relied on instruments, on board probes or orbiters mainly providing data on the atmosphere/ionosphere of the studied bodies (planets, moons, asteroids and comets), the topography and the composition of their surface. Most of the time, when the subsurface was considered, the penetration depth was only a few centimeters. Then came the time of robotic exploration of bodies, starting with our Moon and Mars. Vehicles capable of moving on their surface were obviously equipped with cameras to move safely on the surface and also to analyze rocks for scientific reasons.

In planetary science, the access to the subsurface of bodies, such as planets, moons, asteroids and comets, allows us to collect first-hand information, which is essential to properly document the current environment of these objects, reveal the

---

For a color version of all the figures in this chapter, see www.iste.co.uk/serhir/radar.zip.

Ground Penetrating Radar,
coordinated by Mohammed SERHIR and Dominique LESSELIER. © ISTE Ltd 2024.

processes that shaped them at different stages of their evolution and ultimately provide clues and constraints on the scenarios of the formation of our solar system.

Because electromagnetic waves have the potential to propagate through matter, radars provide a rare opportunity (partially shared with seismometers which need to be accommodated on stationary platforms at the surface and commonly give access to the deeper internal structure) to peer through the surface and reveal buried structures. After some initial radar observations with dedicated radars based on Earth of our neighbor bodies: The Moon in 1946 (Mofensen 1946; Bay 1946), Mars (Goldstein and Carpenter 1963a), Mercury (Carpenter and Goldstein 1963; Dyce et al. 1967), Venus (Goldstein and Carpenter 1963b). The intensive "in situ" exploration of our solar system began in the 1960s (Siddiqi 2002). Our Moon understandably was the first target for exploration and was the target of the first ever sounding radar designed for a space mission. Indeed in 1972, the ASLE (Apollo 17 Lunar and Sounder Experiment) is operated at 5, 15 and 150 MHz center frequencies from lunar orbit (Porcello et al. 1974). A series of challenging missions rapidly followed to explore the outer Solar System (Jupiter, Saturn, Neptune, Pluto). The Pioneer 10 NASA probe, launched in 1972, traveled through the asteroid belt and performed the first flyby of Jupiter. Its companion probe Pioneer 11 explored for the first-time planet Saturn and its main rings.

Since then, most missions designed for solar system exploration carry radars, with most often sounding capacities (Kofman and Safaeinili 2004; Ciarletti 2016). More specifically, all recent missions that have landed a rover on the Lunar and Martian subsurface have one or two GPR onboard their vehicle. It should also be noted that radars which were not initially designed as sounding radars were, under certain favorable conditions, able to access the subsurface and provide unexpected information such as, for example, the bathymetry of the seas of Titan (Mastrogiuseppe et al. 2014).

These radars designed and built to sound the subsurface are, when they are operated from orbit, often called "sounding radars", which is the case for ASLE (Apollo 17 Lunar and Sounder Experiment), MARSIS (Mars Advanced Radar for Subsurface and Ionosphere Sounding) and SHARAD (Shallow Subsurface Radar), while those accommodated on a rover moving on the planet surface are known as GPR (Radar Imager for Mars subsurFAce eXperiment (RIMFAX), Rover Penetrating Radar (RoPeR) and Water Ice Subsurface Desposits Observation on Mars (WISDOM)).

This chapter is specifically devoted to sounding radars and GPRs that are now considered to be essential elements of the instrumental payload in almost all planetary missions. It has been chosen to highlight the specificities of these radars in terms of design, operations and data interpretation. Section 8.2 will give an overview of past, current and future radars developed for planetary exploration and will be illustrated by a limited selection of important results. Then, section 8.3 will show how the spatial context imposes constraints that have an impact on the design and the technical performances of the radar and can have consequences on the data interpretation. Section 8.4 will be a focus on planet Mars. The Martian subsurface has been to date sounded by a series of four radars (two from orbit and two from the surface) operating in different frequency ranges, which provide a range of interesting and complementary results that were able to provide significant inputs to scientific questions. The last section will give some examples of creative radar concepts that have been designed for very specific configurations.

8.2. Overview of the radars designed for planetary missions

In the context of planetary exploration, the variety of scientific questions that can be handled by sounding radars is quite wide: are there reservoirs of liquid water in the Martian deep subsurface? What are the dimensions and depth of lava tubes on the Moon or on Mars? How much water is trapped as ice in permanently shadowed craters on the Moon? What is the internal structure of a comet nucleus? What is the depth of the anticipated oceans beneath the icy shells of Ganymede and Europa? What are the depths of hydrocarbon seas on Titan? What is the chronology of the geological processes that took place in a given site on Mars or on the Moon? etc. These scientific objectives are translated into technical requirements that guide the design of the radar. For sounding radars or GPRs, the main performance indicators are certainly the penetration depth and the vertical resolution, which are defined, in a given environment, by the instrument's center frequency and frequency bandwidth.

Some of the scientific questions mentioned previously require penetration depths larger than a few kilometers (e.g. Martian polar caps, comet nucleus, etc.) or even tens of kilometers (e.g. icy shell of Europa), achievable only in favorable icy environments. As a result, they require radars operating at frequencies of the order of a few MHz or tens of MHz. The direct consequence is the need of antennas which are tens of meters long and which cannot be accommodated on a vehicle moving at the surface of the studied body. They are logically operated from orbit at distances from the surface that can reach thousands of kilometers and their antenna system is deployed once in orbit.

On the other hand, some missions are more focused on the detection of characteristics such as fine stratification in a sedimentary environment in the Martian shallow subsurface, for example. They require radar data with good vertical resolution, which can only be obtained with a broad frequency bandwidth compatible with a high center frequency. The obvious drawback is the limited penetration depth. Note that quite rare exceptions are some unusual radars developed to sound the subsurface from a stationary lander: the GPR of the cancelled CNES Netlander mission (Berthelier et al. 2000; Berthelier et al. 2003), the EISS radar on the original version of the instrumented ExoMars platform (Biancheri-Astier et al. 2010), the LRPR (Lunar Regolith Penetrating Radar) of the Chang'E5 landing platform (Xiao et al. 2019; Ding et al. 2022) or the PRISM radar designed to be operated on Mars under a balloon that unfortunately never flied (Barbin et al. 1995).

Table 8.1 provides a list of the sounding radars and GPRs developed and selected for in situ planetary exploration (some never left Earth).

Radar	Mission	Target	Launch	Platform	Agency	Reference
ASLE–Apollo Lunar Sounder Experiment	Appolo 17	Moon	1972	Orbiter	NASA	Porcello et al. 1974
Cassini Radar	Cassini	Saturn moons	1997	Orbiter	NASA	Elhachi et al. 2004
PRISM–Permafrost Radar Investigation of the Subsurface of Mars	MARS 98	Mars	cancelled	Balloon	Roscosmos	Barbin et al. 1995
GPR	Netlander	Mars	cancelled	Landers	CNES	Berthelier et al. 2000
MARSIS–Mars Advanced Radar for Subsurface et Ionosphere Sounding	MEX (MarsExpress)	Mars	2003	Orbiter	ESA	Picardi et al. 2005
SHARAD–SHAllow RADar	MRO (Mars Reconnaissance Orbiter)	Mars	2005	Orbiter	NASA	Seu et al. 2007a
LRS–Lunar Radar Sounder	Selene/Kaguya	Moon	2007	Orbiter	JAXA	Ono et al. 2000
CONSERT–Comet Nucleus Sounding Experiment by Radiowave Transmission	Rosetta	67P-CG	2004	Orbiter +lander	ESA	Kofman et al. 2007
LPR–Lunar Penetrating Radar	Chang'E 3	Moon	2013	Rover	CNSA	Guang-You et al. 2014

LRPR–Lunar Regolith Penetrating Radar	Chang'E 5	Moon	2014	Lander	CNSA	Shen et al. 2018
EISS–Electromagnetic Investigation of the SubSurface	ExoMars	Mars	Descoped	Lander + Rover	ESA	Biancheri-Astier et al. 2010
LPR–Lunar Penetrating Radar	Chang'E 4	Moon	2018	Rover	CNSA	Guang-You et al. 2014
MOSIR–Mars Orbiter Subsurface Investigation Radar	Tianwen-1	Mars	2020	Orbiter	CNSA	Zou et al. 2021
RoPeR–Mars Rover Penetrating Radar	Tianwen-1/ Zhurong	Mars	2020	Rover	CNSA	Zou et al. 2021
RIMFAX–Radar Imager for Mars Subsurface Experiment	MARS 2020/ perseverance	Mars	2020	Rover	NASA	Hamran et al. 2020
RIME–Radar for Icy Moons Exploration	JUICE (JUpiter ICy moons Explorer)	Europa Ganymede Callisto	2023	Orbiter	ESA	Bruzzone et al. 2013
JuRa–JUventas RAdar	HERA/Juventas	Didymos	2024	cubesat	ESA	Herique et al. 2020
REASON–Radar for Europa Assessment and Sounding: Ocean to Near-surface	Europa Clipper	Europa	2024	Orbiter	NASA	Scanlan et al. 2021
WISDOM–Water Ice Sub-surface Deposits Observation on Mars	EXOMARS/ Rosalind Franklin	Mars	2028	Rover	ESA	Ciarletti et al. 2011
SRS–Subsurface Radar Sounder	Envision	Venus	2030+	Orbiter	ESA	Bruzzone et al. 2020

Table 8.1. *Sounding radars developed for a space mission*

The only exception in this list is the radar of the Cassini mission that was not really designed as a sounding radar but whose data allowed us to estimate the depth of hydrocarbon seas on Titan as previously mentioned.

Figure 8.1 gives an overview of the center frequency (A) and the frequency bandwidth (B) of the radars listed in Table 8.1, where the diamond markers show radars operating from the surface and circle markers show radars operated from orbit.

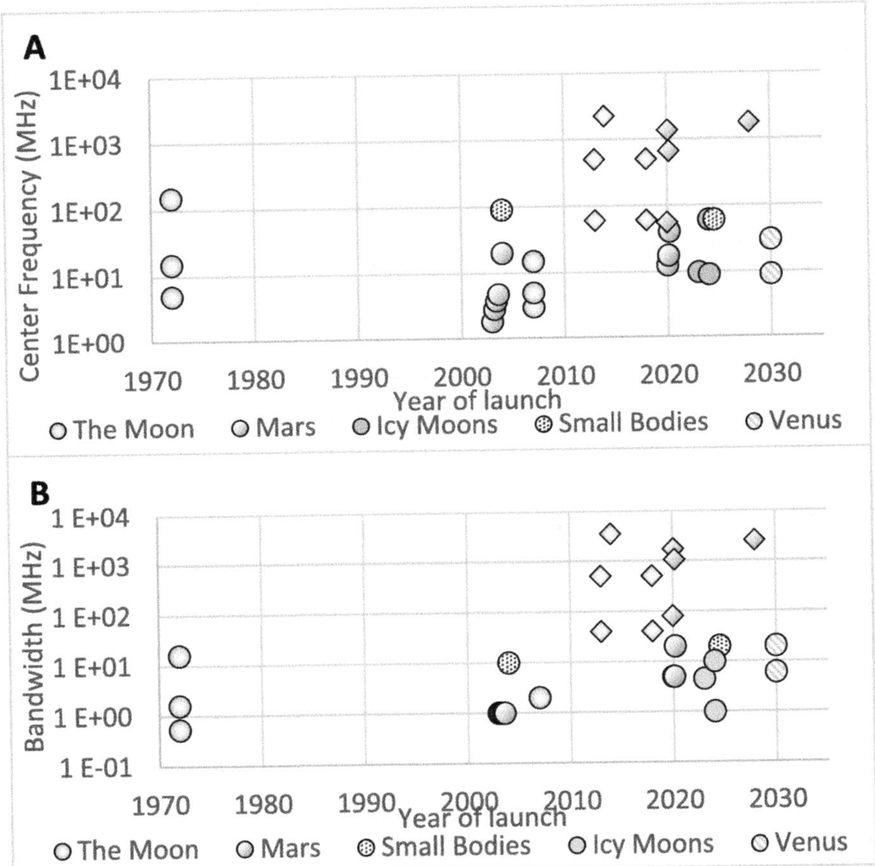

Figure 8.1. *Parameters values for soundings radars designed for space exploration as a function of the year of the mission launch. (A) Center frequency. (B) Frequency bandwidth. The diamonds markers are for radars operating at the surface while circles markers are for sounders in orbit*

Figure 8.2 presents the frequency bandwidth as a function of the center frequency for the radars listed in Table 8.1 with the same markers' convention as in Figure 8.1.

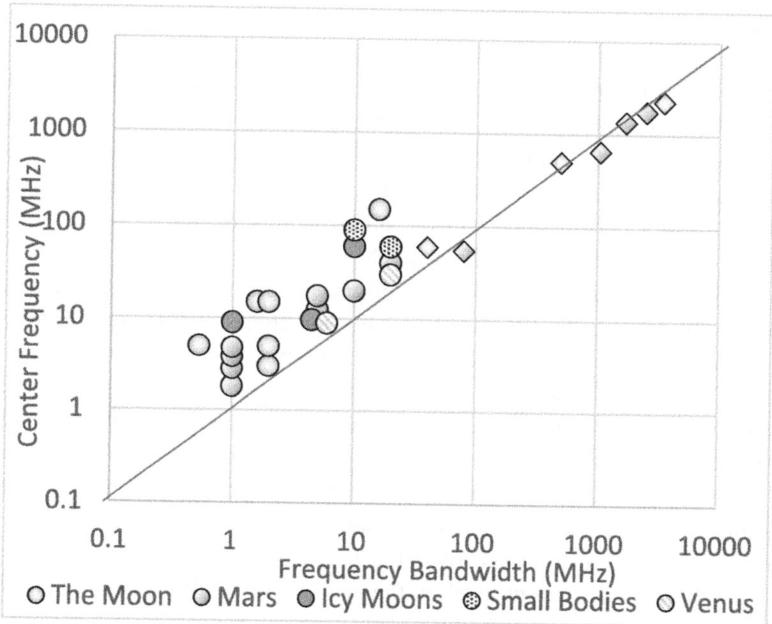

Figure 8.2. *Center frequency values versus frequency bandwidth values of soundings radars designed for space exploration. The diamonds markers are for radars operating at the surface while circles markers are for sounders in orbit*

As expected, the frequency bandwidth increases with the center frequency. The clear trend is that radars designed for rover missions (diamonds markers) tend to be ultrawide GPR (with bandwidth values larger than their center frequency), as they focus primarily on the shallow subsurface and seek detailed stratigraphy.

8.3. Specific constraints on the radar design due to space missions

As mentioned above, the technical objectives of sounding radars are derived from the scientific objectives that are defined and achievable in the frame of a given mission. But then, the actual design of the instrument considers numerous constraints linked to both the mission and the environmental conditions at the body to be explored.

8.3.1. *Building the instrument*

Indeed, Payload instruments developed for space missions must deal with limited resources that strongly constrain some of the choices made for the instrument design. Some of these are rather obvious: they involve mass, volume, power, energy and even data volume to be returned to Earth. They might have a noticeable impact of the instrument's performance. On the other hand, redundancy, which increases the needed resources, is requested as much as possible to limit the risk of point failure in the development of hardware, software or firmware. The instrument also needs to be able to withstand challenging environmental conditions specific to each single mission during the launch (vibrations), cruise (radiations, vacuum and temperature), landing (shock and vibrations) phases as well as through the operations in orbit or at the surface (thermal variations under atmospheric conditions different from those found on Earth, radiations). Each single component or technology used to build the instrument is carefully selected so that it is compatible with the mission's requirements. A delta qualification might even be necessary for components that are not qualified over the needed ranges. Eventually, a series of tests is performed at instrument level on the instrument qualification model (QM) to demonstrate that the instrument flight model (FM) will survive to the stresses of the mission (Hervé et al. 2020; Randy et al. 2020).

In addition to these tests, for missions planning to land in an area considered potentially habitable (due to the presence of liquid water, for example), planetary protection rules apply: the biological load carried by the landing and its payload must be carefully reduced in order to eliminate any risk of contamination by life from Earth. The instruments are built in clean rooms and then sterilized (by Sterad or DHMR) until the level of cleanliness required by the mission and the chosen landing site is reached. The rules on planetary protection are applied from COSPAR (Committee on Space Research), which is an international organization where scientists work together to agree on ways to address common problems in space science (COSPAR 2021).

Ultimately, the constraints imposed on space instruments to limit the risk of failure during their mission mean that the components and technologies chosen must be qualified and therefore cannot be at the state of the art, this point being further aggravated by the timescale of such missions characterized by a long design phase and, sometimes, a cruise of several years to reach the mission's target.

8.3.2. *Accommodating the sounding radar on the mission's platform*

A specific and particularly tricky topic for radars remains the accommodation of the antennas system, the final configuration is the result of a trade-off between the radar requirements corresponding to the scientific objectives and the mission's ones. The size of the antennas, especially when low frequencies are used, may require creative solutions such as deployable antennas sometimes using shape memory materials. In all cases, in the end, the structure of the platform (probe, rover or lander) will modify the radiation pattern of the antennas, be a source of multiple parasitic reflections and make it even difficult to find an accommodation where the antenna's field of view is free. In order to characterize as accurately as possible all of these phenomena, measurements should be performed once the radar is accommodated in its final position on the platform. These measurements are best taken in an anechoic chamber or at least in a configuration where potential obstacles (wall, floor, ceiling, cables, etc.) are identified and far enough so that their potential impact affects the instrument response well after the anticipated impact of the platform. Clearly this kind of measurement is challenging for radar with large antennas. In this case, accurate electromagnetic numerical simulations taking into account the shape of the platform can provide very useful information, complemented, when possible, by experimental tests on Earth which can be performed under a balloon (Pirrotta and Flamini 2011; Hong et al. 2021) or under a drone. Figure 8.3 shows two very different experimental setups.

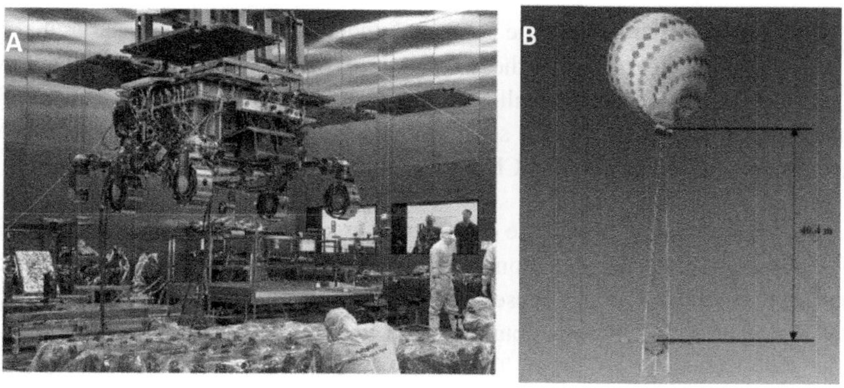

Figure 8.3. *Experimental validation of three radars built for planetary exploration. (A) MOSIR for the Tianwen-1 Orbiter under a balloon (adapted from Hong et al. 2021); (B) WISDOM radar accommodated on the Rosalind Franklin rover lifted off the ground at Airbus Defence and Space Bio Clean Room, Stevenage (adapted from Hervé et al. 2020). Copyright: Max Alexander/Airbus*

The first one (Figure 8.3(A)) is intended for the (Mars Orbiter Subsurface Investigation Radar) MOSIR operating in orbit while the second (Figure 8.3(B)) takes place in a clean room due to the PP rules that apply for WISDOM in the frame of the ExoMars mission, the rover has been lifted off the ground and the ground is covered with absorbent structures to minimize reflections.

Another specificity for GPR designed for space mission rovers is that the antenna system is not in contact with the surface but placed at a distance of about 30–50 cm to reduce the risk of damage to the antennas. These air-launched antennas are not the most favorable configuration for power transmission into the subsurface but provide a clean surface echo, the amplitude of which can be used to estimate the permittivity value of the upper subsurface layer.

Possible interference issues with the communication linked with Earth or with the telecommunication relay in orbit are also to be considered when choosing the frequency band of the radar. Due to the vital importance of having a working communication channel, even if tests show that the radiated power of the radar is too low to significantly interfere with the telecommunications link, the radar will be turned off whenever remote commands are received from Earth and data are transmitted to Earth.

8.3.3. *Some specific difficulties on data interpretation*

GPR data are most of the time presented under the form of 2D radargrams. For GPR accommodated on rovers, these data can easily be displayed along the rover path in 3D environments that allow a better understanding of the context. See Chunlai et al. (2020) for a 3D schematic interpretation of the lunar subsurface structure obtained by LPR of the Chang'E 3 lunar mission.

This is also to some extend the case for sounders operating from orbit, for which 3D views can be generated from a number of 2D radargrams acquired from thousands of straight orbital passes acquired over a given region (e.g. the North polar region of Mars 3D map obtained from SHARAD observations (Putzig et al. 2018, 2022; Xiong et al. 2019). To effectively explore the obtained 3D volumes, tools allowing us to integrate the surface data provided by other instruments and to select cutting planes have been developed (Xiong et al. 2019; Gupta et al. 2020).

For data interpretation, two main levels must be considered. The first one, totally shared with GPRs operating on Earth, results from the ability to unambiguously detect the interfaces between units having different geoelectrical properties. It basically relies on signal and images processing techniques. The difficulty of the task depends on the characteristics of the sounded volume. Smooth interfaces between homogeneous layers significantly thicker than the vertical resolution of the radar will appear very clearly provided that the permittivity contrast at the interface is strong enough. On the other hand, volume scattering in heterogeneous media blurs and sometimes prevents the detection of structures. At this stage, a very good level of knowledge of the instrument is necessary to avoid wrong detections based on potential instrumental artefacts.

The second level of interpretation consists in characterizing the previously detected features (from a geometrical point of view and in terms of permittivity values that provides constraint on the materials) and possibly identifying the physical processes that shaped them. This level of analysis, also necessary on Earth application, is even more essential in planetary exploration, where any information obtained about the subsurface might have a significant contribution to the characterization of the studied body and even for the understanding of our solar system formation. GPRs and sounding radars provide one of the most powerful means of understanding the geological history of the zone studied by revealing, for example, sedimentary deposits involving the lasting presence of liquid water, buried impact craters making it possible to specify the dating of the zone, reservoir of liquid water present at depth that can make the body habitable, old river bed filled with pebbles which makes it possible to estimate the importance and the duration of the flow of the river in question, size of the blocks of potentially pristine material that constitute the nucleus of a comet. This ultimate level of interpretation requires knowledge of the potential processes involved and teamwork that goes well beyond instrumental skills. In the process, the contributions provided by other instruments from the same mission and from past missions are extremely valuable in building a coherent interpretation.

In the fairly recent history of sounding radars for planetary exploration, the difficulty of this exercise in the absence of ground-truth is illustrated by a number of examples of genuine discussions about the interpretation of data. As a first illustration, the published detection of liquid water beneath the South polar cap of Mars (Orosei et al. 2018; Lauro et al. 2021) has been the object of controversial papers highlighting the fact that other interpretations of such a strong echo beneath the ice were possible (Bierson et al. 2021; Lauro et al. 2021; Smith et al. 2021; Grima et al. 2022; Mattei et al. 2022; Pettinelli et al. 2022b; Stillman et al. 2022). A second example is provided in the context of the characterization of lunar subsurface by LPR of the Chang'E 3 mission, the deep interfaces initially detected in radar profiles (Xiao 2015; Zhang et al. 2015) are very likely due to instrumental effects (Chunlai et al. 2017). Eventually, some reviewed and published results have been reinterpreted by the authors of the initial paper themselves who propose a new interpretation of the same data sets. For example, some structures on MARSIS radargrams that were originally interpreted as craters signatures (Picardi et al. 2005; Watters et al. 2006) are now recognized as being due to the Martian ionospheric activity (White et al. 2009).

To try and establish well-founded radar interpretations, scientists rely on complementary approaches. These are experimental measurements carried out in well-documented environments with an instrument model representative of the FM, numerical simulations reproducing as accurately as possible the operations of the sounding radar in the relevant environment, and laboratory permittivity measurements under conditions similar to those encountered during the space mission.

8.3.3.1. *Field tests in analog environments*

The most obvious way to test the end-to-end performance of a GPR in the environment for which it was designed or to validate the algorithms developed to process and interpret its data would be to operate a copy of the FM in a representative analog environment on Earth. However, it is impossible to find sites on Earth with environmental conditions similar to those of the targeted bodies. Only the facilities used for thermal vacuum qualification tests mentioned in section 8.3.1. allow us to reproduce, in a limited volume, the expected temperature variation, under an atmosphere whose composition and pressure are representative of the space mission. It is therefore necessary to work with approximate analogues and to choose sites according to relevant geological characteristics. Because it significantly reduces penetration depth, the presence of liquid water in the Earth's subsurface is a major concern for sounding radars, and often becomes the main driver for selecting a test site. With the Lunar and shallow Martian subsurface being free of liquid water, we will look for either arid environments such as the Atacama Desert (Hall et al. 2019; Dal Bo et al. 2021) or frozen environments (Dorizon et al. 2016; Hamran et al. 2020) to perform the most relevant field campaigns in natural environments, while staying aware of the limitation of the analogy.

8.3.3.2. *Simulated radar data*

Given the variety of configurations possibly encountered on a planet or a moon, and the difficulties to find suitable analog sites, numerical simulations are intensively used to develop and test data processing tools on a variety of realistic environments (Biccardi et al. 2001; Campbell et al. 2003; Ciarletti et al. 2003; Nouvel et al. 2004; Wang et al. 2017; Petersen et al. 2018; Wenmin Lv et al. 2020; Wang et al. 2021; Ling Zhan et al. 2022; Orosei et al. 2022; Oudart et al. 2022).

An obvious illustration of the contribution of simulated data to the interpretation of experimental data is provided by the Martian radars in orbit. Because of the elevation of the radar in orbit (around 300 km for the case of Sharad), the footprint (0.3–1 km along track and 3–7 km across track for the same radar) at the surface is large enough that off-nadir surface echoes due to the Martian topography are received. Depending on the configuration, they can arrive with delays compatible with potential subsurface echoes. To avoid erroneous interpretations, it is thus essential to disentangle reflections coming from the subsurface from unwanted surface echoes (clutter). To date, an efficient way to do this is provided by the comparison between experimental data and synthetic data that correspond to the radar response of the Martian surface topography only. The obtained simulated data are displayed in so-called cluttergrams, in the same way radar experimental data are represented in radargrams (Russo et al. 2008; Choudhary et al. 2016) and used for radargrams interpretation.

The surface topography initially used for these simulations was built with the data collected by the MOLA (Mars Orbiter Laser Altimeter) on board the Mars Global Surveyor (MGS) spacecraft (Smith et al. 2001) between 1998 and 2001. The obtained digital elevation model (DEM) has a horizontal resolution of about 1/256° and an averaged vertical accuracy around 1 m. In 2003, HRSC (high-resolution stereo camera) of the MarsExpress mission started to collect data that allowed us to build DTMs from pairs of images taken from different look angles (Gwinner et al. 2009). To date, more than 70% of the planet's surface is covered by DEM with a resolution better than 20 m (Gwinner et al. 2016). Finally, HiRISE (high-resolution imaging science experiment) which has operated since 2006 onboard the Mars Reconnaissance Orbiter is providing locally very fine DEM (McEwen et al. 2007) with a horizontal resolution of about 0.5 m/pixel and a vertical precision in the tens of centimeters (Ruesch et al. 2021) less than 1% of the planet's surface (Tao et al. 2021) is covered. The surface echo simulation is performed by codes in which the surface is modeled by facets (Nouvel et al. 2004; Berquin et al. 2015; Desage et al. 2022) or a combination of a large-scale topography on the top of which is added a small-scale roughness characterized by statistical descriptors (Campbell et al. 2003; Picardi et al. 2004).

In Peterson et al. (2018), a number of radargrams obtained from SHARAD experimental radar data and corresponding cluttergrams obtained through simulations considering only reflections from the surface are presented. Similarities between the pairs of images are obvious, they correspond to surface echoes. Buried interfaces were detected in some areas. The authors were also able to provide an estimate of the imaginary part of the relative permittivity for the material laying above the detected buried interface, which makes it compatible with a material composed of >80% water ice.

It is very important to understand whether this is happening because the signal is heavily attenuated before it can reach a reflector or because there are no significant buried reflective structures.

8.3.2.3. *Contribution of laboratory permittivity measurements*

GPRs are sensitive to spatial variations of the geoelectrical parameters, namely only the complex effective permittivity is considered here because to the first order, the bodies in our solar system can be considered to be non-magnetic. In favorable conditions, it is thus possible to obtain, from radar data, an estimate of the complex permittivity value of a layer or of a large enough inclusion in the subsurface. Trying to identify a material from the retrieved value is impossible. But given some a priori knowledge of the body is investigated, the estimated permittivity value can provide some constraints about the porosity and composition of the material or can allow it to exclude possible materials. To do this properly, it is essential to have access to data bases that content the permittivity value of all possible materials as a function of the temperature and the frequency. Then mixing laws or extended laboratory measurements provide additional information regarding the permittivity values for porous materials and mixtures of material (e.g. dust/ice). Many studies on sounding radar interpretation indeed rely on effort to characterize relevant samples in well-defined conditions (e.g. Heggy et al. 2006; Paillou et al. 2008; Stillman and Olhoeft 2008; Mattei et al. 2014; Pettinelli, 2015; Brouet et al. 2016; Boivin et al. 2018; Brouet et al. 2019 and many others).

8.4. Focus on Mars

Visible to the naked eye from Earth especially when the planet is closer to the Earth (which happens each ~25.5 months), planet Mars has been the object of interest from very early on.

8.4.1. *A very short summary of the exploration of planet Mars*

The instrumented exploration of Mars started around 1610, when Galileo made the first observation of the planet and its phases with his telescope. A few decades later, Huygens in 1659 obtained a first estimate of the duration of a Martian day. Between 1666 and 1672, Cassini performed a series of observations that allowed him to estimate the distance between the Earth and Mars and draw the first map of

Mars with some large structures, among which is the South polar cap. Scientists seemed to lose interest in the red planet until, in 1784, Hershel established the value of the inclination of Mars, which is responsible for the four Martian seasons and also understands that the bright areas observed at the poles are made of ice and raised the question of possible Martian inhabitants. Almost one century later, Schiaparelli drew a very detailed map of Mars and highlighted some features that looked quite rectilinear. He described them using the world “canali” in Italian which can be mistranslated as “canals” in English. Despite the fact these structures are optical illusions, they gave birth to the golden age of science fiction in literature and later on in cinema.

The observation of Mars underwent a dramatic change when probes were able to leave the Earth’s orbit. In 1965, the Mariner 4 NASA mission made the first fly-by at 10,000 km from Mars and sent to Earth the first images from the Martian surface. The next milestone happened in 1976 with the first successful landing on Mars by the Viking 1 and 2 landers. Eventually, in 1996 the tiny Sojourner rover of the Pathfinder mission was driven for the first time ever on the Martian surface. The most likely future milestones to come will be the deep drilling on Mars to collect samples at depth that should occur in 2029 with the ExoMars rover mission (Vago et al. 2017) and the Mars sample return mission (Muirhead at al. 2020) which was planned in the early 1930s.

To date, 14 orbiters have scrutinized the Martian surface and atmosphere, three of them: MARSIS of the MEX ESA mission, SHARAD of the MRO NASA mission and MOSIR of the Tianwen1 Chinese mission are looking into the subsurface with their sounding radar. Among the 6 rovers that have driven on the Martian surface, two of them have been operating a GPR since 2021: RIMFAX onboard the Perseverance rover of the Mars2020 NASA mission and RoPeR accommodated on the Zhurong rover of the Chinese Tianwen1 mission.

Interest in the subsurface is clearly growing, certainly because it is below the surface that the answer to some of the most important questions about the evolution of the planet Mars can be found: (i) did primitive life arise on Mars at the same time as on Earth, since at that time (~3.8–3.5 billion years ago) the two planets were experiencing fairly similar environmental conditions? Due to the harsh current conditions at the planet’s surface, any potential trace of biosignature is likely to be found in the subsurface. (ii) Did the enormous amount of water that was once present on the surface of Mars leave the planet with the atmospheric escape or are there still reservoirs of liquid water at depths (that may host some primitive life)?

These two questions about the past or present habitability are linked to the presence of liquid water on Mars and call for an accurate study of the subsurface where sounding radars and GPRs have a key role to play.

Table 8.2 summarizes the technical characteristic and science objective of the sounding radars and GPR designed for the Martian exploration.

Radar	Mission	Science objective	Resolution in vacuum	Platform	Depth of exploration
From orbit around Mars					
MARSIS	MEX		150 m	Orbiter	3 km
SHARAD	MRO		15 m	Orbiter	1 km
MOSIR	Tianwen-1		30 m 7.5 m	Orbiter	
At the Martian surface					
RoPeR	Tianwen-1/ Zhurong		2 m 9 cm	Rover	10–100 m 3–10 m
RIMFAX	MARS 2020/ perseverance	FMCW	14 cm	Rover	10–50 m
WISDOM	EXOMARS/ Rosalind Franklin	Step-frequency	6 cm	Rover	3–10 m

Table 8.2. *Present and future sounding radars for Mars*

8.4.2. *Description of the Martian subsurface radars with a few results*

Mars does provide some favorable terrains for sounding radars: the polar caps with their ice-rich deposits are, like on Earth, obviously one of them. The fact that liquid water is not stable at the Martian surface precludes the strong attenuation often encountered on Earth on wet subsoils. Nevertheless, significant volume scattering is to be expected on lithic terrains, especially for GPRs working at wave lengths ranging from a few centimeters to a few tens of centimeters. For radars operating from orbit, during day time, the plasma frequencies of the Martian ionosphere cuts frequencies lower than 3 MHz and also causes distortion, Faraday rotation and attenuation.

8.4.2.1. *MARSIS of the ESA MEX mission*

After the failure of the Mars 96 Russian mission, which never left the orbit of Earth, ESA rapidly decided to use the available spare models of the European instruments and to build the MEX orbiter, taking advantage of the architecture of the orbiter designed for the Rosetta mission. One of the instruments that was not onboard the initial Mars 96 mission is the MARSIS sounding radar designed and built by the Italian Space Agency (ASI) in collaboration with NASA.

The main objective of MARSIS (Picardi et al. 2004, 2005) is to search for potential reservoirs of liquid water in the Martian subsurface. There is no question about the presence of liquid water on Early Mars (Carr 1979; Clifford 1993; Carr 1996; Clifford and Parker 2001; Craddock and Howard 2002; Harrison and Grimm 2004). Some hydrologic models have established that conditions for liquid water could be currently found at a few kilometers below the surface. The MARSIS radar was thus designed to sound as deep as few kilometers into the Martian subsurface. The secondary objectives of MARSIS are the geological study of the Martian subsurface in general and also the characterization of the ionosphere (Picardi et al. 1999).

Figure 8.4. *An artist's impression of ESA MEX orbiter in orbit around Mars. The MARSIS long dipole antenna and the additional monopole are clearly visible. Credit: ESA*

To achieve these objectives, the radar operates at center frequencies of 1.8, 3.0, 4.0 and 5.0 MHz with a bandwidth of 1 MHz, which provides a theoretical vertical resolution in vacuum of 150 m. Two 20-m long monopole antennas (forming a

40-m long dipole) are used for transmission and reception. A shorter 7-m long monopole serves as an additional receiving-only antenna designed to allow a better discrimination of off-nadir echoes (see Figure 8.4), but was never efficiently used. While MEX starts orbiting the planet end of 2003, because of a potential risk for the spacecraft and the other instruments, the deployment of the MARSIS antennas was delayed until mid-2005. Figure 8.4 is an artist's view of the MEX mission at Mars where the MARSIS antennas are visible.

The instrument has been operational since then. The orbit of the MEX orbiter is elliptic with a size of approximatively ~300 km × ~10,000 km and an inclination of 86°. To maximize the chances of detecting deep buried structures, the MARSIS subsurface sounding mode is only triggered when the elevation of the spacecraft is below 800 km. During daytime, when the ionosphere is more active, the signal received by the radar undergoes significant distortions that need to be properly corrected (Mouginot et al. 2008) so that the data collected can be used.

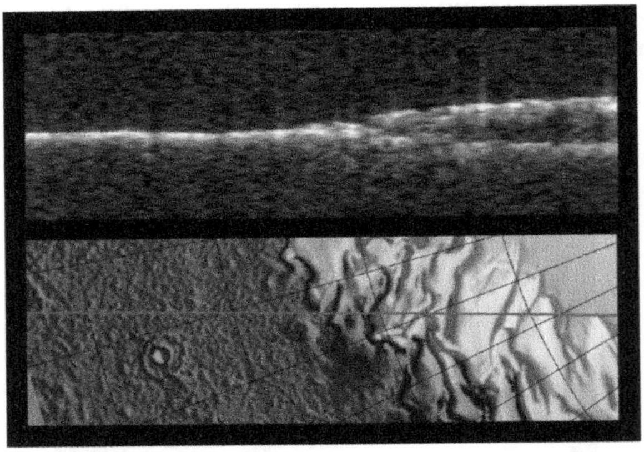

Figure 8.5. *The upper image is a MARSIS radargram acquired around the Martian north pole. The lower image shows the corresponding ground track on a topographic MOLA map. The blue color indicates the lowest terrains, and the orange one indicates the highest terrains. The horizontal extension of the images is 458 kilometers. Credit ASI/NASA/ESA/JPL/Univ. of Rome/MOLA Science Team*

Among the numerous results obtained from MARSIS data (Orosei et al. 2015), we selected:

– The detection of the basal unit beneath the North (Figure 8.5) and South (Plaut et al. 2007; Seu et al. 2007b) polar caps reveals that both polar caps are up to 3 km thick. They are made of water ice and covered by a thin layer of CO_2 ice that sublimated during summer months.

– The reflectivity map of the whole planet surface (Mouginot et al. 2010) is obtained from the radar surface echo and the surface topography provided by MOLA. The map shows a convincing correlation with the hydrogen distribution (assumed to be representative of the water ice content) in the first top meter of the subsurface estimated by the gamma ray spectrometer (GRS) aboard the 2001 Mars Odyssey spacecraft (Boynton et al. 2007). The terrains having a relative permittivity value close to 3 are those where the hydrogen content is high, which concludes that in these areas, the shallow subsurface is probably ice-rich.

– The precise analysis of the radar data collected around the southern polar cap makes it possible to highlight zones in which the amplitude of the echo corresponding to the interface located a few kilometers under the ice is extremely strong (see section 8.3.3). To date, the most recent studies (Pettinelli et al. 2022b) seem to demonstrate that only perchlorates solutions can be responsible for the basal bright reflections detected by MARSIS.

MARSIS is still operating after 20 years in orbit around Mars. The amount of data collected will certainly allow us to make unexpected discoveries about Mars formation and evolution.

8.4.2.2. *SHARAD (SHAllow RADar) of the NASA MRO mission*

SHARAD (Seu et al. 2007a) is the sounding radar aboard the orbiter of the MRO NASA mission (see Figure 8.6). It was designed and built par ASI like the MARSIS radar. MRO launched in 2004 and has continuously operated from the beginning of 2006 until now.

Compared to MARSIS, SHARAD's objectives are focused on the shallower subsurface, namely on the detection, mapping and characterization of interfaces buried in the subsurface over depths of a few hundreds of meters, for a better interpretation of the geologic evolution of selected areas. Thanks to a finer vertical resolution better than 10 m on the subsurface, SHARAD allows for the identification of aqueous deposits and the detailed analysis of the polar cap layered deposits.

To achieve these objectives, a center frequency of 20 MHz was chosen with a bandwidth of 10 MHz (achieved with a frequency modulated chirp between 15 and 25 MHz). The size of the receiving and transmitting antenna is therefore limited to 10 m, which facilitates the accommodation on the spacecraft. The spacecraft's orbit is also polar but almost circular. The MRO elevation above the surface ranges from ~250 to ~310 km.

Figure 8.6. *An artist's impression of NASA's Mars Reconnaissance Orbiter above the Red Planet. The SHARAD dipole antenna is partially visible behind the solar panels. Credit: NASA/JPL*

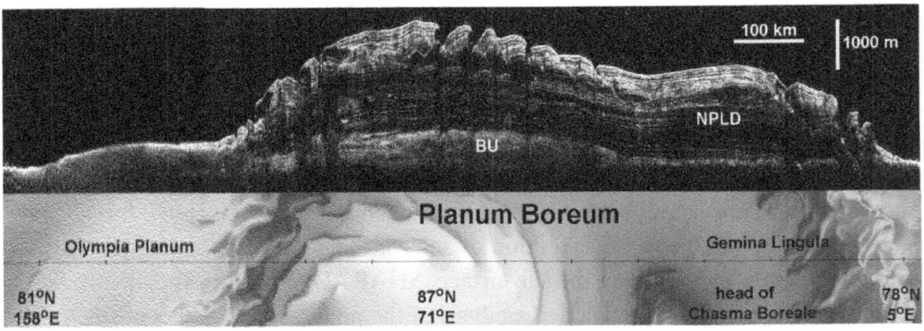

Figure 8.7. *The upper image shows a SHARAD radargram acquired on the north polar cap of Mars. A number of the layers of ice, sand and dust are detected. The image below shows the corresponding ground track on the MOLA topography (red and white represent higher ground, and green and yellow represent lower ground). Credit: NASA/JPL-Caltech/University of Rome/SwRI*

Among the scientific results obtained:

– The detailed layered structure of the polar caps is revealed by SHARAD (see Figure 8.7), which were not resolved by MARSIS. They allow scientists to better constrain theories about the evolution of the planet's climate.

– Lava flows in the Martian volcanic terrains of Tharsis are also successfully characterized by SHARAD: two buried interfaces are detected, estimates of their complex permittivity values are obtained, which are consistent with values measured in laboratory on lunar and terrestrial basalts (Carter et al. 2009; Simon et al. 2014).

– Evidence of buried glaciers in Mars' southern mid-latitudes has been demonstrated (Holt et al. 2008).

– Comparisons of MARSIS and SHARAD radagrams acquired in a same area demonstrate the complementarity of the two radars (Milkovich et al. 2009); the first radar is able to detect deeper structures, and the second radar is able to resolve fine layering (see Figures 8.5 and 8.6 for comparison).

8.4.2.3. *MOSIR of the CNSA Tianwen-1 mission*

China's National Space Administration (CNSA) launched in July 2020 its first mission to Mars, Tianwen-1 (Zou et al. 2021) which was composed of two elements: one orbiter with the MOSIR sounding radar aboard and a rover, called Zhurong, carrying a GPR. In a first mission phase between February and November of 2021, the orbiter was essentially used to image the area planned for Zhurong's landing and serves as a relay for the rover. Then the orbiter's orbit was modified in order to focus on the scientific survey of Mars and data acquisitions. At the end of June 2022, the Chinese space agency announced that the Tianwen-1 space probe has successfully taken images covering the entire planet.

The main objectives of the MOSIR sounding radar on board the orbiter is to build a new Martian topographic map and to sound the Martian ionosphere and the Martian subsurface (Fan et al. 2021). MOSIR operates on two low-frequency (LF) channels centered on 12.5 MHz and 17.5 MHz with a bandwidth of 5 MHZ, and on a higher frequency (HF) channel around 40 MHz with a bandwidth of 20 MHz, which results in resolutions in vacuum of 30 m and 7.5 m. MOSIR transmits a linear frequency modulation chirp signal and uses two perpendicular 10-m long and 9-m long dipoles to perform polarimetric measurements. The achieved horizontal resolution is about several hundred of meters along-track and several thousand of meters across-track.

Like for SHARAD, simulations considering the surface topography and roughness (Hong et al. 2022) are run to identify the echoes actually coming from buried structures. To date, no results have been published about the data collected.

8.4.2.4. *RIMFAX the GPR of the Perseverance rover of the MARS2020 NASA mission*

The MARS 2020 mission (Farley et al. 2020) objective is to explore the area of the Jezero crater. More than 3.5 billion years ago, the Jezero Crater was home to a river-fed lake. An ancient delta clearly visible today indicates that the accumulation of material from the surrounding region has been extensive, making the Jezero crater a very promising place to search for signs of past microbial life. On July 30, 2020, the Perseverance rover, together with its small helicopter Ingenuity, successfully landed on Mars.

The RIMFAX GPR is selected in 2014 to be part of the payload embarked on the rover of the Mars2020 Mission and is, in March 2021, the first GPR to sound the Martian subsurface. The objective of the RIMFAX GPR (Hamran et al. 2014; Hamran et al. 2020) is to contribute to the understanding of the geological history of the Jezero Crater and thus provide the geological context of the samples which will be collected by the rover and sent later to Earth with the Mars Sample Return mission (Muirhead et al. 2020). More precisely, RIMFAX is designed to detect possible buried layers and reveal their relationship with nearby outcrops identified by the rover's surface cameras and to detect a wide range of subsurface materials such as water ice, liquid water or brines. The targeted penetration depth is a few tens of meters, highly depending on the sounded environment.

RIMFAX is a Frequency Modulated Continuous Wave (FMCW) radar. The possible range of frequency lies between 150 and 1,200 MHz, corresponding to a center frequency of 675 MHz. The highest resolution possible with the instrument is therefore about 15 cm in vacuum. The whole frequency bandwidth is used for shallow soundings down to several meters. But since high frequencies are rapidly attenuated with depth, the choice can be made to limit the bandwidth to a lower frequency sub-band when deeper structures are targeted, although this would impact vertical resolution. The operation plan is to interleave shallow and deep soundings along a rover traverse. A single bow-tie antenna located under the rover body is used for both transmission and reception and hard gating is implemented to reduce coupling.

The initial results show that RIMFAX is operating as planned and has allowed us to detect layers in the subsurface of the locations visited by Perseverance in the Jezero Crater (Hamran et al. 2022). Figure 8.8 shows a 90-m long portion of a RIMFAX radargram, acquired with a distance between soundings approximatively equal to 10 cm, that reveals a series of clear dipping interfaces in the Martian subsurface. The radargram highlights the fact that long traverses are needed to optimize the detection of buried structures and that the information about the surface topography provided by cameras in orbit is essential for a proper radargram analysis. The interpretation of the features detected by RIMFAX must be supported by and consistent with data provided by the other instruments of the payload, and is still to be consolidated (Hamran et al. 2022).

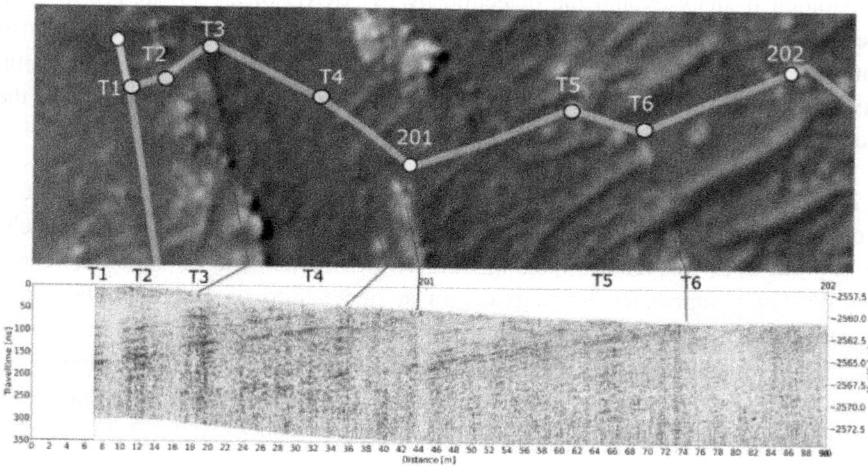

Figure 8.8. *The upper image, provided by MRO's HiRISE instrument, shows the Perseverance path. The image below is the corresponding RIMFAX radargram. Credit: NASA/JPL-Caltech/University of Arizona/USGS/FFI*

While a qualitative understanding of the subsurface configuration is rather easily obtained from processed GPR images (see Figure 8.8), quantitative analyzes are needed to support the attempted identification of subsurface materials and an in-depth comprehension of the local environment. For example, a study based on the amplitude of the signals received by RIMFAX (Eide et al. 2022) makes it possible to estimate, over the whole sounded area, an average attenuation in the subsurface of approximately –2.6 dB/m at 675 MHz, which is compatible with typical values for laboratory measurements of dry magmatic lithologies.

These are very preliminary results from the first GPR that is currently sounding the Martian subsurface.

8.4.2.5. *RoPeR the GPR of the Zhurong rover of the Tanwien-1 CNSA mission.*

As previously mentioned in the section dedicated to the MOSIR, the Tianwen-1 mission (Zou et al. 2021) landed, in May 2021, its rover Zhurong in the south of Utopia Planitia (Mills et al. 2021; Wu et al. 2021). Utopia Planitia is a large impact basin located close to boundary between the northern lowlands and the southern highlands of Mars. This area shows interesting periglacial features such as polygonal terrains (Hiesinger and Head III 2000; Soare et al. 2021). Climate models, surface morphological analyses and the detection by SHARAD of an 80- to 170-m-thick layer of a material with a low permittivity around 2.8 (Stuurman et al. 2016) contribute to the conclusion that a large water ice deposit may be present at depth. This is probably the main motivation for having a GPR probe the subsurface of the area to confirm whether or not this hypothesis is true.

Zhurong carries the RoPeR GPR (Zhou et al. 2020) which is the second GPR to be operated on Mars. RoPeR is a radar operating on two complementary channels: the first one uses frequencies ranging from 15 to 95 MHz and monopole antennas with Wu-King impedance loading to aim at penetration of 100 m in favorable environments (Lu et al. 2021a). The second one operates from 0.45 to 2.15 GHz, and it uses two assembly of Vivaldi antennas accommodated at the rear of the rover, allowing the measurements to be performed in four different polarimetric configurations.

In a little bit longer than 3 months, Zhurong drove over more than 1 km performing soundings along the rover path. Figure 8.9(a) shows a RoPeR LF radargram (Li et al. 2022). Although no clear interface is detected in the radargram, a layered structure is revealed based on the spatial changes in scattering properties observed in the sounded subsurface up to 80 m. An interpretation of the radargram supported by simulated data that shows the same behavior is proposed in terms of scatterers size (Figure 8.9(b)) and bulk permittivity value (Figure 8.9(c)). Hints of sedimentation caused by episodic hydraulic flooding are highlighted by the authors, who also conclude that no evidence of the presence of water was found in the first 80 m of the subsurface.

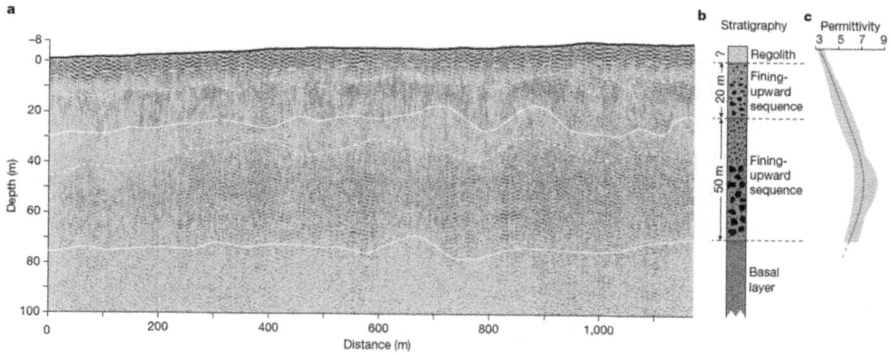

Figure 8.9. *RoPeR low-frequency data and interpretation (Li et al. 2022). (a) RoPeR low-frequency radargram acquired on Utopia Planitia. (b) Interpretation in terms of layers based on heterogeneities. (c) Estimate of the permittivity value vertical profile*

8.4.2.6. *WISDOM the GPR of the Rosalind Franklin rover of the ExoMars mission*

For multiple reasons (programmatic, technical, political, strategical, etc.) partial descoping, successive redesigns of the mission, changes in the international collaborations translated in a series of delays. The ExoMars ESA mission is to date planned for 2028. ExoMars is an astrobiology mission whose aim is to look in the Martian subsurface for traces of past (or present) life (Vago et al. 2017). The specificity of the Rosalind Franklin rover is its unprecedented drilling capacity, which will allow us to take samples up to 2 m below the surface where preserved traces of life are more likely to be found. The selected landing site, about 3.9 billion years old, is located in Oxia Planum (Quantin-Nataf et al. 2021) and presents signs of aqueous alteration. Sedimentary deposits spotted from orbit are favorable places to search for potential traces of ancient life.

In 2003, WISDOM (Ciarletti et al. 2011; Ciarletti et al. 2017a) is the first GPR selected to be part of the payload on a planetary rover mission. It is designed specifically to meet the requirements of the ExoMars mission, that is, to characterize the subsurface along the rover path over a few meters' depth with a fine resolution of around 3 cm (commensurate with the length of the samples that will be collected). WISDOM will provide, together with the other panoramic instruments of the payload, information on the geological context of the landing site. It will also be essential in the selection of the precise location for the collection of samples and will ensure the safety of the drilling operation by detecting possible buried obstacles. Its working frequencies are between 0.5 and 3 GHz to guarantee the high resolution required.

The design selected is a frequency-stepped radar using Vivaldi antennas (visible in Figure 8.3(b)), capable of performing measurements in four polarimetric configurations. The FM of the radar is integrated to the FM of the mission rover since 2019 and has been fully characterized (Hervé et al. 2020). Specific codes have been developed to further improve the vertical resolution by bandwidth extrapolation techniques (Oudart et al. 2021). Figure 8.10 illustrates the result obtained in data collected on Mount Etna, Italy.

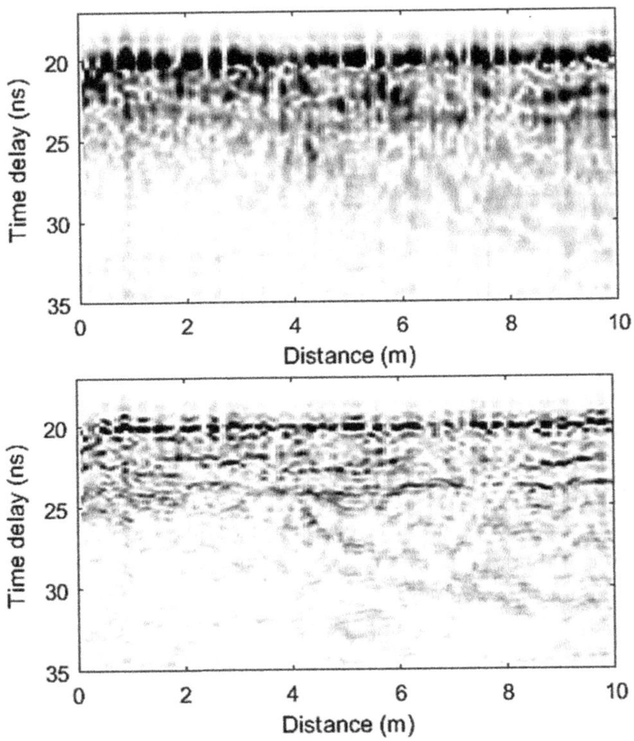

Figure 8.10. *WISDOM radargrams acquired on pyroclastic deposits (Mount Etna). Top: Classical processing; Bottom: Vertical resolution improved by Bandwidth Extrapolation Technique*

8.5. Unusual radar designed

This last section is dedicated to a selection of two quite unusual sounding radars which have been used for a planetary mission.

8.5.1. *The tomographic radar CONSERT (Comet Nucleus Sounding Experiment by Radiowave Transmission) of the ROSETTA mission*

ESA selected in 1993 the challenging Rosetta mission to study a comet nucleus over more than 2 years. The objective of the mission (Glassmeier et al. 2007) is to better understand the processes that led to the formation of our solar system since comets are considered as remnants bodies that were present at the birth of our solar system 4.5 billion years ago, long before the planets (Schulz 2012).

Rosetta consists of two parts: an orbiter called Rosetta and a lander called Philae. Each is equipped with a suite of scientific instruments designed to analyze the comet's plasma, coma, nucleus and their evolution as the nucleus approaches and moves away from its perihelion.

Initially planned in 2003, the launch was postponed because of a launcher failure and the initial target 46P/Wirtanen was replaced by 67P/Churyumov-Gerasimenko. Rosetta arrived close to 67P's nucleus in August 2014 and performed measurements until November 2016. After a phase dedicated to the nucleus observation and the election of an appropriate landing site, Philae successfully landed in November 2014, but ended in an unexpected location, with a non-optimal attitude. As a result, the lander could not recharge its batteries and surface operations were shorter than planned.

The CONSERT radar was selected in 1996 by ESA (Kofman et al. 2007). Its main objective is to characterize the internal structure of the nucleus of comet 67P by tomography, taking advantage of the nucleus rotation and the Rosetta trajectories.

CONSERT is a bistatic sounding radar that uses both Rosetta and Philae as transmitters and receivers. The selected carrier frequency is 90 MHz as a good compromise between the targeted penetration of a few kilometers in the nucleus and the resolution of 20 m achieved with the 10 MHz bandwidth.

Particular attention is paid to the design of the antenna systems. The antennas onboard Rosetta and Philae are used for both reception and transmission. Given the impossibility of predicting the attitude of Philae after its landing hence its antennas orientation with respect to the nucleus' surface and to the Rosetta's antennas, the choice is made to work with circular polarization. In addition, an as broad as possible radiation pattern is needed to receive signals whatever the direction of arrival of the waves are between the orbiter and the lander. The final choice is, as usual, the result of a trade-off between requirements for the instrument and constraints from the mission: the orbiter's antenna consists of two pairs of orthogonal dipoles deployed by a pyrotechnic system after launch. The lander antenna is made of two orthogonal monopoles. This choice of antennas limits budget link problems that are expected in case of a significant modification of polarization during the interaction of the radar waves with the nucleus. Given the non-optimal location and attitude of Philae, this choice, which proves to have taken possible measurements, is a very unfavorable configuration.

Despite the fact that CONSERT data were acquired during only one pass of Rosetta around the nucleus with a non-optimal trajectory (Figure 8.11), interesting and very news results have been obtained.

The retrieved bulk permittivity value (Kofman et al. 2020) for the sounded volume is quite low around 1.27 (Kofman et al. 2015), consistent with a porosity of 75–85% and a volumetric dust/ice ratio of 0.4–2.6. CONSERT's data also suggest that the permittivity gets lower with depth (Ciarletti et al. 2015; Herique et al. 2019) ranging from ~1.7 to 1.95 in the shallow subsurface (<25 m) and about 1.2–1.32 deeper in the nucleus (Kofman et al. 2020). The analysis of the received pulse shape allows us to constrain the size of potential building blocks (Ciarletti et al. 2017b). CONSERT was also used right after the landing to reduce by trilateration the uncertainty about Philae's position and find its final location on the surface (Herique et al. 2015). CONSERT is to date the only sounding radar ever to sound a comet nucleus.

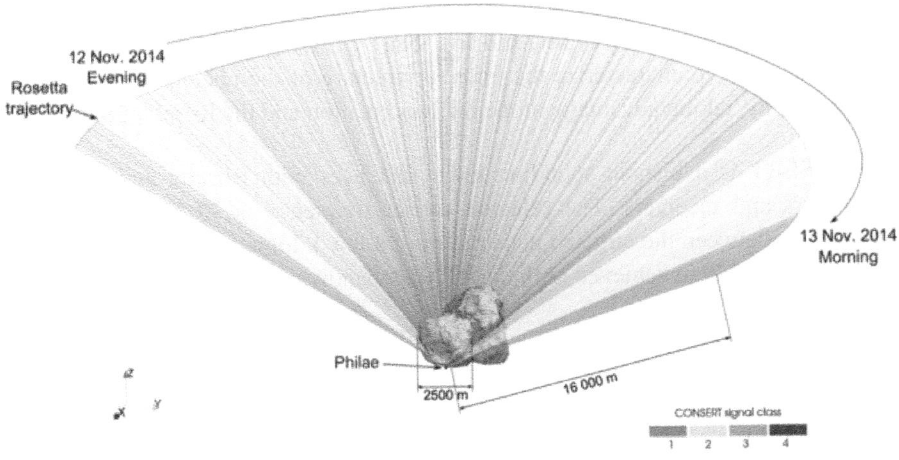

Figure 8.11. *Illustration of the propagation path between Rosetta and Philae through the nucleus of Comet 67P. The green color is for the best signal quality, the red one for no signal. Copyright: ESA/Rosetta/Philae/CONSERT*

8.5.2. *The LRPR on board the lander of the CHANG'E 5 lunar mission*

In December 2020, the Chinese lunar sample return mission Chang'E 5 landed on the moon in the Oceanus Porcellanum region with a stationary platform. The objective of the mission was to take samples from the surface and shallow subsoil with a drill capable of collecting 2 m long samples and sending them to Earth for analysis. The LRPR radar was therefore designed to guide these challenging drilling operations. It is a multi-offset radar using 12 Vivaldi antennas distributed in three groups under the lander body (Lu et al. 2021) as shown in Figure 8.12. This configuration allows us to perform soundings in 11 × 12 different configurations to amp the subsurface beneath the lander.

Given the objectives of LRPR, a high vertical resolution is required hence a large frequency bandwidth. LRPR is an impulse radar using 0.2 ns pulses which corresponds to a frequency range between 0.5 and 4 GHz. The achieved resolution in vacuum was better than 4.5 cm (Shen et al. 2018; Xiao et al. 2019). The expected penetration depth of the instrument was more than 2 m, commensurate with the length of the core sample. The collection of samples was successful and almost 2 kg of lunar rocks were brought back to Earth in December 2022.

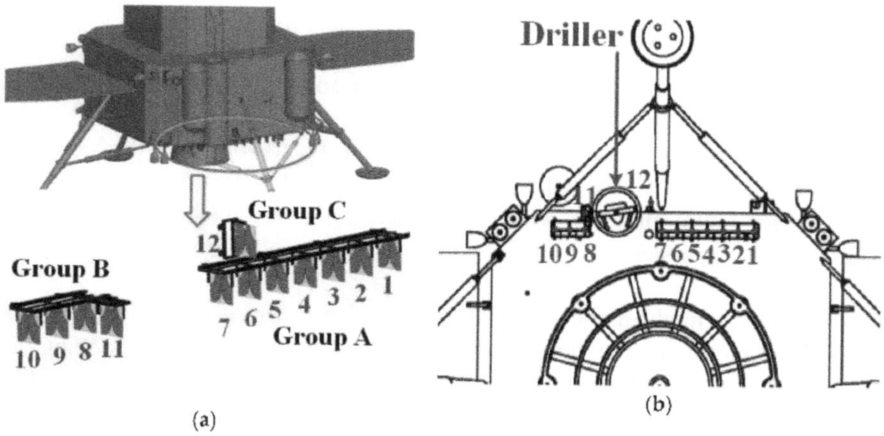

Figure 8.12. *Configuration of the LRPR antennas' array. (a) Oblique view with the lander structure. (b) Bottom view (Lu et al. 2021)*

The first published results show that at the landing site of the platform, the lunar regolith thickness was more than 3 m, and that the GPR was actually able to locate the potential hazards for the drilling and sampling operations. The estimated dielectric constant of the surface lunar regolith is around 2.5, consistent with the bulk density of about 1.4 g/cm^3. The retrieved loss tangent is about 0.013 at 873.6 MHz (Li et al. 2022).

Trying to characterize the subsurface from a stationary platform is challenging, the LRPR was designed specifically for the CHANG'E 5 mission and had to operate at high frequency with a large frequency bandwidth.

On much large spatial scale, aiming at penetration of a few kilometers in the Martian subsurface, the GPR of the Netlander mission was designed to be operated from a small lander at the Martian surface (Berthelier et al. 2000; Berthelier et al. 2003). To achieve the objectives, frequencies of a few MHz were chosen with a bandwidth of 1 MHz. The same design is the basis for the EISS radar on the initial version of the ExoMars mission. EISS was able to operate in a monostatic mode with three 35-m long monopoles antennas deployed on the Martian surface and also in a bistatic mode using the same long electrical antennas for transmission and a 10-cm long magnetic search-coil sensor as a receiver accommodated onboard the rover of the mission (Biancheri-Astier et al. 2010). This part of the ExoMars mission was descoped in 2009.

8.6. Future of soundings radars and GPR in solar system exploration

Thanks to the great results obtained by soundings radars (on Mars, the Moon and Titan) and GPRs (on Mars and the Moon), a series of radars is currently being developed for future missions, which will characterize the subsurface of the other bodies of our solar system.

Because they might harbor oceans of liquid water beneath their ice shell, the Jupiter icy moons are the interest of future astrobiology missions looking for potential habitable worlds in our solar system. In 2012, the ESA selected the JUICE (Jupiter Icy Moons Explorer) mission to study from orbit the icy moons Europa, Ganymede and Callisto of Jupiter. JUICE should be launched in 2023 and arrive in the Jupiter environment in 2031. The main objectives for the radar onboard the orbiter are the search for a buried ocean or shallower reservoirs of liquid water, the study of potential convection layers, the characterization of dust and impurity inside the ice shell, etc. The RIME (Radar for Icy Moon Exploration) radar is logically designed to sound the ice shell of the icy moons over depth of 9 km (Bruzzone et al. 2013). Its design is based on the MARSIS and SHARAD ones with slightly different parameter values. Its central frequency is set at 9 MHz with a maximum value for the bandwidth of 3 MHz, corresponding to a resolution around 50 m in vacuum. The antenna used for both transmission and reception is a 16-m long dipole. The expected cross-track resolution is between 2 and 10 km, and the along-track resolution between 0.3 and 1 km, depending on the phase of the mission.

The Jupiter environment is challenging for radars since the radio noise generated by the planet prevents radar operations below 40 MHz (Bruzzone et al. 2013; Bruzzone and Croci 2019). As a result, RIME will be operated only only on the anti-Jovian side of its moons.

The NASA Europa Clipper (Senske et al. 2019) with its sounding radar REASON (Radar for Europa Assessment and Sounding: Ocean to Near-surface) will be launched in 2024 and arrive at Europa around 2030, with objectives similar and complementary to the JUICE mission (Plaut 2019) but more focused on Europa.

The surface of our neighboring planet Venus is hidden beneath a thick atmosphere of sulfuric acid clouds that prevent the observation by optical remote sensing instruments. It has been revealed by the radars of previous missions: Pioneer Venus (Pettengill et al. 1979, 1980), Venera 15 and 16 (Barsukov et al. 1986) and Magellan (Saunders et al. 1990).

In this context, the aim of the ESA Envision mission to Venus is to investigate the whole planet's environment (atmosphere, surface and subsurface for the first time) and provide a global view of its evolution, present activity and climate.

The SRS (Subsurface Radar Sounder) radar of the Envision mission is designed to sound the first kilometer of the subsurface (Bruzzone et al. 2020). To achieve this science objective, the radar is highly versatile. It can work at wth a central frequency in the range 9–30 MHz. The baseline is to choose the 9 MHz with a 6 MHz bandwidth.

The future Veritas NASA mission to be launched toward Venus in no earlier than 2031 will carry an SAR radar to improve the DEMs of the surface available but no sounding radar is foreseen.

Sounding radars operated from orbit have a typical mass greater than 10 kg, while GPRs designed to be accommodated on a rover are lighter with a mass between 2 kg and 5 kg (for the LPR with its two channels). To date, the smaller and lighter sounding radar developed for a space mission has been built for a nanosatellite. The international AIDA (Asteroid Impact and Deflection Assessment) collaboration targets the asteroid Didymos and its moon. In 2022, in a first phase, NASA's Dart mission (Cheng et al. 2016) successfully impacted the moon of the binary system to study the subsequent trajectory deviation. Secondly, the HERA planetary defense mission will carry a sounding radar called JuRa (Juventas Radar) (Herique et al. 2020). JurRa is hosted in the Juventas 6U-XL cubesat and has objectives of studying the internal structure of the asteroid, the characterization of the crater produced by DART and the detection of potential local changes in the subsoil due to the past impact. The JuRa radar operates between 50 and 70 MHz, uses two dipole antennas 1.5 m long and has a mass which is less than 1.3 kg.

The variety of radars that have been designed for a space mission is impressive in terms of frequency ranges, objectives, platforms from which they are operated, etc. They have paved the way for the exploration of the subsurface and future mission and instrument projects will certainly prove that the creativity of engineers and scientists is not exhausted.

8.7. References

Barbin, Y., Nicollin, F., Kofman, W., Zolotarev, V., Glotov, V. (1995). Mars 96 GPR program. *Journal of Applied Geophysics*, 33(1–3), 27–37.

Barsukov, V.L., Basilevsky, A.T., Burba, G.A., Bobinna, N.N., Kryuchkov, V.P., Kuzmin, R.O., Nikolaeva, O.V., Pronin, A.A., Ronca, L.B., Chernaya, I.M. et al. (1986). The geology and geomorphology of the Venus surface as revealed by the radar images obtained by Veneras 15 and 16. *Journal of Geophysical Research*, 91(B4), D378–D398.

Bay, Z. (1946). Reflection of microwaves from the moon. *Hungarica Acta Physica*, 1, 1–22.

Berquin, Y., Herique, A., Kofman, W., Heggy, E. (2015). Computing low-frequency radar surface echoes for planetary radar using Huygens-Fresnel's principle. *Radio Science*, 50, 1097–1109.

Berthelier, J.J., Ney, R., Costard, F., Hamelin, M., Meyer, A., Martinat, B., Reineix, A., Hansen, T., Bano, M., Kofman, K. et al. (2000). The GPR experiment on NETLANDER. *Planetary and Space Science*, 48(12–14), 1161–1180.

Berthelier, J.J., Ney, R., Ciarletti, V., Reineix, A., Martinat, B., Hamelin, M., Costard, F., Paillou, P., Duvanaud, C., Nevejans, D. et al. (2003). GPR, a ground-penetrating radar for the Netlander mission. *Journal of Geophysical Research Planets*, 108(E4), 8027.

Biancheri-Astier, M., Hassen-Khodja, R., Ciarletti, V., Corbel, C., Simon, Y., Caudoux, C., Faroux, J., Dolon, F., Leray, V., Reineix, A. et al. (2010). EISS: An HF mono and bistatic GPR for terrestrial and planetary deep soundings. In *Proceedings of the XIII International Conference on Ground Penetrating Radar*, Lecce.

Biccari, D., Picardi, G., Seu, R., Melacci, P. (2001). Mars surface models and subsurface detection performance in MARSIS. In *IGARSS 2001. Scanning the Present and Resolving the Future. Proceedings. IEEE 2001 International Geoscience and Remote Sensing Symposium (Cat. No.01CH37217)*, 6, 2560–2562, Sydney.

Bierson, C.J., Tulaczyk, S., Courville, S.W., Putzig, N.E. (2021). Strong MARSIS radar reflections from the base of Martian south polar cap may be due to conductive ice or minerals. *Geophysical Research Letters*, 48, e2021GL093880.

Boivin, A.L., Hickson, D., Tsai, C., Cunje, A., Ghent, R.R., Daly, M.G. (2018). Broadband measurements of the complex permittivity of carbonaceous asteroid regolith analog materials. *Journal of Geophysical Research: Planets*, 123, 3088–3104.

Bruzzone, L. and Croci, R. (2019). Radar for icy moon exploration (RIME). In *2019 IEEE 5th International Workshop on Metrology for AeroSpace (MetroAeroSpace)*, 330–333, Turin.

Bruzzone, L., Plaut, J., Alberti, G., Blankenship, D.D., Bovolo, F., Campbell, B.A., Ferro, A., Gim, Y., Kofman, W., Komatsu, G. et al. (2013). RIME: Radar for Icy Moon Exploration. In *2013 IEEE International Geoscience and Remote Sensing Symposium – IGARSS*, 3907–3910, Melbourne.

Bruzzone, L., Bovolo, F., Thakur, S., Carrer, L., Donini, E., Gerekos, C., Paterna, S., Santoni, M., Sbalchiero, E. (2020). Envision Mission to Venus: Subsurface radar sounding. In *IGARSS 2020 – 2020 IEEE International Geoscience and Remote Sensing Symposium*, 5960–5963, Waikoloa.

Campbell, B.A. and Shepard, M.K. (2003). Coherent and incoherent components in near-nadir radar scattering: Applications to radar sounding of Mars. *Journal of Geophysical Research*, 108, 5132.

Campbell, B.A., Ghent, R.R., Shepard, M.K. (2003). Limits on inference of Mars small-scale topography from MOLA data. *Geophysical Research Letters*, 30(3), 1115.

Carpenter, R.L. and Goldstein, R.M. (1963). Radar observations of Mercury. *Science*, 142(3590), 381–382.

Carr, M.H. (1979). Formation of Martian flood features by release of water from confined aquifers. *Journal of Geophysical Research: Solid Earth*, 84(B6), 2995–3007.

Carr, M.H. (1996). *Water on Mars*. Oxford University Press, New York.

Cheng, A.F., Michel, P., Jutzi, M., Rivkin, A.S., Stickle, A., Barnouin, O., Ernst, C., Atchison, J., Pravec, P., Richardson, D.C. (2016). Asteroid impact & deflection assessment mission: Kinetic impactor. *Planetary and Space Science*, 121, 27–35.

Ciarletti, V. (2016). A variety of radars designed to explore the hidden structures and properties of the Solar System's planets and bodies. *Comptes Rendus Physique*, 17(9), 966–975.

Ciarletti, V., Corbel, C., Plettemeier, D., Cais, P., Clifford, S.M., Hamran, S.-E. (2011). WISDOM GPR designed for shallow and high resolution sounding of the Martian subsurface. *Proceedings of the IEEE*, 99(5), 824–836.

Ciarletti, V., Clifford, S., Plettemeier, D., Le Gall, A., Hervé, Y., Dorizon, S., Quantin-Nataf, C., Benedix, W.S., Schwenzer, S., Pettinelli, E. et al. (2017a). The WISDOM radar: Unveiling the subsurface beneath the ExoMars rover and identifying the best locations for drilling. *Astrobiology*, 17(6–7), 565–584.

Ciarletti, V., Herique, A., Lasue, J., Levasseur-Regourd, A.-C., Plettemeier, D., Lemmonier, F., Guiffaut, C., Pasquero, P., Kofman, W. (2017b). CONSERT constrains the internal structure of 67P at a few-metre size scale. *Monthly Notices of the Royal Astronomical Society*, 469(Suppl_2), S805–S817.

Clifford, S.M. (1993). A model for the hydrologic and climatic behavior of water on Mars. *Journal of Geophysics Research*, 98(E6), 10973–11016.

Clifford, S.M. and Parker, T.J. (2001). The evolution of the Martian hydrosphere: Implications for the fate of a primordial ocean and the current state of the northern plains. *Icarus*, 154, 40–79.

COSPAR (2021). COSPAR policy on planetary protection. *Space Research Today*, 211 12–25.

Craddock, R.A. and Howard, A.D. (2002). The case for rainfall on a warm, wet early Mars. *Journal of Geophysics Research*, 107(E11), 5111.

Dal Bo, I., Klotzsche, A., Bol, R., Moradi, G., Weihermüller, L., Vereecken, H., van der Kruk, J. (2021). GPR and EMI characterization of the hyperarid study site of Yungay, Chile: Implications of applying geophysical methods on Mars. *Earth and Space Science*, 8, e2021EA001790.

Desage, L., Herique, A., Kofman, W., Zine, S. (2022). SHARAD data analysis with high resolution digital terrain models. In *Europlanet Science Congress 2022*, Granada.

Dorizon, S., Ciarletti, V., Plettemeier, D., Benedix, W.-S. (2016). Performance validation of the ExoMars 2018 WISDOM GPR in ice caves, Austria. *Planetary and Space Science*, 120, 1–14.

Dyce, R.B., Pettengill, G.H., Shapiro, I.I. (1967). Radar determination of the rotations of Venus and Mercury. *The Astronomical Journal*, 72(3), 351–359.

Eide, S., Casademont, T.M., Berger, T., Dypvik, H., Shoemaker, E.S., Hamran, S.-E. (2023). Radar attenuation in the shallow Martian subsurface: RIMFAX time-frequency analysis and constant-Q characterization over Jezero Crater Floor. *Geophysical Research Letters*, 50, e2022GL101429.

Fan, M., Lyu, P., Su, Y., Du, K., Zhang, Q., Zhang, Z., Dai, S., Hong, T. (2021). The Mars Orbiter Subsurface Investigation Radar (MOSIR) on China's Tianwen-1 mission. *Space Science Reviews*, 217, 8.

Farley, K.A., Williford, K.H., Stack, K.M., Bharit, R., Chen, A., de la Torre, E., Hand, K., Goreva, Y., Herd, C., Heoso, R. et al. (2020). Mars 2020 mission overview. *Space Science Reviews*, 216(8), 142.

Glassmeier, KH., Boehnhardt, H., Koschny, D., Kürt, E., Richter, I. (2007). The Rosetta mission: Flying towards the origin of the solar system. *Space Science Reviews*, 128, 1–21.

Goldstein, R.M. and Carpenter, R.L. (1963a). Radar observations of Mars. *Science*, 141(3586), 1171–1172.

Goldstein, R.M. and Carpenter, R.L. (1963b). Rotation of Venus: Period estimated from radar measurements. *Science*, 139(3558), 910–911.

Guang-You, F., Bin, Z., Yi-Cai, J., Qun-Ying, Z., Shao-Xiang, S., Yu-Xi, L., Hong-Fei, G., Chuan-Jun, T., Yun-Ze, G., Wei, L. et al. (2014). Lunar penetrating radar onboard the Chang'e-3 mission. *Research in Astronomy and Astrophysics*, 14, 1607.

Gupta, V., Gupta, S.K., Kim, J. (2020). Automated discontinuity detection and reconstruction in subsurface environment of Mars using deep learning: A case study of SHARAD observation. *Applied Sciences*, 10(7), 2279.

Hall, A., Dobke, B., Lisle, M., Shilton, M., Allouis, E., Waugh, L., Carroll, J., Doignon, G., Azkarate, M., van Winnendael, M. et al. (2019). ExoFiT: ExoMars-Like Field Trials – A mission simulation. *15th Symposium on Advanced Space Technologies in Robotics and Automation*. Noordwijk.

Hamran, S.E., Admunsen, H.E.F., Carter, L.M., Ghent, R.R., Kohler, J., Mellon, M.T., Paige, D.A. (2014). The RIMFAX ground penetrating radar on the Mars 2020 Rover. In *Fall Meeting 2014*. American Geophysical Union.

Hamran, S.E., Paige, D.A., Amundsen, H.E.F., Berger., T., Brovoll, S., Carter, L., Damsgard, L., Dypvik, H., Eide, J., Eide, S. et al. (2020). Radar imager for Mars' subsurface experiment – RIMFAX. *Space Science Reviews*, 216(8), 2020.

Hamran, S.E., Paige, D.A., Allwood, A., Amundsen, H.E.F., Berger., T., Brovoll, S., Carter, L., Casademont, T.M., Damsgard, L., Dypvik, H. et al. (2022). Ground penetrating radar observations of subsurface structures in the floor of Jezero crater, Mars. *Science Advances*, 8(34).

Harrison, K.P. and Grimm, R.E. (2004). Tharsis recharge: A source of groundwater for Martian outflow channels. *Geophysical Research Letters*, 31, L14703.

Heggy, E., Clifford, S.M., Younsi, A., Miane, J.L., Carley, R., Morris, R.V. (2006). On the dielectric properties of dust and ice-dust mixtures: Experimental characterization of the Martian polar layered deposits analog materials. In *Fourth Mars Polar Science Conference*, October 2–6.

Herique, A., Roger, Y., Pasquero O.P., Zine, S., Puget, P., Kofman, W. (2015). PHILAE localization from CONSERT/ROSETTA measurement. *Planetary and Space Science*, 117, 475–484.

Herique, A., Plettemeier, D., Goldberg, H., Kofman, W. (2020). JuRa: The Juventas Radar on Hera to fathom Didymoon. In *14th Europlanet Science Congress 2020*, 21 September–9 October.

Hervé, Y., Ciarletti, V., Le Gall, A., Corbel, C., Hassen-Khodja, R., Benedix, W.S., Plettemeier, D., Humeau, O., Vieau, A.J., Lustrement, B. et al. (2020). The WISDOM radar on board the ExoMars 2022 Rover: Characterization and calibration of the flight model. *Planetary and Space Science*, 189, 104939.

Hiesinger, H. and Head III, J.W. (2000). Characteristics and origin of polygonal terrain in southern Utopia Planitia, Mars: Results from Mars Orbiter laser altimeter and Mars Orbiter camera data. *Journal of Geophysical Research: Planets*, 105, 11999–12022.

Holt, J.W., Safaeinili, A., Plaut, J.J., Head, J.W., Phillips, R.J., Seu, R., Kempf, S.D., Choudhary, P., Young, D.A., Putzig, N.E. et al. (2008). Radar sounding evidence for buried glaciers in the southern mid-latitudes. *Science*, 322(5905), 1235–1238.

Hong, T., Su, Y., Dai, S., Zhang, Z., Du, W., Liu, C., Liu, S., Wang, R., Ding, C., Li, C. (2022). An improved method of surface clutter simulation based on orbiting radar in Tianwen-1 Mars exploration. *Radio Science*, 57, e2022RS007491.

Kofman, W. and Safaeinili, A. (2004). Radar techniques applied to subsurface studies in solar system exploration. *Tools and Technologies for Future Planetary Exploration*, 543, 39–50.

Kofman, W., Herique, A., Goutail, JP., Hagfors, T., Wiliams, I.P., Nielsen, E., Barriot, J.-P., Barbin, Y., Elach, C., Edenhofer, P. et al. (2007). The comet nucleus sounding experiment by radiowave transmission (CONSERT): A short description of the instrument and of the commissioning stages. *Space Science Reviews*, 128, 413–432.

Kofman, W., Herique, A., Barbin, Y., Barriot, J.P., Ciarletti, V., Clifford, S., Edenhofer, P., Elachi, C., Eyraud, C., Goutail, J.P. et al. (2015). Internal properties of the nucleus of Comet 67P/ Churyumov – Gerasimenko as measured by the CONSERT instrument. *Science*, 349(6247).

Kofman, W., Zine, S., Herique, A., Yves Rogez, Y., Jorda, L., Levasseur-Regourd, A.-C. (2020). The interior of Comet 67P/C–G; revisiting CONSERT results with the exact position of the Philae lander. *Monthly Notices of the Royal Astronomical Society*, 497(3), 2616–2622.

Lauro, S.E., Pettinelli, E., Caprarelli, G., Guallini, L., Rossi, A.P., Mattei, E., Cosciotti, B., Cicchetti, A., Soldovieri, F., Cartacci, M. et al. (2021). Multiple subglacial water bodies below the south pole of Mars unveiled by new MARSIS data. *Nature Astronomy*, 5(1), 63–70.

Li, C., Xing, S., Lauro, S., Su, Y., Dai, S., Feng, J., Cosciotti, B. (2018a). Pitfalls in GPR data interpretation: False reflectors detected in lunar radar cross sections by Chang'e–3. *IEEE Transactions on Geoscience and Remote Sensing*, 56, 1325–1335.

Li, C., Xing, S., Lauro, S., Su, Y., Dai, S., Feng, J., Cosciotti, B., Di Paolo, F., Mattei, E., Xiao, Y. (2018b). Pitfalls in GPR data interpretation: False reflectors detected in lunar radar cross sections by Chang'e-3. *IEEE Transactions on Geoscience and Remote Sensing*, 56(3), 1325–1335.

Li, C., Zheng, Y., Wang, X., Zhang, J., Wang, Y., Chen, L., Zhang, L., Zhao, P., Liu, Y., Lv, W. et al. (2022a). Layered subsurface in Utopia Basin of Mars revealed by Zhurong rover radar. *Nature*, 610, 308–312.

Li, Y., Zhou, B., Shen, S., Lu, W., Tang, C., Li, S., Su, Y., Dai, S., Fang, G. (2022b). The lunar regolith structure and electromagnetic properties of Chang'E-5 Landing Site. *Remote Sensing*, 14(18), 4539.

Lu, W., Li, Y., Ji, Y., Tang, C., Zhou, B., Fang, G. (2021). Ultra-wideband MIMO array for penetrating lunar regolith structures on the Chang'e-5 lander. *Electronics*, 10, 1–12.

Mastrogiuseppe, M., Poggiali, V., Hayes, A., Lorenz, R., Lunine, J., Picardi, G., Seu, R., Flamini, E., Mitri, G., Notarnicola, C. et al. (2014). The bathymetry of a Titan sea. *Geophysical Research Letters*, 41, 1432–1437.

Mattei, E., Pettinelli, E., Lauro, S., Stillman, D.E., Cosciotti, B., Marinangeli, L., Tangari, A.C., Soldovieri, F., Orosei, R., Caprarelli, G. (2022) Assessing the role of clay and salts on the origin of MARSIS basal bright reflections. *Earth and Planetary Science Letters*, 579, 117370.

Milkovich, S.M., Plaut, J.J., Safaeinili, A., Picardi, G., Seu, R., Phillips, R.J. (2009). Stratigraphy of Promethei Lingula, south polar layered deposits, Mars, in radar and imaging data sets. *Journal of Geophysical Research*, 114, E03002.

Mills, M.M., McEwen, A.S., Okubo, C.H. (2021). A preliminary regional geomorphologic map in Utopia Planitia of the Tianwen-1 Zhurong landing region. *Geophysical Research Letters*, 48, e2021GL094629.

Mofensen, J. (1946). Radar echoes from the Moon. *Electronics*, 19, 92–98.

Mouginot, J., Kofman, W., Safaeinili, A., Herique, A. (2008). Correction of the ionospheric distortion on the MARSIS surface sounding echoes. *Planetary and Space Science*, 56(7), 917–926.

Mouginot, J., Pommerol, A., Kofman, W., Beck, P., Schmitt, B., Herique, A., Grima, C., Safaeinili, A., Plaut, J.J. (2010). The 3–5 MHz global reflectivity map of Mars by MARSIS/Mars Express: Implications for the current inventory of subsurface H_2O. *Icarus*, 210(2), 612–625.

Muirhead, B.K., Nicholas, A., Umland, J. (2020). Mars sample return mission concept status. In *2020 IEEE Aerospace Conference*, 1–8, Big Sky.

Ono, T. and Oya, H. (2000). Lunar radar sounder (LRS) experiment on-board the SELENE spacecraft. *Earth Planets Space*, 52, 629–637.

Orosei, R., Lauro, S.E., Pettinelli, E., Cicchetti, A., Coradini, M., Cosciotti, B., Di Paolo, F., Flamini, E., Mattei, E., Pajola, M. et al. (2018). Radar evidence of subglacial liquid water on Mars. *Science*, 361(6401), 490–493.

Oudart, N., Ciarletti, V., Le Gall, A., Mastrogiuseppe, M., Herve, Y., Benedix, W.-S., Plettemeier, D., Tranier, V., Hassen-Khodja, R., Statz, C., Lu, Y. (2021). Range resolution enhancement of WISDOM/ExoMars radar soundings by the bandwidth extrapolation technique: Validation and application to field campaign measurements. *Planetary and Space Science*, 197, 105173.

Paillou, P., Lunine, J., Ruffié, G., Encrenaz, P., Wall, S., Lorenz, R., Janssen, M. (2008). Microwave dielectric constant of Titan-relevant materials. *Geophysical Research Letters*, 35, L18202.

Petersen, E.I., Holt, J.W., Levy, J.S. (2018). High ice purity of Martian lobate debris aprons at the regional scale: Evidence from an orbital radar sounding survey in Deuteronilus and Protonilus Mensae. *Geophysical Research Letters*, 45, 11595–11604.

Pettengill, G.H., Ford, P.G., Brown, W.E., Kaula, W.M., Keller, C.H., Masursky, H., McGill, G.E. (1979). Pioneer Venus radar mapper experiment. *Science*, 203(4382), 806–808.

Pettengill, G.H., Eliason, E., Ford, P.G., Loriot, G.B., Masursky, H., McGill, G.E. (1980). Pioneer Venus radar results altimetry and surface properties. *Journal of Geophysical Research*, 85(A13), 8261–8270.

Pettinelli, E., Lauro, S., Mattei, E., Cosciotti, B., Stillman, D., Caprarelli, G., Orosei, R. (2022). The search for liquid water below the South Polar Layer Deposits: Where we stand? In *Europlanet Science Congress 2022*, EPSC2022-1024, Granada.

Picardi, G., Biccari, D., Seu, R., Marinangeli, L., Johnson, W.T.K., Jordan, R.L., Plaut, J., Safaenili, A., Gurnett, D.A., Ori, G.G. (2004). Performance and surface scattering models for the Mars Advanced Radar for Subsurface and Ionosphere Sounding (MARSIS). *Planetary and Space Science*, 52(1–3), 149–156.

Picardi, G., Plaut, J., Biccari, D., Bombaci, O., Calabrese, D., Cartacci, M., Cicchetti, A., Clifford, S.M., Edenhofer, P., Farrell, W.M. et al. (2005). Radar soundings of the subsurface of Mars. *Science*, 310(5756), 1925–1928.

Picardi, G., Biccari, D., Seu, R., Plaut, J., Johnson, W., Jordan, R., Safaeinili, A., Gurnett, D., Huff, R., Orosei, R. et al. (2015). Mars Advanced Radar for Subsurface and Ionospheric Sounding (MARSIS) after nine years of operation: A summary. *Planetary and Space Science*, 112, 98–114.

Pirrotta, S. and Flamini, E. (2011). SoRa first flight. Summer 2009. *Memorie della Societa Astronomica Italiana Supplementi*, 16, 145.

Plaut, J.J., Picardi, G., Safaeinili, A., Ivanov, A.B., Milkovich, S.M., Cicchetti, A., Kofman, W., Mouginot, J., Farrell, W.M., Phillips, R.J. et al. (2007). Subsurface radar sounding of the south polar layered deposits of Mars. *Science*, 316, 92–95.

Porcello, L.J., Jordan, R.L., Zelenka, J.S., Adams, G.F., Phillips, R.J., Brown, W.E., Ward, S.H., Jackson, P.L. (1974). The Apollo lunar sounder radar system. *Proceedings of the IEEE*, 62(6), 769–783.

Putzig, N.E., Smith, I.B., Perry, M.R., Foss II, F.J., Campbell, B.A., Phillips, R.J., Seu, R. (2018). Three-dimensional radar imaging of structures and craters in the Martian polar caps. *Icarus*, 308, 138–147.

Putzig, N.E., Foss, F.J., Campbell, B.A., Holt, J.W., Perry, M.R., Smith, I.B., Russell, A.T., Nerozzi, S., Christoffersen, M.S., Mueller, I.H., Sava, P.C. (2022). New views of the internal structure of Planum Boreum from enhanced 3D imaging of Mars Reconnaissance Orbiter Shallow Radar data. *The Planetary Science Journal*, 3(11), 259(14).

Randy, D., Boyan, K., Gabriella L.S. (2020). Highlights of the Mars 2020 environmental test program [Online]. Available at: https://hdl.handle.net/2014/52312.

Saunders, R.S., Pettengill, G.H., Arvidson, R.E., Sjogren, W.L., Johnson, W.T.K., Pieri, L. (1990). The Magellan Venus radar mapping mission. *Journal of Geophysical Research*, 95(B6), 8339–8355.

Scanlan, K.M., Young, D.A., Steinbrügge, G., Kempf, S.D., Grima, C., Blankenship, D.D. (2021). Delay Doppler SAR focusing and quantitative quality control of the Radar for Europa Assessment and Sounding: Ocean to Near-Surface (REASON) sounding data product. *IEEE Journal of Selected Topics in Applied Earth Observations and Remote Sensing*, 14, 4352–4369.

Schulz, R. (2012). The Rosetta mission – Exploring solar system formation. *Planetary and Space Science*, 66(1). doi:10.1016/j.pss.2012.04.001.

Senske, D., Pappalardo, R., Korth, H., Blaney, D., Blankenship, D., Christensen, P., Retherford, K., Turtle, E. (2019). Science of the Europa Clipper Mission: Comprehensive remote sensing to investigate the ice shell EPSC. In *EPSC-DPS2019-731-1*, 13, 15–20 September, Geneva.

Seu, R., Phillips, R.J., Biccari, D., Orosei, R., Masdea, A., Picardi, G., Safaeinili, A., Campbell, B.A., Plaut, J.J., Marinangeli, L. (2007a). SHARAD sounding radar on the Mars Reconnaissance Orbiter. *Journal of Geophysical Research*, 112, E05S05.

Seu, R., Phillips, R.J., Alberti, G., Biccari, D., Bonaventura, F., Bortone, M., Calabrese, D., Campbell, B.A., Cartacci, M., Carter, L.M. et al. (2007b). Accumulation and erosion of Mars' south polar layered deposits. *Science*, 317, 1715–1718.

Shen, S.X., Hua, X.L., Zhou, B., Li, Y.X., Lu, W., Liu, Q., Ji, Y.C., Fang, G.Y., Wang, L. (2018). Lunar regolith penetrating radar on the Lander for Chang'E-5 Mission. In *17th International Conference on Ground Penetrating Radar (GPR)*, 1–4, Rapperswil.

Siddiqi, A.A. (2002). *Deep Space Chronicle: A Chronology of Deep Space and Planetary Probes 1958–2000*. NASA, Washington, DC.

Smith, I.B., Lalich, D.E., Rezza, C., Horgan, B.H.N., Whitten, J.L., Nerozzi, S., Holt, J.W. (2021). A solid interpretation of bright radar reflectors under the Mars south polar ice. *Geophysical Research Letters*, 48, e2021GL093618.

Soare, R.J., Conway, S.J., Williams, J.-P., Philippe, M., McKeown, L.E., Godin, E., Hawkswell, J. (2021). Possible ice-wedge polygonisation in Utopia Planitia, Mars and its latitudinal gradient of distribution. *Icarus*, 358, 114208.

Stillman, D.E., Pettinelli, E., Lauro, S.E., Mattei, E., Caprarelli, G., Cosciotti, B., Primm, K.M., Orosei, R. (2022). Partially-saturated brines within basal ice or sediments can explain the bright basal reflections in the south polar layered deposits. *Journal of Geophysical Research: Planets*, 127, e2022JE007398.

Vago, J.L., Westall, F., Coates, A.J., Jaumann, R., Korablev, O., Ciarletti, V., Mitrofanov, I., Josset J.-L., De Sanctis, M.-C., Bibring, J.-P. et al. (2017). Habitability on early Mars and the search for biosignatures with the ExoMars Rover. *Astrobiology*, 17(6–7), 471–510.

Wang, Y., Feng, X., Zhou, H., Dong, Z., Liang, W., Xue, C., Li, X. (2021). Water ice detection research in Utopia Planitia based on simulation of Mars rover full-polarimetric subsurface penetrating radar. *Remote Sensing*, 13(14), 2685.

Watters, T.R., Leuschen, C., Plaut, J., Picardi, G., Safaneili, A., Clifford, S.M., Farrell, W.M., Ivanov, A.B., Philips, R.J., Stofan, E.R. (2006). MARSIS radar sounder evidence of buried basins in the northern lowlands of Mars. *Nature*, 444, 905–908.

White, O.L., Safaeinili, A., Plaut, J.J., Stofan, E.R., Clifford, S.M., Farrell, W.M., Heggy, E., Picardi, G. (2009). MARSIS radar sounder observations in the vicinity of Ma'adim Vallis, Mars. *Icarus*, 201(2), 460.

Wlodek Kofman, K., Herique, A., Barbin, Y., Barriot, J.P., Ciarletti, V., Clifford, S., Edenhofer, P., Elachi, C., Eyraud, C., Goutail, J.P. et al. (2015). Internal properties of the nucleus of Comet 67P/ Churyumov – Gerasimenko as measured by the CONSERT instrument. *Science*, 349(6247).

Wu, B., Dong, J., Wang, Y., Rao, W., Sun, Z., Li, Z., Tan, Z., Chen, Z., Wang, C., Liu, W.C. et al. (2022). Landing site selection and characterization of Tianwen-1 (Zhurong rover) on Mars. *Journal of Geophysical Research: Planets*, 127, e2021JE007137.

Xiao, Y., Su, Y., Dai, S., Feng, J., Xing, S., Ding, C., Li, C. (2019). Ground experiments of Chang'e-5 lunar regolith penetrating radar. *Advances in Space Research*, 63, 3404–3419.

Xiong, S. and Muller, J.-P. (2019). Automated reconstruction of subsurface interfaces in Promethei Lingula near the Martian south pole by using SHARAD data. *Planetary and Space Science*, 166, 59–69.

Zhang, J., Yang, W., Hu, S., Lin, Y., Fang, G., Li, C., Peng, W., Zhu, S., He, Z., Zhou, B. et al. (2015). Volcanic history of the Imbrium basin: A close-up view from the lunar rover Yutu. *Proceedings of the National Academy of Sciences of the United States of America*, 112(17), 5342–5347.

Zhang, L., Xu, Y., Zeng, Z., Li, J., Zhang, D. (2022). Simulation of Martian near-surface structure and imaging of future GPR data from Mars. *IEEE Transactions on Geoscience and Remote Sensing*, 60(5101812), 1–12.

Zhou, B., Li, Y., Li, D., Lu, W., Shen, S., Fang, G.-Y., Su, Y., Dai, S. (2018). Initial laboratory field tests of the rover-mounted GPR for China's first mission to Mars. In *17th International Conference on Ground Penetrating Radar (GPR)*, 1–4, Rapperswil.

Zhou, B., Shen, S.X., Lu, W., Liu, Q., Tang, C.J., Li, S.D., Fang, G.Y. (2020). The Mars rover subsurface penetrating radar onboard China's Mars 2020 mission. *Earth and Planetary Physics*, 4(4), 345–354.

Zou, Y., Zhu, Y., Bai, Y., Wang, L., Jia, Y., Shen, W., Fan, Y., Liu, Y., Wang, C., Zhang, A. et al. (2021). Scientific objectives and payloads of Tianwen-1, China's first Mars exploration mission. *Advances in Space Research*, 67(2), 812–823.

List of Authors

Abderrahmane ABOUDOURIB
Université Paris-Saclay
CNRS
CentraleSupélec
Laboratoire de Génie Electrique
et Electronique de Paris
Gif-sur-Yvette
France

Michele AMBROSANIO
Dipartimento di Ingegneria
Università degli Studi di Napoli
"Parthenope"
Naples
and
Consorzio Nazionale Interuniversitario
per le Telecomunicazioni (CNIT)
Parma
Italy

Grégory ANDREOLI
Cerema
ENDSUM Research Team
France

Vincent BALTAZART
Université Gustave Eiffel
Nantes Campus
France

Martina Teresa BEVACQUA
Dipartimento di Ingegneria
dell'Informazione delle Infrastrutture e
dell'Energia Sostenibile
Università Mediterranea di
Reggio Calabria
and
Consorzio Nazionale Interuniversitario
per le Telecomunicazioni (CNIT)
Parma
Italy

Ilaria CATAPANO
National Research Council of Italy
Institute for Electromagnetic Sensing
of the Environment
Naples
Italy

Valérie CIARLETTI
LATMOS/IPSL
UVSQ/Univeristé Paris-Saclay
Sorbonne Université
CNRS
Guyancourt
France

Lorenzo CROCCO
National Research Council of Italy
Institute for Electromagnetic Sensing of the Environment
Naples
Italy

Antonio CUCCARO
Department of Computer, Modeling, Electronic and Systems Engineering
University of Calabria
Rende
Italy

Angela DELL'AVERSANO
Dipartimento di Ingegneria
Università della Campania
Luigi Vanvitelli
Caserta
Italy

Xavier DÉROBERT
Université Gustave Eiffel
Nantes Campus
France

Giuseppe ESPOSITO
National Research Council of Italy
Institute for Electromagnetic Sensing of the Environment
Naples
Italy

Cyrille FAUCHARD
Cerema
ENDSUM Research Team
France

Alessandro FEDELI
Department of Electrical, Electronic, Telecommunications Engineering and Naval Architecture
Università di Genova
Genoa
Italy

Gianluca GENNARELLI
National Research Council of Italy
Institute for Electromagnetic Sensing of the Environment
Naples
Italy

David GUILBERT
Cerema
ENDSUM Research Team
France

Amine IHAMOUTEN
Université Gustave Eiffel
Nantes Campus
France

Tommaso ISERNIA
Dipartimento di Ingegneria dell'Informazione delle Infrastrutture e dell'Energia Sostenibile
Università Mediterranea di Reggio Calabria
and
Consorzio Nazionale Interuniversitario per le Telecomunicazioni (CNIT)
Parma
Italy

Sébastien LAMBOT
Earth and Life Institute
Université catholique de Louvain
Louvain-la-Neuve
Belgium

Dominique LESSELIER
Université Paris-Saclay
CNRS
CentraleSupélec
Laboratoire des Signaux et Systèmes
Gif-sur-Yvette
France

Giovanni LUDENO
National Research Council of Italy
Institute for Electromagnetic Sensing
of the Environment
Naples
Italy

Maria Antonia MAISTO
Dipartimento di Ingegneria
Università della Campania
Luigi Vanvitelli
Caserta
Italy

Vito PASCAZIO
Dipartimento di Ingegneria
Università degli Studi di Napoli
"Parthenope"
Naples
and
Consorzio Nazionale Interuniversitario
per le Telecomunicazioni (CNIT)
Parma
Italy

Matteo PASTORINO†
Department of Electrical, Electronic,
Telecommunications Engineering and
Naval Architecture
Università di Genova
Genoa
Italy

Andrea RANDAZZO
Department of Electrical, Electronic,
Telecommunications Engineering and
Naval Architecture
Università di Genova
Genoa
Italy

Valentina SCHENONE
Department of Electrical, Electronic,
Telecommunications Engineering and
Naval Architecture
Università di Genova
Genoa
Italy

Mohammed SERHIR
Université Paris-Saclay
CentraleSupélec
SONDRA
Gif-sur-Yvette
France

Jean-Michel SIMONIN
Université Gustave Eiffel
Nantes Campus
France

Arthur SLUŸTERS
Louvain Research Institute in Management and Organizations (LouRIM)
Institute of Information and Communication Technologies, Electronics and Applied Mathematics (ICTEAM)
Université catholique de Louvain
Louvain-la-Neuve
Belgium

Francesco SOLDOVIERI
National Research Council of Italy
Institute for Electromagnetic Sensing of the Environment
Naples
Italy

Raffaele SOLIMENE
Dipartimento di Ingegneria
Università della Campania Luigi Vanvitelli
Caserta
Italy

Bachir TCHANA-TANKEU
Cerema
ENDSUM Research Team
France

Shreedhar Savant TODKAR
Université Gustave Eiffel
Nantes Campus
France

Jean VANDERDONCKT
Louvain Research Institute in Management and Organizations (LouRIM)
Institute of Information and Communication Technologies, Electronics and Applied Mathematics (ICTEAM)
Université catholique de Louvain
Louvain-la-Neuve
Belgium

Kaijun WU
Earth and Life Institute
Université catholique de Louvain
Louvain-la-Neuve
Belgium

Index

A, B, C

D, E, F

R, S, T

U, V, W

Printed and bound by CPI Group (UK) Ltd, Croydon, CR0 4YY

13/09/2026

14961069-0002